The Space Shuttle: An Experimental Flying Machine

Thirty Years of Challenges

Ben Evans

The Space Shuttle: An Experimental Flying Machine

Thirty Years of Challenges

 Springer

Published in association with
Praxis Publishing
Chichester, UK

Ben Evans
Space Writer
Atherstone, Warwickshire, UK

SPRINGER-PRAXIS BOOKS IN SPACE EXPLORATION

Springer Praxis Books
Space Exploration
ISBN 978-3-030-70776-7   ISBN 978-3-030-70777-4   (eBook)
https://doi.org/10.1007/978-3-030-70777-4

Cover Design: Jim Wilkie
Project Editor: David M. Harland

This Springer imprint is published by the registered company Springer Nature Switzerland AG
The registered company address is: Gewerbestrasse 11, 6330 Cham, Switzerland

To my wife, Michelle, for everything

Foreword

Experimental Flying Machine is a meticulously researched and beautifully written narrative of the Space Shuttle Experience. Ben Evans weaves together the technology trades, the political influences and the real human beings who created and flew this amazing vehicle. If you ever wondered why the Shuttle looked the way it did, Ben explains it in a way everyone can understand. Most engineering efforts are a compromise and the Space Shuttle is the prime example.

Ben Evans reminds me of Stephen Ambrose. Ambrose was able to get World War II veterans to open up about their experiences in books like The Band of Brothers. In the same way Ben Evans has been able to get the normally tight-lipped cadre of astronauts to open up about their personal experiences in the high risk, high stress environment flying the Shuttle. Their excitement, frustrations and fears. And our fears were well founded. The author describes the many close calls – red flags – that exposed the vulnerabilities and razor thin margins that foretold the Challenger and Columbia accidents. Much of what you will learn has only been available to those who lived the experience – passed on by word of mouth or buried in obscure government reports. But Ben ties it all together in a narrative that is fascinating, thought provoking and informative. I was surprised at how much I learned and I lived it.

Experimental Flying Machine should be a required read for every engineering and management student. It is not an engineering book, but it explores the decisions engineers and managers must make and exposes the consequence of those decisions including their human impact. Ben Evans presents a perspective on the Space Shuttle that is far different than the finely honed narrative you have heard from NASA Public Affairs. It is a book I could not put down.

Sidney M. Gutierrez (Colonel, U.S. Air Force, Ret.)
NASA Astronaut 1984-1994
Pilot, STS-40 (June 1991)
Commander, STS-59 (April 1994)

Acknowledgments

This book would not have been possible without the support of a number of individuals, to whom I am enormously indebted. I must first thank my wife, Michelle, for her love, support and encouragement during the time it has taken to plan, research and write the manuscript. As always, she has been uncomplaining during the weekends and holidays when I sat up late, typing on the laptop or poring through piles of books, old newspaper cuttings, magazines, interview transcripts, press kits or websites. It is to her, with my love, that *Experimental Flying Machine* is dedicated. My thanks also go to Clive Horwood of Praxis for his enthusiastic support, to David M. Harland for reviewing the manuscript and offering a wealth of advice and guidance, to Jim Hillhouse of AmericaSpace.com for a decade of friendship, and to former Space Shuttle commander Sid Gutierrez for kindly contributing the Foreword. I deeply appreciate not only their support, but also their patience in what has been an overdue project and one which proved far more difficult to write than I had ever imagined. Additional thanks go to Ed Hengeveld, who has been gracious with his time in identifying and supplying suitable illustrations for this book, including many 'unfamiliar' ones which will hopefully bolster the text. Others to whom I owe a debt of gratitude are my parents, Marilyn and Tim Evans, and to Sandie Dearn and Malcolm and Helen Chawner. To those friends who have encouraged my fascination with all things 'space' over the years, many thanks: to the late and much-missed Andy Salmon, to Andy Rowlands and Dave Evetts and Mike Bryce of GoSpaceWatch, and to Rob and Jill Wood. Our golden retriever, Milly, also provided a ready source of light relief as she seized any available opportunity to drag me away from the laptop and play with her or take her for a walk.

Author's Preface

Astronaut Steve Hawley once remarked that the Space Shuttle was a feat of such technical enormity that simply launching it was nothing less than a minor miracle. It required thousands of people to work together to ensure that millions of discrete mechanical and electronic parts operated perfectly and in tandem. Several hundred of those parts, indeed, were so critical to the safety of the crew and the spacecraft that their single-point failure could spell disaster. The Shuttle required a winged vehicle, awkwardly bolted onto an immense fuel tank and a pair of boosters, to launch like a rocket, fly like a spacecraft, then descend back to Earth like an aircraft to a smooth runway landing…and then, after a period of refurbishment, repeat the achievement again and again. It required perfect weather conditions at its launch site and at several emergency landing sites around the world. It required precise control over huge amounts of volatile propellants and a computing power smaller than can be found in one of today's mobile phones. And with survivability and emergency escape provisions which many astronauts found questionable at best, it required steely-eyed bravery on the part of the men and women who dared to fly this most experimental of experimental flying machines.

I grew up with the Space Shuttle. It made its first flight over the California desert when I was a few months old, it first launched into space as I attended nursery, it triumphantly repaired the Hubble Space Telescope as I finished high school and it began building the International Space Station when I graduated from university. Over its 30 years of active operational service – first launched on 12 April 1981 and last landed on 21 July 2011 – this fleet of five winged orbiters flew an impressive 135 times. Fleet leader Discovery completed 39 missions, Atlantis made 33, Columbia made 28, Endeavour made 25 and Challenger made ten. Crews as small as two and as large as eight launched and repaired satellites, carried out scientific research and built and maintained space stations. Across that bundle of years, the Shuttle achieved unimaginable dreams and opened space exploration to more walks of life than ever before. The descendants of its technology continue to live and breathe as America prepares for its next great step into deep space. Modified versions of its engines, its boosters and its pressure suits will power the next generation of human explorers to the Moon and back.

But the Shuttle was from the outset, and remained to the very end, a highly dangerous vehicle to fly. Even late in life, with substantial safety improvements having been made, the risk of a launch disaster was predicted to be one-in-500; by the end of the program, that figure was refined to no better than around one in a hundred. In its original form, the Shuttle was intended to be the spacefaring equivalent of a commercial airliner, capable of carrying passengers and payloads into space regularly, reliably and cheaply and flying dozens of times each year. But as the vehicle's design morphed in response to external political and military pressures, it grew gradually more complex. When it eventually flew, it required far too much attention after each mission to ever come close to achieving those early dreams. And in any case, few people intimately involved in the program ever considered its politically mandated goals to be realistic.

The Shuttle's fallibility was tragically exposed on two occasions. In January 1986, a failure of one of its boosters resulted in the deaths of seven astronauts. And in February 2003, a failure of its fuel tank and critical damage to its thermal protection system claimed another seven lives. Both failures traced their cause to human as well as technical shortfalls. And betwixt those two dates, numerous other missions came within a hair's breadth of disaster: from aborted launch attempts on the pad to in-flight engine failures and from heat shields suffering severe damage to maddening computer malfunctions.

This book was written on the eve of the 40th anniversary of the first Shuttle flight, by which time the surviving members of the fleet had been retired for almost a decade. It offers a glimpse at the inherent dangers of this reusable spacecraft, whilst at the same time offering the reader a perspective of some of its successes. Of course, to trace every single problem faced by the Shuttle – every single weather delay, every time a Reaction Control System (RCS) thruster failed, every time a computer conked out, every time the toilet broke – would require a book far larger than this, and as such the troubles described in these pages are far from exhaustive. But *Experimental Flying Machine* seeks to offer a snapshot of some of the Shuttle's most visible technical and human challenges over its three astonishing decades of service.

Contents

1

A Troubled Childhood

FIRST MOTIONS

As clocks across Florida struck eight on the last Monday of 1980, a new era began. In the early morning gloom, the doors of the Vehicle Assembly Building (VAB) – a 53-story sugar-cube of a structure which stood dominant over the marshy flatness of the Kennedy Space Center (KSC) on Merritt Island – clanked ajar to reveal a spacecraft like no other. It shimmered in the rays of the newly risen sun as the warming fingers of dawn gently caressed its flight surfaces and picked out its textures and contours in stark relief. The first Space Shuttle, named 'Columbia' in honor of the female personification of the United States, was on the move, bound for the launch pad. And after a troubled decade in development, the greatest experimental flying machine in history was about to spread its wings.

Clearing the confines of the VAB and emerging into the salty Florida air, this 56-meter behemoth cannot have failed to impress those who were there to see it on 29 December 1980. It surely caused many a jaw to drop in astonishment. A tight knot of engineers and managers attired in suits and ties, shirts and jeans, watched agape, some with hands stuffed into pockets, others conversing in hushed tones as Columbia made her creaking departure. When they saw her next in this building, she would be a 'used' spacecraft, having circled the globe 36 times, spent two days in orbit and deep into processing for another mission. Indeed, the central tenet of the Shuttle concept was that its fleet of winged 'orbiters' (of which Columbia was one) were fully reusable, capable of lifting more people into space more frequently than ever before and drawing down the immense cost of doing so.

Yet cost was not the Shuttle's only immense feature; so too was the spacecraft itself. Everything about it shrieked 'huge'. And whilst we dwellers of the 21st century may scratch our heads over the physical dimensions of this Great Pyramid of

© Springer Nature Switzerland AG 2021
B. Evans, *The Space Shuttle: An Experimental Flying Machine*,
Springer Praxis Books, https://doi.org/10.1007/978-3-030-70777-4_1

our age, it is difficult to grasp in a few sentences its sheer scale and monumentality. Even the tracked 'crawlers' which hauled the 2-million-kilogram Shuttle 'stack' along a roadway of Alabama and Tennessee river-gravels from the VAB to one of two pads at KSC's historic Launch Complex 39 – a distance of some 5.6 kilometers – were a marvel of industrial engineering. To this day, they remain the largest self-powered land vehicles in the world, tipping the scales at 2.7 million kilograms. With a Mobile Launch Platform (MLP) and fully laden Shuttle atop them, the crawlers lumbered along at a glacial pace of just 1.6 kilometers per hour.

Unsurprisingly, not all went according to plan and between October 1983 and December 2010 a total of 18 missions were not only 'rolled out' to the launch pad, but also 'rolled back' to the VAB. The finger of blame often pointed squarely at Florida's intractable weather, with the Shuttle ordered off the pad and back to the safety of the assembly building in response to the ravages of Tropical Storm Klaus in October 1990, Hurricane Erin in August 1995 and Hurricanes Bertha and Fran in July and September 1996. On one occasion in August 2006, threatened by severe approaching weather from Tropical Storm Ernesto, one Shuttle was rolled off the pad and was partway back to the VAB when conditions began to improve. The rollback was halted on the crawler roadway and the stack returned to the pad. But most problems which enforced rollbacks over the years were in response to technical difficulties not easily accessible or fixable in a vertical configuration at the pad: a suspect booster nozzle, the replacement of engines following a pair of aborted launch attempts, an issue with a primary payload, hydrogen leaks, cracked fuel-line door hinges and even – as will be discussed in Chapter 4 – attacks by hailstones and woodpeckers. Nor was the MLP and crawler hardware itself immune to difficulties. During one rollout in January 1997, a crack 7.3 meters in length was spotted on the MLP deck, whilst faulty bearings in the steering linkage on one of the crawler's four trucks and hydrogen leaks conspired against other missions. These problems provided a stark and constant reminder that nothing about this endeavor was routine.

As Columbia moved, the onlookers continued to watch and chat. A helicopter fluttered overhead, filming the proceedings for posterity. Technicians in hardhats, wearing headsets and clutching walkie-talkies, plodded alongside the crawler's huge tracks, their eyes and ears attuned to the slightest hint of damage or structural wear. Minutes ticked into hours as the morning chill gave way to a fine Florida afternoon. Eventually, the "big bird" (as Shuttle commander Jack Lousma once described it) reached its destination and set about navigating the upward slope to the concrete surface of Pad 39A. This was hallowed ground in America's space program, having already been added to the National Register of Historic Places (NRHP) in 1973. A decade before the arrival of Columbia, it had seen off most of the mighty Saturn V rockets on their Apollo expeditions to the Moon and it was also from here that America's first space station, Skylab, speared into orbit. Steeped in history, Pad 39A now stood proud for a new and very different chapter in its life.

Fig. 1.1 Displaying its unusual "butterfly-and-bullet" configuration, Columbia reaches Pad 39A atop the Mobile Launch Platform (MLP) and crawler on 29 December 1980. Clearly visible are the white-painted External Tank (ET) and twin Solid Rocket Boosters (SRBs).

Sophisticated levelling machines kept the Shuttle perfectly upright as it inched its way up the incline. And having set its precious cargo securely onto the concrete pedestals of the pad surface at three in the afternoon, the crawler withdrew. It left behind an ungainly machine utterly at odds with how a rocket 'should' look. For unlike all rockets before it, the Shuttle was no pencil. Two reusable Solid Rocket Boosters (SRBs), each 45.5 meters tall, flanked the stack like a pair of great Roman candles, providing the lion's share of the thrust (about 80 percent, or 2.5 million kilograms) at liftoff. A bulbous External Tank (ET), measuring 47 meters long and discarded after each flight, carried the liquid oxygen and hydrogen propellants to feed the Shuttle's cluster of three main engines which delivered a total of 535,000 kilograms of thrust. As for the orbiter itself, in size and shape it was not dissimilar to a DC-3 jet airliner: 37.2 meters long and spanning 27.8 meters between the tips of its delta-shaped wings. These constituent parts forged a flying machine which astronaut Story Musgrave likened to a butterfly that was awkwardly bolted onto a bullet.

"The Shuttle is an asymmetric vehicle," remembered Neil Hutchinson, ascent flight director in Mission Control for Columbia's maiden voyage. "It doesn't look like it ought to launch right because it's not a pencil. Some of us in the early days wondered how that was going to work, not being an aerodynamicist. In fact, the Shuttle…it's a very tricky vehicle to launch. It has to be pointed carefully in the right direction at certain times or you'll tear the wings off or tear it off the External Tank. It is not a casual launch process." The Shuttle, therefore, was highly problematic in terms of its aerodynamic behavior and although hundreds of engineering tests and launch abort simulations and computational fluid dynamics models were conducted before its first flight, the unknowns remained. "I'm not sure that we have risk-takers in NASA these days," Hutchinson added, "that would take that kind of risk."

A REUSABLE SPACESHIP

For three decades, from April 1981 to July 2011, five Shuttle orbiters – Columbia, Challenger, Discovery, Atlantis and Endeavour – launched 135 times, traveled 873 million kilometers, spent 1,323 days circling the Earth a total of 21,030 times, and carried 357 individual men and women from 16 sovereign nations. More than two-thirds of that number launched and landed on the Shuttle at least twice, with a handful doing so as many as six and even seven times. All told, Discovery flew 39 times, Atlantis 33 times, Columbia 28 times, Endeavour 25 times and Challenger ten times. Across that bundle of years, the Shuttle attained the loftiest of heights and achieved the most unimaginable of dreams: launching and retrieving satellites, performing cutting-edge research across multiple scientific disciplines, revealing the home planet in unprecedented detail and building and restocking the International Space Station (ISS). But it did so at the cost of two appalling human tragedies and many other brushes with misfortune and near-disaster. Right from the start, in fact, the Shuttle was a direct consequence of political and military compromise.

Indeed, Richard Nixon harbored no innate love of space. He entered office as the 37th U.S. president in January 1969 with a full plate of political priorities upon which America's space aspirations held comparatively little sway. Ending an unpopular war in Vietnam and solving problems of civil unrest, student protests and racial division at home were in no shape or form dependent upon a multibillion-dollar space program, which Nixon felt contributed little to the average man or woman in the street. At its genesis, the stated intent of Project Apollo – the national drive to land a man on the Moon – was purely to beat the Soviet Union. Nixon's predecessors in the White House, John Kennedy and Lyndon Johnson, had pushed through this aggressive goal to demonstrate technological and ideological might over a Cold War foe. And in July 1969, when Neil Armstrong and Buzz Aldrin walked on the Moon's Sea of Tranquility, Nixon had little interest in further space spectaculars.

Fig. 1.2 The sheer size of the crawler is visible as it withdraws in June 2011, after depositing STS-135, the last Space Shuttle, on the launch pad.

But he did acquiesce that America needed a future in space. In February 1969, the Space Task Group (STG) convened within the remit of the National Aeronautics and Space Council (NASC). Chaired by U.S. vice president Spiro Agnew, its mandate was to chart the United States' space-flying course after Apollo. The group proposed four options to the White House: an advanced base on the Moon, a manned voyage to Mars, an Earth-circling space station or a reusable winged 'shuttle' which could visit space more reliably, frequently and cheaply than ever before. Only the latter piqued Nixon's interest. A machine capable of launching and landing again and again, like an airliner, and with similar levels of frequency and low cost, might give it the chance to compete with (and perhaps replace) the United States' fleet of eye-wateringly-expensive disposable rockets. The Space Shuttle was glowingly endorsed by the American Institute of Aeronautics and

Astronautics (AIAA) and the President's Science Advisory Committee (PSAC), which pointed out that its "early goal of replacing all existing launch vehicles" not only promised to deploy and recover satellites and build a space station, but also achieve a "radical reduction in unit-cost of space transportation".

Ideas for a reusable winged spacecraft, even at this historical juncture, were far from new. As early as the 1930s, the German aerospace engineer Eugen Sänger conceptualized a rocket-propelled aircraft capable of velocities over ten times the speed of sound and altitudes as high as 70 kilometers. He found that adding wings enhanced the potential of such machines and the 'lift' thereby generated during re-entry allowed them to 'skip' off the atmosphere, adding the capability to circle the globe and land back at their launch site. Within a few years, dreams morphed into reality when Chuck Yeager piloted the rocket-powered Bell X-1 'Glamorous Glennis' aircraft through the sound barrier in level flight in October 1947. At those velocities, aerodynamic heating was not yet a substantial obstacle, but as the U.S. military focused upon the practicalities of sending heavy warheads across intercontinental distances, speed and stability attained new heights of importance. In 1952, the National Advisory Committee for Aeronautics (NACA) – forerunner of today's National Aeronautics and Space Administration (NASA) – began work on an aircraft capable of exceeding Mach 5 (the generally recognized lowermost velocity for 'hypersonic' flight) and set about considering how to achieve enhanced stability and better thermal protection.

If the aircraft re-entered the atmosphere with its nose oriented in the direction of flight, its streamlined shape could induce catastrophic overheating and destructive aerodynamic loads. But re-entering with the nose positioned at a slightly higher angle-of-attack (and with its flat belly presented to the hypersonic airflow) offered a more manageable approach, permitting it to bleed off speed in the rarefied high atmosphere, minimizing overheating and reducing aerodynamic stress. Yet temperatures remained far more extreme than anything previously experienced in aviation. Bell was already working on a chrome-nickel alloy called 'Inconel-X', which, when combined with stainless steel 'shingles', could radiate heat away from the airframe in conjunction with water-cooling at the leading edges of the wings.

In 1954, the Aircraft Panel of the Scientific Advisory Board advocated the field of hypersonic flows to be a principal research and development goal for the next decade, telling U.S. Air Force chief of staff Nathan Twining that "much of the necessary physical knowledge still remains unknown at present" and cautioning that "an ingenious and clever application of existing laws of mechanics is probably not adequate". The board considered the time to be ripe for a new aircraft to surpass Mach 5 and reach altitudes up to 150 kilometers. The result was the single-seat North American X-15, which flew 199 times out of Edwards Air Force Base in California in 1959–1968, reaching the highest speed ever recorded by a crewed, powered aircraft of 7,274 kilometers per hour (almost Mach 5.9) and an altitude of 107.8 kilometers. Two of its missions exceeded 100 kilometers, surpassing the 'Kármán Line' which is accepted by the Fédération Aéronautique Internationale (FAI) as the edge of space. Not only did it push velocity and altitude to new heights, the X-15 also trialed a throttleable rocket engine (the XLR-99), which laid the groundwork for the Shuttle's main engines.

Elsewhere, significant strides were being made in the adjunct field of manned lifting bodies and in 1963–1967 the U.S. Air Force flew a series of small unmanned hypersonic gliders – the Aerothermodynamic/elastic Structural Systems Environment Test (ASSET) and Precision Recovery Including Maneuverable Entry (PRIME) – at speeds of up to 25,000 kilometers per hour and achieved a cross-range capability of over 1,000 kilometers. Neither ASSET or PRIME could land on runways and instead parachuted back to Earth, but they showcased supreme maneuverability and could endure the furnace-like temperatures of atmospheric re-entry at hypersonic velocities. Their findings paved the way for the M2-F1, M2-F2, HL-10 and X-24 lifting bodies, flown out of Edwards in 1963–1975. These lifting bodies proved that human pilots (including future Shuttle commander Dick Scobee) could land wingless vehicles whose very shape provided the same aerodynamic lift as wings.

Even as Agnew and the STG labored on a 'road map' for space exploration, NASA was considering an "integral launch and re-entry vehicle" and initiated a four-phase solicitation process for U.S. industry to analyze, define, design, produce and operate a new spacecraft after Apollo. Study contracts were awarded to Lockheed, General Dynamics/Convair, McDonnell Douglas and North American Rockwell in February 1969 for what became 'Phase A' of the Shuttle. STG board member Robert Seamans (a former NASA deputy administrator and since January 1969 the secretary of the Air Force) noted that this attracted "considerable military interest".

It was an interest which grew (for good or ill) to define the size, shape and capabilities of the new vehicle. Concurrently, NASA formed a Space Station Task Group (SSTG) in May 1969, led by the agency's associate administrator for spaceflight George Mueller, which raised the Shuttle's payload-to-orbit capability from 11,300 kilograms to 22,600 kilograms to suit growing calls for it to be a 'truck' for a future space station and a satellite launcher. Its 'payload bay' was set at 6.7 meters in length and discussions with the Department of Defense led to an August 1969 decision that it would not use expendable boosters, but would instead be a fully reusable, two-stage system. Although 'partial' reusability was considered an appropriate means of minimizing development costs for a space truck, it was not anticipated to work as well if the Shuttle were to take on a broader role with a correspondingly higher flight rate. And whilst full reusability would increase development costs, like an airliner it promised lower operational running costs in the longer term.

Fig. 1.3 The eventual Shuttle system adopted a 'parallel-burn' architecture, with the three main engines (seen here in the process of ignition on STS-51) and twin Solid Rocket Boosters (SRBs) both started and verified as healthy whilst on the launch pad.

This multi-purpose role for the Shuttle, as the SSTG told the STG in June 1969, now encompassed several mission types. As well as supporting a future space station, it could also deploy, retrieve, refuel and repair satellites and perform research. NASA recognized that building a vehicle flexible enough for these applications (whilst remaining sufficiently economical to fly) put serious demands on a fully reusable system. Potential 'trades' included piloted flyback boosters, off-the-shelf engines, a vertical rather than horizontal takeoff profile and igniting the engines 'sequentially' during ascent, rather than in 'parallel' on the ground. Very soon, however, the Air Force weighed in with its own demands for a cross-range capability of 2,780 kilometers, enabling the Shuttle to rapidly return to secure military airfields. "The military was interested in a great cross-range," said Caldwell Johnson, chief of NASA's spacecraft design division at the time, "so they could land where they wanted to in one orbit at any given time and get back." It also wanted a cargo capacity of 29,500 kilograms and payload bay length of 18.2 meters to accommodate its large reconnaissance/intelligence satellites. This was significantly greater than the 6,800-kilogram cargo capacity that NASA desired.

"What the Air Force had in mind for the Shuttle and what NASA had in mind for the Shuttle were two different animals," said former North American Rockwell chief program engineer Alan Kehlet. "One was an elephant and the other one was a gigantic elephant. The Air Force wanted a big payload bay. NASA wanted a small vehicle to refuel the space station and the two of them can't meet. You had these conflicting requirements which determined the size of the vehicle and the booster."

But the Shuttle was in for a rude awakening. In September 1969, Agnew's STG submitted its report to Nixon. It pointed out that the reusable spacecraft offered substantial improvements over NASA's current way of doing business through reduced costs and higher operational flexibility, as well as supporting "a broad spectrum" of missions. The STG offered three long-range plans for America. The first, costing $10 billion per year, envisaged a space station in Earth orbit (serviced by the Shuttle), plus a lunar-orbiting complex and manned voyage to Mars. A second, at $8 billion per year, deleted the lunar complex. And a third at $5 billion per year kept the space station and the Shuttle.

Nixon rejected all three.

WINNING THE PRESIDENT'S EAR

NASA now sat in the unenviable position of having to build political support to get the Shuttle approved. In October 1969, the Phase A industrial teams submitted their orbiter/booster concepts. North American Rockwell (which ultimately won the contract to build the Shuttle) proposed a straight-winged orbiter, 61.5 meters long with a 44.5-meter wingspan, mounted slightly forward of the booster. In keeping with Air Force requirements, its payload bay was 18.2 meters long and 4.5 meters wide but could still only lift 6,800 kilograms. Powered by a pair of boost engines and four turbojet engines, it carried two pilots and up to ten passengers. The booster was 85.3 meters long with a 73.4-meter wingspan. The orbiter/booster would take off vertically and execute a 90-degree roll at high altitude, before parting company at 70 kilometers. The orbiter would then propel itself firstly into a 185-kilometer 'parking' orbit and then a 500-kilometer 'phasing' orbit. Meanwhile, the booster would land on a runway, as would the orbiter at the end of its mission. North American Rockwell expected to build six vehicles and fly 50 missions per year.

In May 1970, the four companies were winnowed down to two, with North American Rockwell and McDonnell Douglas selected for definition and preliminary design studies in 'Phase B'. By this stage, the Shuttle's maiden voyage was scheduled for 1977 with an expectation that two orbiters would be built. The first was a straight-winged vehicle with a low cross-range of 370 kilometers, the other a delta-winged vehicle with a high cross-range of 2,780 kilometers. Inclusion of a delta-wing in the design of the high-cross-range orbiter afforded it better aerodynamic lift over the broad Mach range as it decelerated through hypersonic and subsonic flight regimes. Added to the complexity was the need to use different control modes to manage the vehicle through conditions of minimal atmospheric drag into full aerodynamic flight with the support of rudders and ailerons and from orbital velocities of 28,200 kilometers per hour to land at about 370 kilometers per hour. "You had to transition your control systems as they became functional and transition 'out' the ones that were no longer functional," said NASA aerospace engineer Emery Smith. "The guidance had to make sure you were targeted so you didn't burn the vehicle up. We had contractors all over the country at that time working on the same problem."

Under the terms of the design requirements, each orbiter could remain in space for a week at a time and fly 25–75 missions per year, with landing-to-launch turn-around times of only 14 days. But the payload capacity of 6,800 kilograms still fell far short of Air Force requirements. Moreover, the booster – which would lift the orbiter not only to 70 kilometers but also to a speed of 11,200 kilometers per hour at 'staging' – would be the biggest, heaviest and fastest aircraft ever made. Twenty-five percent larger than a Boeing 747 and ten times faster, it would weigh 1.4 million kilograms when fully fueled. And although clever design of its cryogenic tanks would permit them to carry some of these weights and aerodynamic loads, there remained an acute risk of stress-related fractures, leaks and a potentially catastrophic build-up of gaseous hydrogen under its skin during re-entry.

However, a multitude of other troubles faced NASA when it became obvious that the agency's funding outlook through the mid-1970s was effectively flat. The

Fig. 1.4 The Shuttle's delta-shaped wing, pictured here as Discovery stood poised to launch STS-96 in May 1999, permitted greater aerodynamic 'lift' over a much broader Mach-range during the passage from the hypersonic through subsonic flight regimes.

picture for Fiscal Year 1971 was barely $3.2 billion and the Office and Management and Budget (OMB) informed NASA that this was unlikely to change for five years. The result was that there would be sufficient funding to build a reusable orbiter, but not a reusable booster. Coupled with internal analyses that showed a fully reusable Shuttle was simply not competitive against expendable rockets, NASA's emphasis shifted once again to a partially reusable design. In June 1971, an earlier idea of the External Tank now re-entered consideration. This called for the liquid oxygen and hydrogen propellants for the Space Shuttle Main Engines (SSMEs) to be shifted outside the body of the orbiter and into a disposable piece of hardware.

Despite the estimated $740,000 cost of building a new tank for each mission, the design eliminated the need to refurbish the booster's thermal protection system and allowed the orbiter to be smaller and lighter, with a substantial reduction in its development cost. North American Rockwell's orbiter correspondingly shrank from 61.5 meters in Phase A to 58.5 meters in Phase B and eventually came in at 37.2 meters. Since the ET would not be reused, its own need for thermal protection was minimal. And the overall cost of producing tank after tank for mission after mission would gradually decline as flight rates increased and manufacturing efficiencies matured. The number of SSMEs on the Shuttle itself increased from two to three for added margins of safety. This reduced the risk of an outright 50-percent power loss in the event of a single engine failure. Furthermore, because the SSMEs could be throttled up in an emergency, three engines afforded greater flexibility in abort situations. NASA now began to consider a 'phased' approach, whereby the reusable orbiter would initially be tested alongside an interim expendable booster, with and expectation that "full-scale hardware development of a reusable booster" would commence later.

"The preferred configuration, which is emerging from the studies is a two-stage, delta-wing reusable system in which the orbiter has external propellant tanks that can be jettisoned," NASA administrator James Fletcher explained in June 1971. "Although our studies to date have mostly been based on a 'concurrent approach', in which development and testing of both the orbiter and the booster stages would proceed at the same time, we have been studying in parallel the idea of sequencing the development, test and verification of critical new technology features of the system. We now believe that a 'phased approach' is feasible and may offer significant advantages."

In July, four industry teams – North American Rockwell, McDonnell Douglas, Grumman Aerospace and Lockheed, together with Martin Marietta, General Dynamics and Boeing as major subcontractors – were selected to examine this phased approach. The boosters under consideration included the S-IC first stage of the Saturn V, an 'outgrowth' of the Titan III rocket, a single solid-fueled motor with a diameter of 6.6 meters or a cluster of solids, each with a diameter of 3–4 meters. Liquid-fueled boosters were flexible and 'throttleable' during flight, whereas solids delivered a harsher impulse and could not be turned off once ignited but were simpler to design and less risky to develop. Contracts were later extended to April 1972 as emphasis gravitated towards building the entire Shuttle in parallel. And hopes for a fully-reusable, two-stage vehicle vanished.

As the development process moved into 'Phase B Prime' in October 1971, the decision of exactly what type of booster to use remained an open question. So too did the issue of whether the Shuttle would adopt a 'sequential-staging burn' (in which the booster would ignite on the ground and the orbiter's engines would ignite in flight) or a 'parallel-burn' (with ignition of all engines on the ground). Advantages of the latter allowed the SSMEs to be verified as 'startable' and healthy before the stack went airborne. As a result, the Thrust Assisted Orbiter Shuttle (TAOS) concept had gained broad acceptance by the end of 1971. Its orbiter/ET combination would function in parallel from liftoff through to insertion into orbit, with the boosters – whether reusable or otherwise – providing added thrust for the first two minutes of ascent, prior to being jettisoned. The parallel-burn TAOS architecture was priced at approximately $6 billion over six years.

Fig. 1.5 From June 1971, the propellants for the Shuttle's three main engines were shifted outside the orbiter and into the External Tank (ET), the rust-colored base of which can be seen during installation onto the twin Solid Rocket Boosters (SRBs) in the Vehicle Assembly Building (VAB).

Still, it remained a problematic solution. "Before the Space Shuttle, we would have generally built launch vehicles…stacked like telephone poles; they were all in line," said former Space Shuttle program office manager Bob Thompson. "You've got a good thrust vector that goes right up the center of the backbone of the vehicle. You can drop off parts of the vehicle as you 'stage'. People wanted to put the booster rockets behind the orbiter and push it along the axis, but they weren't very practical if you wanted to burn the orbiter engines at liftoff. We very much wanted to turn those engines on before we lifted off to make sure they were working, so we wanted the orbiter down in the 'firepit'…where we could light off those engines." That requirement prompted additional decisions to use the ET itself as part of the structural backbone of the stack, onto which the boosters could be mounted, and played an integral role in creating its peculiar butterfly-and-bullet appearance.

But to gain Nixon's approval, Fletcher had to make concerted overtures to the Department of Defense to commit to using the Shuttle for all its launch needs; an estimated one-third of all future space traffic. This made it imperative that the spacecraft be able to fully accommodate the military requirements. The often-touted payload bay length of 18.2 meters and the capacity to put 29,500 kilograms into low-Earth orbit or 18,150 kilograms into polar orbit were non-negotiable. Moreover, the Air Force wanted to reach polar inclinations from Vandenberg Air Force Base in California to reach and service its classified satellites, then return to Earth after a single 90-minute orbit. On such missions, the landing site would appear to 'move' eastwards as the Earth rotated, necessitating a cross-range capability of up to 2,780 kilometers. This was greater than NASA's desire to land back at its launch site after 24 hours. Still, there were some corollary benefits for NASA in terms of launch aborts, including the ability to land the Shuttle at a downrange site or return home quickly after a single orbit.

As early as September 1970, NASA increased the Shuttle's payload capacity from 6,800 kilograms to 11,340 kilograms, although even this still represented less than half of the Air Force's requirement. The size of the payload bay remained critically important and in mid-1971 the Air Force's assistant secretary for research and development Grant Hansen told NASA's associate administrator for space-flight Dale Myers that anything less than 18.2 meters meant that nearly half of all military cargoes would not fit the Shuttle. It is unsurprising that the Air Force's lack of total confidence prompted it to declare that it would continue developing its own Titan and Atlas expendable rockets. Its position softened in late 1971 with an agreement to continue purchasing existing designs alongside the Shuttle.

By New Year 1972, analysis fell in favor of two payload bay sizes. NASA refined its requirement to 13.7 meters and 18,150 kilograms, but the Air Force remained unflinching in its insistence for 18.2 meters and 29,500 kilograms. On 5 January, at the Western White House on California's picturesque San Clemente coast, Fletcher and NASA deputy administrator George Low presented Nixon with a model of the TAOS design and the president was visibly fascinated. He liked the fact that it would carry ordinary people into space, but other factors prevailed equally on his mind. Fletcher pledged that starting the Shuttle in 1972 would generate direct employment on the order of 8,800 people by the end of the year and 24,000 by December 1973. And as Project Apollo wound down and the Soviet Union's manned space program appeared on the ascendancy, the prospect of America having no capability to put people into space was unconscionable. Nixon therefore formally instructed NASA to "proceed at once" with building the Shuttle. He cared not a jot about whether the orbiter had a 13.7-meter payload bay or an 18.2-meter one, nor about its cargo-lifting credentials. His principal concern was that the United States retained a capacity to put people into space and that NASA did not exceed the fiscal ceiling of approximately $5 billion mandated by the OMB.

COMPETING PRIORITIES

With Nixon's approval in hand, decisions could at last be made about what form the Shuttle would physically take. Aluminum was selected for its airframe, based upon Air Force preferences and the fact that very few aerospace contractors possessed the requisite expertise with handling titanium. But this choice also made it likely that a complex and expensive patchwork of silica tiles would be needed for its Thermal Protection System (TPS) to protect it from the searing temperatures of hypersonic re-entry. A parallel-burn architecture was selected, in which all engines – those of the orbiter and the booster – would be ignited on the ground before liftoff. And the nature of that booster had also shifted definitively in favor of solids. Boeing's plan to modify its S-IC first stage could not be done cheaply enough to fit within OMB guidelines and, in any case, solids promised significant cost savings over the others. NASA's budgetary outlook raised the stature of solids even further. Their cost savings would allow any 'difference' to be kept in reserve for unexpected development problems. The decision to adopt solids came on 15 March 1972. Fletcher announced that they would have a diameter of 4 meters and could be made faster and $700 million cheaper than liquid-fueled boosters, bringing the Shuttle's overall cost down from $5.5 billion to $5.15 billion. It was noted that the boosters would be fully reusable, detaching from the stack at an altitude of 45.7 kilometers and parachuting to an oceanic splashdown, after which the orbiter would continue into space under the power of its SSMEs. Unsurprisingly, NASA agreed to meet all Air Force requirements. The payload bay would be 18.2 meters long and 4.5 meters wide and capable of lifting 29,500 kilograms into a 185-kilometer 'due-east' orbit.

Fig. 1.6 The sheer enormity of the External Tank (ET) and twin Solid Rocket Boosters (SRBs) is illustrated in this view of the STS-110 crew posing in front of 'their' vehicle in early 2002. The attachment points linking the boosters to the tank are readily apparent.

Two days later, on 17 March, NASA released its request for proposals to North American Rockwell, McDonnell Douglas, Grumman Aerospace and Lockheed, together with their major subcontractors Martin Marietta and Boeing. This 'Phase C/D' contractual element required each orbiter to have a 'useful' lifetime of ten years and a capacity to fly a hundred missions before major refurbishment. The Shuttle had to be able to return to its launch site after a single orbit, although its cross-range capability was (initially) left unspecified. North American Rockwell's revised design for the orbiter was 38 meters long with a wingspan of 24.3 meters. A bubble-like canopy over its flight deck improved the astronauts' visibility of the payload bay and its landing gear retracted into 'wheel wells' in the wings, rather than the fuselage.

On either side of the 'pyramid' of SSMEs were a pair of Orbital Maneuvering System (OMS) engines for use in space and two Abort Solid Rocket Motors (ASRMs) for use in the first 30 seconds of ascent to facilitate a 'meaningful' crew-escape capability in an emergency. Air-breathing engines were housed in the Shuttle's rear payload bay, with an air intake just underneath the vertical stabilizer. Both the air-breathing engines and the ASRMs were subsequently deleted to recover Air Force payload requirements. Additionally, the ASRM elimination saved $300 million and their usefulness, in any case, would only have been effective in the opening few seconds of a mission. The orbiter's wheels, brakes and tires were drawn in design from the B-1A Lancer bomber and a 'drag chute' to assist with deceleration on the runway came from the heritage of the B-52 Stratofortress. (Ironically, the drag chute, too, was deleted, although it would be reintroduced later in the Shuttle era.) North American Rockwell's concept for the ET envisaged a cylindrical structure, 64 meters long and 10 meters in diameter, with a retro-rocket 'pod' at its tip. Two solid-fueled boosters – equipped with fins for greater stability after staging – sat 18.3 meters aft of the ET's nose, their nozzles exhausting behind the trailing edges of the Shuttle's wings.

The proposals for the Shuttle were submitted by May and on 9 August 1972 NASA awarded North American Rockwell a $2.6 billion letter contract to begin developing two Shuttles. One of them (originally intended to be called 'Constitution' but eventually named 'Enterprise') would make a series of Approach and Landing Tests (ALTs) in the low atmosphere, before being modified for space missions, whilst the other ('Columbia') would be built for orbital flights from the outset. Of the four teams, North American Rockwell's proposal gained the highest scores in terms of mission suitability and the most lightweight design. NASA's Source Evaluation Board paid glowing tribute to its guidance, navigation and control architecture, which it regarded as clean and simplistic with minimal interfaces. Additionally, the company had excellent analyses of maintainability and turnaround, as well the lowest costs. Two space-rated orbiters, a full-scale Structural Test Article (STA) and a Main Propulsion Test Article (MPTA) were incorporated into the contract. And the Shuttle would fulfill the final round of Air Force needs with a cross-range capability of 2,035 kilometers. Frustratingly, this broad cross-range went largely unused. "Cussed a lot," remembered Bruce Jackson, former chief of NASA's engineering analysis division. "Their requirement was to have a certain cross-range and they never used it. The military requirement is what dictated the Shuttle configuration, not NASA's requirement. Configuration looks as it does simply because of the long cross-range requirement the Air Force placed on the configuration. That was a big expense."

From the outset, the contract required 50 percent in dollar-value to be subcontracted to other U.S. companies. In November 1972, North American Rockwell issued a solicitation for proposals to design and fabricate the spacecraft's wings, mid-fuselage and vertical stabilizer. Unfortunately, the Shuttle contract award came under immediate and sustained fire, not least because North American Rockwell was headquartered at Downey, in Nixon's home state of California (with its 55-vote monopoly in the Electoral College), and five members of its board had contributed thousands of dollars to his 1972 presidential re-election campaign. Jean Westwood, chair of the Democratic National Committee (DNC), harshly criticized the president's "calculated use of the American taxpayers' dollars for his own pre-election purposes". Moreover, NASA's Dale Myers had spent much of his early career with the company and hand-picked the members of the Source Evaluation Board. Thus began an ugly period of recrimination and not until April 1973 did 'Rockwell International Corporation' – the result of North American Rockwell's merger with Rockwell Manufacturing – sign the definitive Shuttle production contract with NASA.

Nevertheless, there were solid engineering reasons for not selecting the others. Lockheed's craft, for example, was considered heavy, "unnecessarily complex" (according to Fletcher) and left a minute-long duration in the ascent phase with no provision for an abort. In retrospect, this proved an ironic criticism, given that the eventual Shuttle design imposed two full minutes (during first-stage flight burning the solids) with no viable means of crew escape. McDonnell Douglas' proposal was deemed to be technically deficient and weak, whilst Grumman came second. Its presentation was impressive, identifying fundamental problems and offering good solutions, but struggled with costs and management. "I can't verify what happened in the final selection," said former Grumman president Joe Gavin, "but the gossip has it that Mr. Nixon put it in California and that's all I should say about it." Even North American Rockwell had several shortfalls, including a difficult-to-build crew cabin. One intriguing footnote is that its approach to hiring minorities garnered it something of an edge. By 1972, the company had more Black, Hispanic and Asian workers than the others. High scores and low costs, therefore, were the principal rationale behind the decision to pick North American Rockwell to build the Shuttle. In fact, when George Low asked the unsuccessful bidders to comment on the overall fairness of the contract award, all three considered it to have been one of the best and fairest competitions that they had participated in.

Fig. 1.7 The Orbital Maneuvering System (OMS) provided the capability to conduct orbital insertion and deorbit 'burns', as well as maneuvers in space. Two large OMS engines (one of which is visible here) were located in the Shuttle's aft fuselage, on opposite sides of the main engine 'pyramid'. Also visible at the center of this image are the aft-mounted Reaction Control System (RCS) thrusters.

For the three contenders vying to build the SSMEs, the result of the competition was far less pleasant. In late 1968, Aerojet General, Pratt & Whitney and North American Rockwell's Rocketdyne division were chosen for Phase A studies into an engine which would burn liquid oxygen and hydrogen. Throttleable in increments from 73 percent to 100 percent of rated performance, it would limit maximum aerodynamic stresses and G-loads on the spacecraft during ascent. In its Phase A incarnation, it could also operate at a lower power setting for on-orbit maneuvers, although this functionality was deleted in the Phase B design and the OMS concept was introduced instead. As the Shuttle's payload-lifting requirement increased in line with Air Force demands, so the power requirement for the SSME also increased. This suitably positioned Rocketdyne's engine as the front-runner, as it sat much closer to mandated performance requirements than the others. A digital controller was added to allow the engine to throttle across a range from 50 percent to 115 percent, although the Shuttle's airframe contractors were requested to go no higher than 105 percent in abort situations and no lower than 65 percent in the most aerodynamically challenging phases of flight.

In July 1971, NASA selected Rocketdyne to design, fabricate and deliver the engines. But Pratt & Whitney angrily protested to the General Accounting Office (GAO), accusing NASA of unfair favoritism. The company had confidently expected to win the contract, even taking out adverts in major aerospace magazines and declaring its readiness to get started. Pratt & Whitney alleged that the selection was "manifestly illegal, arbitrary and capricious and based upon unsound, imprudent procurement decisions". But in March 1972 comptroller-general Elmer Staats ruled in NASA's and Rocketdyne's favor. Staats slammed Pratt & Whitney's case. He told their lawyers that the decision was equally unfair on NASA, which had helped Pratt & Whitney "to bring [an] original, inadequate proposal up to the level of other, adequate proposals, by pointing out those weaknesses which were the result of [Pratt & Whitney's] own lack of diligence, competence or inventiveness". The engine contract between NASA and Rocketdyne was finally signed in August 1972. As circumstances transpired, the SSME was in for a painfully complex development process.

"The main engine is very high performance, with a very high chamber pressure for that day and time and very lightweight for the thrust that they were producing," former NASA director of engineering Henry Pohl remembered. "We came out with that at the only time when it would have been successful. If we had waited another two years before starting development on the Shuttle, we probably would not have been able to do it, because the people that designed the main engine were the same that designed previous rocket engines. That group had designed and built seven different engines before they started the Shuttle development. A lot of them retired and so if we'd waited another two years, those people would have been gone and we would have had to learn all over again on the engine development."

Each SSME stood 4.2 meters tall and weighed 3,400 kilograms. It was throttleable in incremental steps from 65 percent to 104 percent of rated thrust. "How one can run an engine at more than 100 percent never made much sense to me, either, but the bottom line is that the Shuttle's main engines turned out to be more powerful than the designers thought they would be," wrote astronaut Jerry Linenger in his memoir, *Off the Planet*. "Consequently, we can actually run the main engines at 4 percent greater thrust than what was originally thought to be full-speed-ahead." Certainly, the second orbiter, Challenger, benefitted from redesigned SSME components and for each percentage-point incremental hike she could lift an additional 450 kilograms of payload to orbit, compared to her sister Columbia.

Fig. 1.8 Three Space Shuttle Main Engines (SSMEs), configured in a pyramidal shape in the aft fuselage, provided about 20 percent of the liftoff thrust for each mission.

As the orbiters and their engines took shape, so too did the ET and solids. In the wake of the June 1971 decision to pursue partially-reusable systems and lighter external tankage for the Shuttle, the required 'staging velocity' – the point in the flight at which the solids would be discarded – was reduced from 10,950 kilometers per hour to 7,600 kilometers per hour. And the cost of building a new tank for every mission would be offset by the smaller Shuttle and the smaller amount of propellants carried, which in turn allowed for a lighter structure and less beefy TPS. In May 1973, Boeing, Chrysler, McDonnell Douglas and Martin Marietta submitted proposals for the ET. Three months later, Martin Marietta won the contract to build three ground-test units and six development flight tanks. Each tank would stand 57.4 meters tall and 8.1 meters in diameter. To deorbit it safely into the ocean after each mission, a retrorocket 'spike' was incorporated into its nose. However, as the design matured, the spike was deleted and the tank's nose took on an ogive shape to better reduce parasitic air drag. Its length was correspondingly reduced to 47 meters.

The twin Solid Rocket Boosters also grew smaller in size and weight as the anticipated staging velocity dropped to 7,600 kilometers per hour. In fact, the solids would be much closer to the launch site at the point of being jettisoned from the Shuttle and this factor contributed a 40-percent weight reduction to their design. The lower staging altitude necessitated less thermal shielding and allowed parts of the boosters' structure to be made from cheaper and lighter aluminum, rather than titanium or Inconel-X. One contractor even proposed bringing the staging velocity down to 6,750 kilometers per hour, which permitted 80 percent of the airframe to be made from aluminum. In October 1973, Aerojet General, Lockheed Propulsion, Thiokol and United Technology submitted proposals to build the SRBs and in June 1974 Thiokol won the contract. In their final form, the SRBs shrank in length from 56 meters to 45.5 meters. Each booster consisted of a reusable Solid Rocket Motor (SRM) and four stainless steel 'segments', each measuring 3.7 meters in diameter.

THE SHUTTLE TAKES SHAPE

In the months which followed the contract award to Rockwell International, progress on the Shuttle accelerated. The company offered important subcontracts to its rivals, with Grumman building the orbiter's delta-shaped wings – "the consolation prize", according to Joe Gavin – at its Bethpage facility in Long Island (the first of which were delivered in April 1975) and a pair of modified Gulfstream II aircraft for training Shuttle pilots in approach and landing techniques. An unsuccessful attempt was made to interest Pratt & Whitney in sharing the SSME development, whilst McDonnell Douglas received contracts to fabricate the OMS 'pods' and

Aerojet General was selected to build the OMS engines. Capable of producing 2,700 kilograms of thrust and intended to circularize the Shuttle's orbit, execute the re-entry 'burn' and perform major maneuvers whilst in space, the OMS was fueled by nitrogen tetroxide and monomethyl hydrazine and test-fired for the first time at White Sands Test Facility, near Las Cruces in New Mexico, in September 1978. This test campaign was concluded in February 1980. Additionally, 44 primary and 'vernier' Reaction Control System (RCS) thrusters (16 mounted in the orbiter's nose and 28 in the aft fuselage) would furnish attitude control in space and during the onset of re-entry. They were built by the Marquardt Corporation, with the primary thrusters test-fired over 14,000 times and the verniers over 100,000 times, prior to certification in February 1980.

Fig. 1.9 View of the Forward Reaction Control System (FRCS) in the Shuttle's nose during pre-flight processing. The thruster manifolds are secured with red covers. Visible at center-right is the orbiter's Reinforced Carbon Carbon (RCC) nosecap.

The performance of these components over time would exhibit their own problems. On more than a quarter of all Shuttle missions, at least one RCS thruster would be declared 'failed' in flight, with others exhibiting fuel or oxidizer leaks, heater failures, low injector temperatures and issues with microswitches and other functionality. Prior to Columbia's STS-4 mission in June 1982, local storms even deposited rainwater into the thrusters. On several flights, the lightweight Tyvek rain-covers – installed over the RCS manifolds before launch and detached early in ascent – were found to have only partially separated, leaving debris around the 'lips' of their respective thrusters. Many of these problems turned out to be inconsequential to overall mission success, but several flights were directly impacted by RCS failures. Notably, in February 1995, Discovery was tasked with the Shuttle's first rendezvous with the Russian Mir space station. However, one aft-mounted RCS thruster failed and another began leaking. This was followed by a leak in a nose-mounted jet. The result was not only a degraded maneuvering capability, but also a threat to Mir and her crew on a high-stakes mission. The aft-mounted thrusters, as dictated by the flight rules, needed to be fully functional for the rendezvous to go ahead. Over a couple of days, the leak gradually diminished in severity and Discovery was able to complete a perfect rendezvous, but the incident served as an acute reminder of the fallibility of the criticality of the RCS.

The bulbous OMS pods and their powerful engines, on the other hand, would go on to perform near-flawlessly throughout the Shuttle's career. Indeed, beginning in April 1998 the OMS was used to 'assist' the SSMEs during ascent, enabling an additional 1,800 kilograms of payload to be lifted into orbit. Occasional issues cropped up in the form of gaseous nitrogen accumulator leaks, nozzle cracks, fuel probe and heater failures, stuck fuel quantity gauges and troublesome regulators, but the main concern about the pods was the frequency of damage to their TPS tiles and loosened or torn protective blankets during ascent or in flight. Late in 2001, fleet-wide inspections were conducted into improperly drilled bolt-holes used to affix the OMS pods to the orbiters and in May of the following year a leaking nitrogen valve contributed directly to delay a launch.

The 74.3-cubic-meter habitable area of the Shuttle consisted of a two-tiered cockpit, with the upper 'flight deck' for operations, connected via a hatch in the floor to the lower 'middeck' for working, eating and sleeping. The cockpit backed onto the 18.2-meter-long payload bay and aft fuselage which housed the three SSMEs, twin OMS pods and vertical stabilizer. The graphite-epoxy payload bay doors were, for their time, the largest aerospace structures ever made from composite materials and had to be opened within a couple of hours of reaching orbit to allow radiators lining their interior faces to shed excess heat from electrical systems.

The five-piece, clamshell-shaped doors were hinged at opposite sides of the orbiter's mid-fuselage, mechanically latched at the forward and aft bulkheads and thermally sealed at the centerline. Ordinarily, they were driven 'open' and 'closed' by electromechanical power, but if they could not be opened the Shuttle was required to return to Earth at the earliest opportunity. If they did not close properly at mission's end, two crew members would operate the mechanism manually on a spacewalk. In most cases, the doors performed without incident, although on occasion actuators stalled, latches proved stubborn to secure and closure indicators misbehaved. During pre-flight processing for a flight in March 1985 a work access platform was accidentally dropped onto one of the doors, producing a two-week launch delay. In June 1991 a section of door-seal became detached, along with some thermal blanketing around the aft bulkhead but the doors were successfully closed without incident at the end of the mission and the Shuttle landed safely. And in November 1994, a wastewater dump left a sizeable deposit of ice on a payload bay door.

As the hardware took shape, so too did the infrastructure. In June 1973, modifications got underway on NASA's Industrial Plant at Downey in California to build the orbiters. That September, KSC took center-stage as its principal launch and landing site, with the huge VAB and the two pads of Launch Complex 39 already in place from Project Apollo. All told, more than a billion dollars had been invested in launch facilities at the Cape, which made it a logical front-runner as the Shuttle's home base. However, Vandenberg in California was also proposed as a candidate and the state's Democratic Senator Alan Cranston set up a task force to champion its cause. Others included White Sands in New Mexico, which possessed the requisite telemetry and support facilities, as well as a suitable location, 1,200 meters above sea level.

In April 1972, KSC was selected as the primary site for 'due-east' and NASA launches, with Vandenberg picked for polar-inclination missions and military flights. Having two launch facilities carried unique benefits. Polar-orbiting missions out of KSC risked overflying Cuba and posed safety concerns for Mexico and the southern United States. And although missions of 57 degrees were achievable from Florida (with higher inclinations of up to 62 degrees also feasible by performing a 'dog-leg' maneuver during ascent), the additional energy required to 'turn' the orbiter/ET stack at high velocities yielded a drop in payload capacity. In February 1990, the only Shuttle mission ever to reach 62 degrees did so by following a 'normal' flight azimuth downrange from KSC, then maneuvering to a higher azimuth whilst high over the Atlantic Ocean. But in doing so, its ascent profile carried it over heavily populated areas of Cape Hatteras, Cape Cod and parts of Canada, creating a highly undesirable safety risk which was only permitted on account of the importance of its payload.

Fig. 1.10 From the outset, the Kennedy Space Center (KSC) and its twin pads at Launch Complex 39 was envisaged as the Shuttle's principal launch site. Pictured here in the foreground on Pad 39A is Columbia, having just experienced a last-second shutdown of her main engines on 22 March 1993. Visible in the distance on Pad 39B is her sister Discovery, being readied for her own mission a few weeks later.

Vandenberg, too, had its own limitations. Due-east missions out of the West Coast site risked launching directly over the heartland of the United States and would have nowhere to safety jettison their SRBs and ETs. As such, the KSC/Vandenberg duopoly was explicitly chosen for its "cost, operational and safety advantages over any possible single site or any other pair of sites in the United States". It was estimated that $150 million would be required to bring KSC up to Shuttle specifications, with the Air Force contributing $500 million to upgrade its facilities at the old Space Launch Complex (SLC)-6 at Vandenberg. And it was expected that at least one orbiter would be semi-permanently detailed to Vandenberg, as part of the payoff for Air Force political support during the Shuttle's evolution.

Planning was also at an advanced stage to build Runway 15/33, also known as the Shuttle Landing Facility (SLF) – an enormous concrete strip at KSC measuring 4,500 meters long and 90 meters wide, with 300-meter paved overruns at each end – to support not only the vehicle's return from space, but also to serve as a Return to Launch Site (RTLS) 'contingency abort' option, should an emergency occur during the first few minutes of a mission, demanding an immediate landing. Even today, the SLF remains one of the longest runways in the world. However, in October 1974, NASA determined that at least "the first few" Shuttle flights would terminate on either the vast expanses of dry lakebed which formed Runways 17/35, 05/23 or 15/33 or the all-concrete Runway 04/22 at Edwards Air Force Base in California, "for the added safety margins and good weather conditions" afforded there. Edwards would also remain a 'secondary' landing site behind KSC during operational Shuttle flights. The great compacted gypsum flats of White Sands were picked in March 1979 as a backup site for due-east missions. As well as offering (almost) year-round dry weather conditions, White Sands possessed twin runways, each 4,500 meters long with 3,000-meter overruns at each end, crossing one another in an 'X' pattern. As well as being an excellent training site for astronauts, White Sands was a close analogue for conditions they would encounter at the SLF, Edwards and the Transoceanic Abort Landing (TAL) runways. White Sands also lay directly beneath the Shuttle's flight path during its critical first orbit, making it ideally suited to support an immediate return to Earth in the event of a serious malfunction. Meanwhile, at Vandenberg's north base, the existing 1,700-meter strip and overruns of Runway 12/30 were correspondingly lengthened to 4,500 meters to accommodate West Coast polar-orbiting flights.

QUALIFYING THE SYSTEM

But in the years between the contract awards and the Shuttle's maiden voyage in April 1981, the program was repeatedly set back, most visibly due to frustrating problems with the patchwork of TPS silica tiles and blankets needed to shield it during its blistering descent through the atmosphere at the end of each mission. A conventional aluminum airframe for the orbiter had already gained preference and NASA and Lockheed worked to refine a ceramic Reusable Surface Insulation (RSI) thermal insulator. Its thickness allowed it to protect the Shuttle against re-entry temperatures as high as 1,370 degrees Celsius, whilst keeping the underlying aluminum skin cool. Moreover, the lightweight nature and temperature durability of the tiles meant that up to 4,500 kilograms could be shaved from the gross weight of the orbiter. And that, in turn, translated directly into an estimated

$80 million reduction in development costs. But the protective material was brittle and, far from covering the entire vehicle, it had to be installed in the form of over 20,000 tiles, each of which was individually designed. Since the airframe expanded when heated, the tiles – which could not be permitted to open any gaps – were fitted to a 'dynamic' base and Nomex felt pads were sandwiched between them and the aluminum skin.

Other areas of the spacecraft (specifically its nose cap and the leading edges of its wings) were predicted to get considerably hotter than 1,370 degrees Celsius during re-entry. As such, a Reinforced Carbon Carbon (RCC) material was implemented to guard against extremes well over 1,500 degrees Celsius whilst keeping the underlying airframe cool. Black-colored High Temperature Reusable Surface Insulation (HRSI) tiles, capable of withstanding up to 1,260 degrees Celsius, would coat the Shuttle's belly and parts of her OMS pods and fuselage. White-colored Low Temperature Reusable Surface Insulation (LRSI) tiles on the vertical stabilizer and main fuselage would guard against temperatures of up to 650 degrees Celsius. And a series of lightweight, durable Advanced Flexible Reusable Surface Insulation (AFRSI) 'blankets' would be used where the heating was expected not to exceed a relatively benign 400 degrees Celsius.

Late in the 1970s, a process of 'densification' was implemented in which engineers applied a silica solution to the base of the tiles to improve their adhesion to the Nomex pads. "You always still have this question about when you look at a vehicle, you'll have flows that are different, depending on little vortices that form here or there," remembered former NASA chief engineer Milton Silveira. "The pressure on certain of the tiles are different. Some of them have higher density than the other ones and what you look for is the flow-field around the orbiter to maybe put these higher-density tiles in those areas where the flow is stronger. It takes a certain amount of testing and analysis to determine where these higher heating areas are and to be able to handle the application of the tile. It's always a worry." But even after Columbia was delivered from Rockwell International's Palmdale facility in California to KSC in March 1979, it took engineers nearly two years to test, install, remove, re-test and re-install the pesky tiles. With technicians installing an average of 1.3 tiles per man, per week, at one stage (and with students drafted in to help during summer breaks), there were times when it seemed that hardly any progress was being made at all.

This already complex picture was complicated yet further by lingering worries that the tiles were simply not up to the task. Dozens fell away from Columbia during her transit from California to Florida and analysis indicated that tiles on the OMS pods might themselves fracture and detach in flight. And even though thermal and pressure tests on the tiles had been successfully completed as early as

Fig. 1.11 The process of installing Thermal Protection System (TPS) tiles onto the orbiter for each mission was a long and laborious process. More than 20,000 individually designed tiles protected the spacecraft from the extreme thermal stress of re-entry.

April 1976, doubts remained. One nightmarish possibility centered upon the so-called 'zipper' effect: if a single tile somehow became de-bonded from the airframe during ascent or re-entry, there were fears that it might pull hundreds more away in its wake. Such a situation would inevitably result in a catastrophe of the highest order: a Loss of Crew and Vehicle (LOCV). By September 1979, plans were afoot for inspections and tile repairs by astronauts. In one scenario, an extendable boom could provide television images of any damage to the crew. Another option was to send astronauts outside in pressurized space suits – wearing a unique jet-propelled backpack, called the Manned Maneuvering Unit (MMU) – to apply either a cure-in-place filler to replace missing tiles with a modified caulking gun or install pre-cured ablative 'blocks' to fix areas of more substantial damage. In January 1980, NASA contracted with Martin Marietta to develop a TPS repair kit, which would "not be flown on the first test flight but will be held in reserve for possible use on later flights". These issues with the TPS and the need for a reliable means of inspecting and repairing it would return again and again during the Shuttle's lifetime and the agency's failure to decisively resolve the issue would eventually kill a crew.

The vulnerability of the tiles came to the fore in October 1981, shortly before Columbia's second launch, when highly toxic nitrogen tetroxide being loaded into the RCS accidentally spilled onto the nose of the Shuttle. "When they got ready to disconnect the ground support equipment from the flight hardware, the quick disconnect failed," said Chester Vaughan of NASA's propulsion and power division. "It continued to flow propellant that had no place to go except to the outside." The failure of the quick disconnect was traced to concentrations of iron nitrate in its 'head' and hardened lubricants. Although the situation was detected within minutes, the spillage deposited around 80 liters of nitrogen tetroxide onto Columbia's nose and 378 protective tiles had to be removed, decontaminated and replaced. "The overall event would have been almost a non-event had it not been that the tiles and their…attach materials were not compatible," continued Vaughan, "so in an hour or so, the tiles started sliding off the vehicle." It was a worrisome indicator, as will be discussed in Chapter 6, that the TPS malady was here to stay.

If the tiles were problematic to fabricate and keep affixed to the airframe, then bringing the SSMEs up from the drawing board to full certification was truly maddening. Following the contract award to Rocketdyne, their development got underway at the National Space Technology Laboratory (NSTL) in Mississippi in 1974. A full-thrust chamber test of an integrated subsystem demonstrator took place in the summer of the following year and in March 1976 the first engine was test-fired for 42.5 seconds at 65 percent of its rated performance. However, several serious technical defects were uncovered. Cracks in the turbine blades of the high-pressure fuel and oxidizer turbopumps created a potential for failure, which prompted a decision to incorporate further modifications and monitoring instrumentation. An extensive campaign of 25 SSME tests began in April 1977 and encountered no significant difficulties, although the troublesome turbopumps continued to blight the process. In July 1979, the main fuel valve experienced a major fracture, which allowed hydrogen gas to leak into a mockup of the Shuttle's aft fuselage. The system shut itself down correctly, but not before sustaining major structural damage. Four months later, a test firing of a three-engine 'cluster' – scheduled to run for 510 seconds, equivalent to a full flight duration – was halted shortly after ignition when the oxidizer turbopump failed again. Success followed hard on the heels of failure. A perfect test in December 1979 was followed by a premature SSME shutdown in April 1980, then another success, then another agonizing shutdown. In November 1980, a weak brazing section on a nozzle failed and left a sizeable hole in the engine. It became almost a running joke: "What engine did we lose *this* week?" quipped Neil Hutchinson.

Ironically, although the SRBs represented new technology for NASA, their development was surprisingly unwrinkled. The first 'hydroburst' test of an empty booster segment casing occurred in September 1977 at Thiokol's facility in Utah to validate its structural integrity and fracture mechanics for better analysis of the growth and evolution of damage. Four static firings of the SRMs took place between July 1977 and February 1979, followed by three qualification tests between June 1979 and February 1980 and a second hydroburst test in September 1980. Six air drops with full-scale pilot, drogue and main parachutes were conducted near El Centro in California between June 1977 and September 1978 to evaluate the boosters' decelerator subsystems under actual flight conditions. In fact, the verification of the SRBs in just seven test firings was dwarfed by no fewer than 726 tests – totaling over 62,000 seconds' worth of burn-time – needed to qualify the SSMEs.

TRAGEDY AT THE FINISH LINE

The pre-dawn rollout of Columbia to Pad 39A on 29 December 1980, therefore, was a watershed moment after a tough decade in which the Shuttle had morphed into something entirely at odds with what NASA originally desired. Work to build Columbia had begun in June 1974 and she was finally rolled out of Palmdale on 8 March 1979. Four days later, to the amazement of motorists in the sweltering heat of a late California spring, she was towed overland to Edwards Air Force Base, before being bolted to the top of a Boeing 747 – specially modified as a Shuttle Carrier Aircraft (SCA) – and flown 'piggyback' to Florida. After traveling 4,100 kilometers in several stages across the continental United States, on 25 March this astonishing combo reached Florida and the wheels of the SCA kissed the concrete runway surface of the SLF. Columbia was removed from the Boeing and transferred to the nearby Orbiter Processing Facility (OPF). Eventually, on 24 November 1980, she was rolled over to the VAB for 'stacking' onto her ET and boosters.

Significantly, before her first launch, Columbia was put through a Wet Countdown Demonstration Test (WCDT), which culminated on 20 February 1981 in a 20-second firing of her three SSMEs. This Flight Readiness Firing (FRF) demonstrated their ability to throttle between 94 and 100 percent of rated performance and 'gimbal' as they would during flight. Similar 'wet' (or fully fueled) tests had been performed before Saturn V launches, but on none of those occasions were the engines fired. Preparations for the FRF proceeded in a manner not unlike a 'real' countdown. Launch controllers started the clock at T-53 hours and powered-up the SRBs, ground support equipment and systems aboard Columbia. Four seconds before the simulated launch, the engines roared perfectly to life at 120-millisecond intervals, producing a sheet of orange flame, a trio of Mach-diamonds and ramping up to 100 percent of their rated thrust precisely at T-0. Three seconds later, engineers simulated the retraction of the ET umbilical and SRB hold-down posts. After 15 seconds of stable, continuous thrust, shutdown commands were issued to all three engines.

Fig. 1.12   Columbia resides on Pad 39A before her maiden voyage, STS-1.

FRFs were performed before the maiden voyages of all five Shuttles, plus an additional test on Discovery in August 1988 before STS-26, the return-to-flight mission after the loss of Challenger. However, not all FRFs ran according to plan. During Challenger's FRF on 18 December 1982, engineers detected levels of gaseous hydrogen in the aft compartment which significantly exceeded allowable limits. When it became impossible to pinpoint the cause or location of the leak, a second FRF was ordered. New instrumentation was installed both inside and outside the aft compartment to determine whether the leaking hydrogen gas was from an internal or an external source. Suspicion focused initially on the second possibility, because vibration and current had found their way into the aft compartment, behind the engines' heat shields. Extra sensors and a higher than ambient pressurization level was installed to preclude penetration by 'external' hydrogen sources. A second FRF on 25 January 1983, during which Challenger's SSMEs were run at 100 percent for 23 seconds, again revealed leaking hydrogen gas. Several more days of troubleshooting identified a cracked weld in tubing leading to the upper engine, which was removed. A replacement arrived from the NSTL, but inspections in the VAB uncovered a leak in an inlet line to its liquid oxygen heat exchanger. Before it could even be installed onto Challenger, the 'replacement' was itself replaced by a third engine. And after further checks at the NSTL (including a full-flight-duration test firing), it was dispatched to Florida and fitted.

Whilst this work was ongoing, painstaking efforts were underway to ensure that the other two 'original' engines did not exhibit any leaks. Unfortunately for Challenger, they did. Tiny hairline cracks were found in one of the fuel lines to the left-hand engine and borescope observations of the right-hand engine revealed a similar problem. Both were removed, returned to the VAB and repaired. With the arrival of the replacement SSME from the NSTL, all three were installed by mid-March 1983 and verified as ready for flight. The leaks were apparently caused by a generic 'seepage' in a 45-centimeter Inconel-625 tube in the ignition system. It apparently occurred underneath a protective sleeve brazed onto a small hydrogen line which sent fuel to the augmented spark igniters that lit the engines. The sleeves were meant to stop potential chafing. After practicing cutting off the sleeve on the NSTL test stand, Rocketdyne technicians proceeded to Florida and replaced it with a non-sleeved Inconel-625 tube on each of Challenger's engines.

By blissful contrast, the FRFs of the last three Shuttles – Discovery on 2 June 1984, Atlantis on 12 September 1985, Discovery again on 10 August 1988 and most recently Endeavour on 6 April 1992 – proved relatively uneventful. However, during her first FRF, Discovery suffered some de-bonding of one of her SSME heat shields and her second FRF was scrubbed on 4 August 1988, due to a sluggish fuel valve, before being satisfactorily completed a few days later. Despite these challenges, FRFs cleared the final major milestone on the road towards the maiden voyage of each orbiter. And in Columbia's case, it removed another obstacle on the road to 'STS-1': the first orbital flight of the 'Space Transportation System', the official name of the Shuttle. According to the STS-1 press kit, the two-day mission was targeted to begin "no earlier than" 17 March 1981, but several technical issues and a human tragedy had another card to play in the evolution of this experimental flying machine. Following the FRF, engineers had to repair a section of super-light ablator material, which had become de-bonded from the ET during a test of its cryogenic propellants in January. This delayed the launch to no sooner than 5 April. A strike against Boeing by machinists and aerospace workers pushed this date back to 10 April. In the meantime, on 19 March STS-1 commander John Young and pilot Bob Crippen participated in a Terminal Countdown Demonstration Test (TCDT), during which they donned their pressure suits, rode out by bus to Pad 39A and boarded Columbia. These integrated tests at the pad with the astronauts were performed before each mission throughout the Shuttle's career. Working hand in glove with an integrated team at the Launch Control Center (LCC) at KSC and the Mission Control Center (MCC) at the Johnson Space Center (JSC) in Houston, Texas, Young and Crippen rehearsed each step of their final countdown milestones, prior to a simulated failure and shutdown of the SSMEs. They then practiced escaping from the orbiter and rode an emergency slidewire basket from the launch tower to the ground. If such a hair-raising scenario happened for real (as it almost did on more than one mission), the astronauts might find themselves running through invisible hydrogen flames, dicing with the foul breath of Death itself.

Although Death did not visit Columbia on 20 February 1981, its presence was forever near. And it would remain an ominous specter before each flight. Not long after Young and Crippen completed their TCDT, the Shuttle claimed its first lives following a routine gaseous nitrogen purge test. A group of Rockwell International technicians had been cleared to enter Columbia's aft engine compartment for routine closeout tasks. Believing the small space to have been flushed with breathable air and thinking conditions inside to be safe, they did so without having donned air packs or masks. However, traces of the odorless, colorless nitrogen gas lingered, and the men lost consciousness even without realizing anything was amiss. Another technician donned an air pack and dragged them out, but efforts to get them to hospital were delayed by a tragic breakdown of communications.

In the confusion, it was incorrectly assumed that a deadly ammonia leak had occurred and security guards at the Pad 39A gate accordingly refused to allow the first ambulance access as it did not carry the requisite breathing apparatus. At length, when the enormity of the situation became apparent, the injured men were airlifted to safety, but after several minutes of exposure to gaseous

Fig. 1.13 Clad in their pressure suits, the backup crewmen for STS-1 inspect the slide-wire baskets that would carry them away from the launch pad to a fortified bunker at ground level in the event of a pre-launch emergency. At left is STS-1 commander John Young, with pilot Bob Crippen and backup pilot Dick Truly and commander Joe Engle (both with their backs to the camera) at right.

nitrogen the damage was done. John Bjornstad died on the way to hospital. Forrest Cole slipped into a coma and died several days later. Another technician endured relentless migraines, which his wife described as "non-stop, 24 hours a day, for three and a half years". And Nick Mullon, who saved two other men's lives that day, ultimately paid with his own. Left brain-damaged by the effects of hypoxia and nitrogen ingestion, Mullon was plagued by sleep disorders, anxiety and personality changes associated with survivor's guilt for the rest of his life. He died in April 1995, three weeks before his son's high school graduation.

When Columbia finally reached space, the astronauts of STS-1 paid a touching tribute to the victims. "They believed in the space program and it meant a lot to them," Young said. "I am sure they would be thrilled to see where we have the vehicle now." But as time would tragically tell, John Bjornstad and Forrest Cole and Nick Mullon were only the first of many lives prematurely lost in the name of the Space Shuttle.

2

A Machine Flown By People

CHANGES

It was around nine o'clock on the fine Saturday morning of 17 June 1989 when the clamor of a vintage AT-6D Texan aircraft first caught the attention of eyewitnesses on the outskirts of the small city of Earle, right on the Arkansas-Tennessee border, about 45 kilometers west of Memphis. As they watched, mesmerized by the reverberating buzz of its propeller and the incessant drone of Old Growler's radial piston engine, they may have noticed the lone pilot – a gruff, mustachioed Vietnam War veteran, named Dave Griggs – performing a series of aileron rolls over the golden wheat fields of eastern Arkansas. It was a small aircraft, but its heritage was impressive: for the Texan originated in the 1930s to prepare American fighter pilots for aerial combat, earning for itself the apt moniker of 'Pilot Maker'. But half a century later, it was more often found in warbird museums or air shows. And it was for that reason that Griggs was now taking it for a spin. But with more than 9,500 hours in 45 different types of aircraft, he could never have known on that bright June morning that the tiny Texan would be the last machine he would ever fly.

Griggs was preparing for a weekend air show in Clarksville, a couple of hundred kilometers west of Earle, and had been in the air for only about 40 minutes. Eyewitnesses recalled seeing the Texan approach McNeely Airport from the west, in what they described as a "low and slow" approach, as if the pilot were preparing to land. "Then it was maneuvered to an inverted attitude over Runway 8," the National Transportation Safety Board (NTSB) noted in its official accident report. At this stage, Griggs was flying barely 20 meters above the ground. "After remaining inverted for a short time, the aircraft was rolled back to an upright attitude," the report continued. "However, as the aircraft rolled upright, it angled to the right side of the runway, descended and crashed." One of the Texan's wings touched the ground and the aircraft seemed to 'slip' sideways into a nearby field, bursting into flames. Forty-nine-year-old Griggs – retired U.S. Navy rear-admiral, engineer, husband, father and seasoned Space Shuttle astronaut – was killed instantly.

At the time of the accident, he had been training to pilot STS-33 aboard the orbiter Discovery, scheduled for launch only five months later. Although the accident happened whilst he was off-duty and not in a NASA aircraft, his untimely death sent shockwaves reverberating through the close-knit astronaut corps. After the funeral, his STS-33 crewmate Kathy Thornton walked into the Outpost tavern in Houston, Texas, her cheeks soaked with tears, and placed a wreath of flowers onto the bar. It was neither the first nor the last time that astronauts would lose their lives in the line of duty, but of all Shuttle missions STS-33 seemed to retain a dark strain of tragedy at its heart. With Griggs gone, a new pilot had to be announced, and a first-time flier was hardly advisable in view of the short training time available before launch. Many of the experienced Shuttle pilots had already been assigned to other crews. At length, STS-33 commander Fred Gregory suggested a seasoned pilot named John Blaha. He had flown in space only a few months earlier and, although he was already training for another mission (STS-40), that flight was far enough into the future for a new pilot to be selected in his stead. At the end of June, Blaha formally joined Gregory, Thornton, Manley 'Sonny' Carter and Story Musgrave to train for STS-33.

Taking Blaha's place on STS-40 was a 'rookie' pilot named Sid Gutierrez and he had no idea that a mission assignment was coming his way. "It was summer and we headed off on an extended family vacation, driving to the East Coast, visiting relatives in Ocean City and touring sites around D.C.," Gutierrez told this author. "We then planned to stay at a friend's cabin in West Virginia before starting the drive home. It was the days before cell phones or even personal pagers, so I left my assistant with good phone numbers to reach me up until leaving the cabin in West Virginia. From there on back to Houston, I told her we would just drive along and stop when and where we felt like it, so there were a few days where she could not reach me. While reading a paper in D.C., we learned of the tragic accident involving Dave." On returning home that weekend, Gutierrez wandered into his front yard and was startled by a congratulatory call from his next-door neighbor.

Gutierrez thought the neighbor was being sarcastic. Perhaps making it through an extended family vacation in one piece *was* cause for celebration.

"I was talking about your flight assignment."

"What do you mean?"

"It's in the paper." Then he spotted the one in Gutierrez's hands. "Not that one. It's old news. It happened days ago." Gutierrez hurried indoors and began rummaging through old newspapers and found the article announcing his selection to STS-40. It was a moment of pure joy for the Air Force pilot to receive news of his first flight assignment, tempered with sadness at it coming in the shadow of Griggs' death.

Fig. 2.1 A group of astronauts and training instructors practice boarding the launch pad slidewire escape baskets to simulate a Mode 1 Egress from the Shuttle.

STS-33 flew the following November. The astronauts memorialized Griggs through their formal crew patch, which included a gold star in honor of their fallen colleague. After the flight, all five of them moved into other duties and by the end of the following year all had been named to new missions. But on the calm afternoon of 5 April 1991, Atlantic Southeast Airlines Flight 2311 was approaching Glynco Airport (today's Brunswick Golden Isles Airport) in Brunswick, Georgia, after an hour-long flight from Atlanta. As it approached the runway, in near-perfect weather conditions, observers on the ground noticed that the jet was flying at a much lower altitude than normal. All at once, it rolled sharply to the left, then descended in a nose-down dive and crashed into a patch of trees. All 20 passengers and three crew aboard the aircraft were killed. Among the dead were two small children, a Texas senator and astronaut Sonny Carter.

On the very day that Carter died, Space Shuttle Atlantis had just launched on STS-37. In Mission Control that day was Carter's STS-33 comrade, Story Musgrave, and it would be his task to notify the crew over the radio. But first

Musgrave himself had to be told the devastating news. "I remember going and getting Story, but I wasn't in a position to tell him what was about to happen and what he was about to be told," reflected NASA flight director Phil Engelauf in an interview for Musgrave's biography, *The Way of Water*. "And as I brought him out to the front room, he stood and he talked with a couple of the managers and I just saw all the color go out of his face. He sat down on a step in the control center and put his face in his hands. He sat there for several minutes and didn't speak. And the emotion of that moment has kind of stayed with me ever since."

Although not many astronauts lost their lives whilst training to fly the Shuttle, the dual tragedies of Griggs and Carter underscored a reality about this group of people that we outsiders tend to forget. Astronauts were and are in many ways inured to the risks involved in the spaceflight business and the hazards posed by the Shuttle, many of them having come from military careers, and in spite of this (or perhaps because of it) their closeness as a unit was often tighter than any flying squadron; closer and more intimate, in some ways, even than family. Bob Crippen, who commanded three Shuttle missions, considered it a fundamental part of his role to know how his crewmates would react to certain malfunctions and emergencies; to him, it was imperative that a bond of mutual trust should develop between them all, for every astronaut knew that one day they might depend on each other for their lives.

And deaths, illnesses, family tragedies and other issues which resulted in crew changes added a further layer of complexity to the already hefty challenge of preparing to fly the Shuttle. The same was true for the crew's spouses. In his memoir *Sky Walking*, astronaut Tom Jones recalled a conversation between a group of astronauts' wives with Carter's widow, Dana.

"Are you all close to each other?" she asked.

"Not especially."

"You might want to develop those relationships," replied Mrs. Carter. "You might find you'll need them someday."

SCHEDULE PRESSURE, TURNAROUND TIMES AND FLIGHT RATES

On the morning of 12 April 1985, seven astronauts boarded Space Shuttle Discovery, convinced that they would not launch that day. Thick grey overcast clouds hung ominously over the Kennedy Space Center (KSC), to such an extent that pilot Don Williams jokingly remarked that perhaps he should give the weather officer a cloud 'tops' report. Behind Williams, equally certain that they would be going nowhere that gloomy day, mission specialist Dave Griggs unstrapped and sat on the back of his seat, chatting to his crewmates downstairs on the middeck. The first of two launch 'windows' for the morning had been missed when a ship inadvertently strayed into the danger area. A second opportunity to get Discovery off the ground existed an hour later, but Mother Nature intoned otherwise. "We didn't think we were going to go and we were kinda just chatting around," Williams remembered, "because we figured we were going to do a scrub turnaround for 24 hours and come out the next day."

Then, as engineers and managers in the Launch Control Center (LCC) considered calling a scrub, NASA administrator James Beggs entered the firing room. Shortly thereafter, in consultation with the payload customer, the NASA Test Director (NTD) granted an extension to the launch window and the countdown clock was manually restarted. "All of a sudden, there's this big scramble to tighten up your harnesses, get your helmets back where they're supposed to be," added Williams. "We couldn't believe they were going to launch us, because there was still an overcast." Hastily, crewmate Jeff Hoffman strapped Griggs back into his seat, then returned to his own seat, next to Discovery's side hatch. Less than a minute before the closure of the second window, Discovery roared smoothly into space. But the events of that morning demonstrated that, whether intentionally or not, managerial pressures to meet launch schedules and a heady mix of 'Go Fever' remained a driving factor in how the Shuttle was operated in that timeframe. It was a toxic perversion of common sense.

Before the loss of Challenger in January 1986, hope sprang eternal that this fleet of reusable orbiters would make the arduous voyage into low-Earth orbit increasingly 'routine', with launches every week or two, as the Shuttle transformed human spaceflight from the preserve of the few into one of the many. This pervasive line of rhetoric was espoused by NASA's political puppet-masters, who argued that the Shuttle would deliver commercial satellites into orbit far more frequently and cheaply than expendable rockets, with the unique bonus of 'free' astronaut seats for customers' representatives which other launch services could not provide. "There was a constant push from outside the program," remembered flight director Neil Hutchinson, "to declare the Shuttle operational."

The peculiar 'groupthink' which characterized the orbiters as the spacefaring equivalents of a fleet of Boeing 747s was diametrically opposed to the reality of what was a highly labor-intensive machine. "You do a 747 and you get it to a certain point where it's all checked out and now you're selling hundreds of them to different customers and so they're usable airplanes; they're production craft," said Robert Wren, a former structural division manager for Shuttle payloads systems. "The Shuttle system never got to that and it really wasn't intended to get to that. It's so complicated and so detailed that the only way you could ever classify it is as an experimental vehicle." After each mission, the level of refurbishment was so intricate that each vehicle was virtually taken apart and reassembled, producing an almost brand-new craft in advance of its next flight. Difficulties with brakes and tires, maddening failures of the Space Shuttle Main Engines (SSMEs), unresolved problems with the Thermal Protection System (TPS) and fatal flaws in the design of the Solid Rocket Boosters (SRBs) were only the most publicly visible of hundreds of potentially disastrous failure points. "The Shuttle never really was an operational vehicle," admitted former NASA branch chief John 'Denny' Holt. "It's always had that kind of developmental nature."

Fig. 2.2 Payload specialist Christa McAuliffe listens as STS-51L commander Dick Scobee briefs her on flight deck systems in the Shuttle simulator.

But in the months and years prior to January 1986, an outward veneer of success as mission after mission launched successfully created a belief that the Shuttle was indeed operational and, by extension, sufficiently 'safe' to carry not only professional astronauts, but ordinary civilians from other walks of life. It was a catastrophic fallacy. Right from the start, three main 'markets' had been identified for the vehicle: commercial, scientific and military. Europe's Ariane 3 rocket, first flown in August 1984, began to prove its worth as a reliable launch vehicle, whilst NASA's promise of flying the Shuttle every week or two seemed increasingly hollow. Only days before Discovery's 12 April 1985 launch, it was revealed that 'spare' Delta expendable rockets were being fabricated to hedge against further Shuttle delays; hardly a glowing endorsement of how seriously the reusable spacecraft's schedule-keeping abilities were being taken. Thirteen missions were planned for 1985, but the year would end after just nine. NASA associate administrator for spaceflight Jesse Moore insisted that the Shuttle was quite dissimilar to Ariane 3. The Shuttle was human rated, for starters, and "schedule is a secondary priority; mission safety and success are top priority". Whilst his sentiment was undoubtedly sincere and reflected the majority view within NASA at the time, this stance proved bitterly ironic after the loss of Challenger, when investigators uncovered evidence of cannibalism of Shuttle parts, a widespread acceptance of critical safety-of-flight issues, and a "normalization of deviance" in working practices and how the orbiters were handled.

The reality was that the driving factor propelling the Shuttle in 1985 was not the liquid oxygen and hydrogen of the SSMEs or the ammonium perchlorate in the SRBs, or even the blood, sweat, tears and incessant training of her astronauts and instructors. Rather, it was schedule pressure. And as mission after mission flew safely and 'routinely', that pressure intensified. Three Shuttle missions in 1982 rose to four in 1983, then five in 1984 and nine in 1985. Fifteen were scheduled for 1986 and 24 the year after. The two simulators at the Johnson Space Center (JSC) in Houston, Texas, could only support training for (at best) around a dozen flights per annum. "That was for political PR," remembered former NASA guidance and propulsion systems manager David Whittle. "It became immediately obvious that you were not going to fly 12 times a year and you were *not* going to fly that thing a hundred times. Other than making us work real hard, that was not a big player, but I don't think we ever believed we were going to fly 12 times a year. I don't think that the control center could support that, given the amount of sims and preparation and people."

Columbia successfully launched the Shuttle's first pair of commercial communications satellites on STS-5 in November 1982, netting NASA a negotiated fee of $18 million. It was expected that eventually the reusable spacecraft would outcompete Ariane 3, but few expected it to work in the longer term. "No one ever felt like that was going to be a mission for the Shuttle for long, because they filled squares," remembered astronaut James 'Ox' van Hoften. "They're all done on unmanned things now, as they probably should be, because it was just too expensive a way to launch a satellite like that."

Fig. 2.3 One of the three Space Shuttle Main Engines (SSMEs) is installed into Discovery, prior to STS-42 in January 1992. Preparing each of these infinitely complex vehicles for each mission required a timescale on the order of several months.

Expensive, indeed, in terms of both financial cost, technological challenge and risk to human life. Between the first Shuttle mission and the loss of Challenger on the program's 25th mission, although the flight rate grew year-on-year, the multitude (and magnitude) of problems slowed the anticipated cadence of launches from a sprint to a crawl. Plans to fly every two weeks with four operational vehicles necessitated rapid turnaround times. "It wasn't until after the Shuttle was flying that the engineers who were responsible for all the subsystems started to get cold feet," said Hoffman. He even recalled an engineer telling him, with a perfectly straight face, that individual orbiters could be turned around in as little as 16 days. "Which again is totally outrageous," Hoffman scoffed. "It takes a minimum of three months to turn a Shuttle around and that's pretty quick! But it's just typical of the way people were thinking back then." In correspondence with this author, one former Shuttle engineer expressed serious reservations that nine or ten missions per year was barely achievable and stretched available resources to their limits, even counting overtime and around-the-clock refurbishment operations in the Orbiter Processing Facility (OPF) at KSC. Numerous contractors routinely worked 72-hour weeks in the months preceding the loss of Challenger and frequently logged 12-hour shifts.

"The potential implications of such overtime for safety were made apparent during the attempted launch of Mission 61C on 6 January 1986," read the report issued by the Rogers Commission, the presidential inquiry into the loss of Challenger, "when fatigue and shift-work were cited as major contributing factors to a serious incident involving a liquid oxygen depletion that occurred less than five minutes before scheduled liftoff." STS-61C, the last successful Shuttle mission before the destruction of Challenger, is an interesting case, not least because it was one of only two Shuttle flights to have scrubbed no fewer than six launch attempts. Originally targeted to fly on 18 December 1985, it was routinely postponed by 24 hours to give engineers additional time to close out Columbia's aft fuselage. Next day, the countdown was halted at T-14 seconds, when the Hydraulic Power Unit (HPU) in the right-hand SRB indicated that turbine speeds were above allowable limits. The signal turned out to be erroneous, but it proved a moot issue because NASA ran out of time in the launch window and the attempt was called off. Two other tries to get Columbia off the ground were frustrated by poor weather.

But the two attempts which really raised eyebrows – and made the pages of the Rogers Commission's damning report into the failings which caused the loss of Challenger and the deaths of her seven astronauts – could have proven catastrophic. On 6 January 1986, the countdown clock was automatically halted at T-31 seconds, due to the accidental draining of 1,800 kilograms of liquid oxygen from the ET by an overworked and exhausted technician. The great tank's fill-and-drain valve, it seemed, had not properly closed when commanded to do so. Three days later, Columbia again brushed with disaster, when a liquid oxygen sensor broke away from the launch pad and lodged itself into the prevalve of one of the three SSMEs. Had Columbia launched that day, the affected engine would have ingested the sensor into its High Pressure Oxidizer Turbopump (HPOT). With its turbine blades spinning 30,000 times per minute, the pump would have torn itself to pieces, destroying the Shuttle and killing the crew. "That would have been a bad day," noted STS-61C pilot Charlie Bolden with understated test-pilot gallows humor. "It would have been catastrophic, because the engine would have exploded, had we launched."

Furthermore, the Rogers Commission found disturbing evidence that NASA's provisions to support expanded flight rates and meet tight schedules were woefully inadequate. Spare parts for individual orbiters were in notoriously short supply before the destruction of Challenger, with only 65 percent of the required inventory in place by January 1986. This had already led to an increasingly reckless practice of 'cannibalizing' parts from one orbiter to equip the next and resources tended to focus upon 'near-term' priorities, rather than longer-term problems. An $83.3 million budget cut in October 1985 had forced additional major deferrals of spare parts purchases. Moreover, taking from one vehicle to fix another "increases the exposure of both orbiters to intrusion by people," observed Shuttle commander Paul Weitz in his Rogers testimony. "Every time you get people inside and around the orbiter, you stand a chance of inadvertent damage of whatever type: whether you leave a tool behind or, without knowing it, step on a wire bundle or a tube."

Prior to the disaster, the shortage of spare parts had little serious impact on the flight rate, but cannibalism was "possible only so long as orbiters from which to borrow are available". In previous years, one vehicle was either being constructed or was undergoing maintenance and refurbishment, allowing such cannibalism to occur. But 1986 was expected to mark the first year in which all four vehicles had a full plate of missions and there would be insufficient time to robustly investigate problems. "There was never time," said former Shuttle director Arnold Aldrich. "And once we started flying, the flights came so quick, one after another, you couldn't stop and fix anything." Another senior processing director predicted that even if Challenger had flown safely, the flight schedule for 1986 would have come to its knees later that spring owing to the spare parts issue alone. "Compounding the problem was the fact that NASA had difficulty evolving from its 'single-flight' focus to a system that could efficiently support the projected flight rate," noted the Rogers Commission. "It was slow in developing a hardware maintenance plan for its reusable fleet and slow in developing the capabilities that would allow it to handle the higher volume of work and training associated with the increased flight frequency." Certainly, the progressive ramping-up of the Shuttle rate in the months preceding Challenger precipitated disaster, threatening to saturate the capacity of the system to safely prepare the vehicles and (as we shall see later in this chapter) adequately train the crews.

Fig. 2.4 Just as each Shuttle crew prepared for almost every eventuality, including fire-fighting at the launch pad, in the pre-Challenger period 'firefighting' of a different nature took place in the processing facility, as teams struggled to fix problems against the demands of an overwhelming flight schedule.

In the years after the loss of Challenger, more than 250 hardware and software changes were implemented and the Shuttle's flight rate was dramatically reduced to more manageable levels. This afforded additional time for the analysis of pertinent data from one mission in advance of proceeding with the next. Following the resumption of operations in September 1988, co-ordinated plans were cemented in place to remove each orbiter from service after every seven or eight missions for an extended period of maintenance, inspection and refurbishment, either at KSC or at prime contractor Rockwell International's plant in Palmdale, California. This permitted not only the implementation of upgrades – carbon brakes, drag chutes, external airlocks, improved computers and better avionics – but also gave engineers time to shave away unnecessary weight, conduct observations of wiring deficiencies and structural corrosion and investigate gremlins which lurked unseen in complex systems. "There is no 'off-ramp' for an orbiter," said engineer Kevin Templin, who worked on Endeavour during her first maintenance period undertaken in 1996–1997. "When it's flying, and something's failing, it usually happens quickly. It's a very dynamic environment; you're moving very fast. We plan to not fail. We put redundancy in there. We put systems that fail safely on board, so that we don't have catastrophic failures where we can envision something going wrong." Templin's words served as another reminder that the sheer dynamism of the myriad environments inhabited by the Shuttle during an 'average' mission (whether ascent, on-orbit or re-entry) could not be more distantly removed from the commercial airliner that it was intended to replicate.

Another facet of the fleet which went away after Challenger were commercial satellites as 'primary' payloads, following the logic that NASA should only risk astronauts' lives on missions which demanded the Shuttle's unique abilities. Indeed, when Arabsat-1B flew as a payload on STS-51G in June 1985, it had reportedly failed all its pre-launch safety reviews. "The crew recommended that it not be flown," said STS-51G astronaut John Fabian, "and the safety office recommended that it not be flown, but NASA management decided to fly it." Political embarrassment, it seemed, in the United States not flying a cargo for a major ally in the form of Saudi Arabia was too much to bear. "It was a good experience to work with the commercial side," remembered former chief flight director Randy Stone, "though it was kind of cumbersome from the standpoint that we couldn't do all the things as rapidly as the commercial people would like you to do them." But former NASA Administrator James Beggs was unwavering in his conviction that removing commercial entities from the fleet was a mistake. "The argument was that you shouldn't risk lives, he reflected, "but you're risking lives anyway when you fly the Shuttle." A deviation from that policy came in May 1992, when Endeavour, was tasked on her maiden mission to retrieve the stranded Intelsat-603 commercial communications satellite, attach a new motor and put it back into orbit. Although STS-49 and the $90 million salvage operation was successful, it indicated that some of the lessons of Challenger were gradually fading from NASA's consciousness. The military also backed away from using the highly visible Shuttle for its clandestine payloads and returned to expendable rockets. Together with an increased conservatism in testing and maintenance, this brought the Shuttle's flight rate down in its post-Challenger heyday to no more than seven launches per year.

CREW DYNAMICS AND REAL LIFE

In the halcyon days of the Shuttle's youth, 'core' NASA-trained crews – commander, pilot and mission specialists – were typically assigned as complete units, although their payloads were frequently subjected to change as technical difficulties delayed some flights whilst others moved forward on the flight manifest. One such flight was STS-41E, originally planned for August 1984 to deploy a pair of commercial satellites and a free-flying solar observatory. But after a harrowing launch pad abort a few weeks earlier, STS-41E was canceled and rescheduled for February 1985. (That flight, too, was also canceled and the unfortunate crew eventually launched with a totally different payload the following April.) Delays and cancellations of this nature were intensely disappointing for the astronauts, who had trained for many months and mentally and physically steeled themselves to fly, only for the rug of the ever-changing mission manifest to be pulled from beneath them. STS-41E commander Karol 'Bo' Bobko considered it his responsibility to keep his crew's spirits up. "I remember Bo had all us all over for dinner," recalled crewmate Jeff Hoffman. "He recognized that he had to do something for morale."

The fate of the STS-41E crew was far from unique in the pre-Challenger era. Some astronauts found themselves assigned to two missions simultaneously, which made it difficult for crews to effectively gel together and complete their requisite training. "In those days, the Shuttle was going to fly 60 times a year," said Carl Shelley, former deputy chief of the training division of the Flight Operations Directorate. "We had sold that thing on the idea that it was going to be very routine to fly this thing. My job was, well, how in the world are you going to get enough astronauts through the pipe to fly that many flights? How many flights a year can an astronaut fly? We finally concluded, under the best conditions, maybe we could get him through four." In some astronauts' minds, flying rapidly in this back-to-back manner was the proper approach. "There are no more-trained astronauts than the people who have just landed," Shuttle commander John Blaha

Fig. 2.5 After landing aboard Discovery in July 1995, STS-70 commander Tom Henricks (bending down to inspect the Shuttle's landing gear) and pilot Kevin Kregel (left) were turned around to fly a second mission together, less than a year later in June 1996.

once observed. "If they turned right around and flew, I've always said I think you could have about a month and a half of training and they could launch. Because the ascent and entry consume so much of the training, you're way up on a stump, and with a few little proficiencies of ascent and entry that's how you could do that." Fellow astronaut Kathy Sullivan also emphasized the importance of turning around recently flown crew members. "If you come back on a second flight within a year or two, the whole first tranche of basic building blocks will still be current, so you can jump into the pipeline somewhere downstream." Others, including Jeff Hoffman, found their families asking if they would ever get vacation time.

As with so many best-laid plans, it never happened. Only 27 Shuttle astronauts (about seven percent of the total) launched more than once in a single calendar year; none flew more frequently than that. (One complete crew launched two missions within 12 weeks in April and July 1997, but theirs was a reflight of the same orbiter and the same payload.) A notable example of the difficulties caused by flying too often was Steve Nagel, who in February 1984 was named to pilot a mission planned for launch in October 1985. At the time, he was training for another flight in October 1984. Had both flown as intended, the spacing between them – roughly a year – would have given him a reasonable period of preparation. But problems arose when Nagel's first flight was postponed into June 1985 and his second flight did not move. As a result, he found himself in the unenviable situation of training for two complex missions, spaced only four months apart. The commander of his first flight noticed the problem and successfully negotiated for Nagel to remain on both crews, but it proved far from ideal.

Another astronaut upon whose shoulders mission after mission fell with a high degree of regularity was Bob Crippen, who commanded three flights between June 1983 and October 1984. His last two commands occurred six months apart. The rationale was that as the Shuttle's flight rate quickened, astronauts needed to rotate rapidly from one mission to the next. But this practice uncovered serious operational concerns. When he was assigned to his final command, STS-41G, Crippen was still midway through training for a previous flight and unavailable to join the rest of his crew for several months. Worse yet, that crew was relatively inexperienced and forced to train in his absence. "There was a lot of pressure on us to know what we were doing and not screw up, because Mission Operations were looking at us carefully to see if this is something that could be done or not," said STS-41G astronaut Dave Leestma. "Can you train without one of the crew members, who is doing another flight?"

Fortunately, the only other veteran crew member, Sally Ride, had flown with Crippen before. She knew how he 'ran' a Shuttle flight and was able to act as a surrogate commander for a few months in his stead. "Sally basically was a continuity link in a number of respects, both for Crippen and in very important ways for all of us rookies," said STS-41G's Kathy Sullivan. "We were new to training, we were going to be new to flying and we hadn't worked with Crippen before. Sally could be a stable point and give us some perspective on all three of those things, which would help us know as we went through all the detailed process development and habit-pattern development that comes with training, that we had minimal risk of having to undo some things that we would then have learned deeply when Crip showed up." Yet even when Crippen did join them and brought his wealth of experience to bear on the mission, the crew still found themselves putting in punishing 90-hour working weeks in the Shuttle simulator to get ready. From NASA's perspective, it afforded valuable data to show that astronauts could fly multiple missions each year. But from the perspective of outsiders, including journalist Henry S.F. Cooper, it worked counter to the very grain that it was seeking to create. "It went against NASA's policy of building up a pool of experienced astronauts, essential…to achieve a rate of one flight a month," Cooper wrote in his book *Before Liftoff*. "And using [Crippen] again seemed even more unusual, in light of his late arrival."

Crew dynamics, therefore, proved a critical enabler for every Shuttle crew to function at its best under the most challenging of circumstances. And the central figure in determining the 'tone' of each flight (and the buck-stops-here responsibility for its success or failure) was the mission commander. "There's some amount of loneliness at the top…and with it comes the responsibility for accomplishing the mission," said former Shuttle commander Don Williams. "If something goes wrong, it's not somebody else's fault, it's the person in command's fault. In the Navy, if you run a ship aground, if it happens in the middle of the night and you're the captain and you're in your bunk, it's not the helmsman or the officer of the deck that gets relieved; it's the commanding officer who's accountable. The same thing is true when you command a mission. Any landing you walk away from is a good one."

Over the course of a year or more training together in confined and highly stressful environments, the role of the Shuttle commander in engendering a harmonious crew dynamic was essential. "The commander is responsible for the success of the mission; that's a given," said former Shuttle commander and NASA deputy administrator Fred Gregory. "A secondary role for the commander is to ensure that the crew has had fun. The crew becomes a family. It's not a dysfunctional family. It's a family that accepts the strengths and weaknesses. It's a family that one person does not become so headstrong that he or she believes that the success of the crew is only dependent on one. I always said that a good crew is like a ballet: it's musical in some ways and it's very co-ordinated and beautiful. You stand back from it and you watch it and it appears perfect, like an oil painting, even though there are lots of little flaws."

The success of any given mission was further underpinned by the leadership style of its commander. John Blaha, who served as Gregory's pilot on STS-33 before going on to lead two flights of his own, was particularly impressed by commanders who avoided 'micromanaging' their crews but allowed them to make mistakes as a way of ensuring that they were thoroughly prepared by launch day. "They're all sharp people," pointed out Blaha. "They don't need to be mothered to death. You have to tell them what their responsibility is and then don't look over their shoulder." Fellow Shuttle commander and chief astronaut Robert 'Hoot' Gibson agreed. "My leadership style is not to be barking out orders and telling people what to do," he admitted. "Don't over-supervise. Nobody likes you looking over their shoulder, nitpicking every little thing they do and ready to jump on them if they don't do it right. You don't have to pound your chest. Everybody knows who the commander is. You don't have to try to emphasize it."

Fig. 2.6 The commander of each flight set the tone in terms of expectations for the crew and carried overall responsibility for mission success. STS-107 commander Rick Husband offers a thumbs-up to a crewmate during water survival training.

On several occasions, Shuttle crews were changed midway during training as astronauts were replaced for reasons ranging from illness and injury to disciplinary actions and family tragedy. In July 1990, Gibson and another veteran Shuttle commander, Dave Walker, were reprimanded for infringements of NASA flying rules. Gibson had been involved in a collision with another aircraft whilst participating in a weekend air show. Walker inadvertently flew his T-38 jet trainer too close to a commercial airliner. Both were suspended and lost their respective Shuttle commands but went on to fly subsequent missions. One pilot stepped down from a mission to care for his cancer-stricken wife. In May 1993, Story Musgrave almost lost his spot as chief spacewalker on the first Hubble Space Telescope servicing mission when he suffered severe frostbite during an equipment test in the thermal vacuum chamber; only the timely intervention of a specialist in Alaska saved his fingers. Others lost flight assignments for more obscure reasons, including Mark Lee, who trained for more than two years as chief spacewalker for a major International Space Station (ISS) construction mission, only to be grounded and replaced with no explanation. The seeming unfairness of the incident, wrote Lee's STS-98 comrade Tom Jones in his memoir *Sky Walking*, so disgusted the rest of the crew that they were tempted to resign themselves. But they knew that by throwing their badges on the table, another less experienced crew would be appointed "and our colleagues would have less than a year to prepare for the mission".

Flights were also delayed in response to crew illness. In February 1990, STS-36 commander John Creighton came down with a severe cough and upper respiratory infection only days before launch. He was moved out of the astronaut quarters to limit his contact with his four crewmates, who couldn't resist the opportunity to prank. They served him breakfast, lunch and dinner by pushing a meal tray to his bedroom door using a long-handled push-broom, then retreating, all the while wearing plastic bags over their heads. But the serious impact that illness could have on the successful execution of a multi-million-dollar Shuttle mission was lost on no one. In October of that same year, the Ulysses solar probe needed to be launched within a critically short, three-week 'window'. To hedge against the possibility that any of the Shuttle crew members might fall ill, the unprecedented decision was made to assign a backup team to 'shadow' their training. As circumstances transpired, the phantom crew was stood down. Late in 1990, deep into training for STS-39, astronaut Guy Bluford suffered a herniated disk in his back. Although physical therapy worked well, the malady required an operation. This raised the possibility that Bluford – one of only two veterans on the seven-man crew – might have to be replaced. Fortunately, judicious rescheduling of the crew's training workload by mission commander Mike Coats allowed Bluford to undergo his operation, recover and rejoin the flight. In total, he was out of work for only two weeks and within a month was fit and healthy.

However, these episodes underlined the impact that illness or injury (as well as sheer exhaustion from relentless training and 'sleep-shifting') could have on the preparedness of crews to fly missions safely and successfully. In September 1991, six months before he was scheduled to fly an important Earth sciences mission, astronaut Mike Lampton was replaced on STS-45 following a cancer diagnosis. And in January 1997, eight weeks before he was set to launch, Don Thomas slipped down a flight of stairs and broke his ankle. A replacement was assigned in his stead, but Thomas recovered and was able to fly as planned. Less fortunate were astronauts Christopher 'Gus' Loria and Tim Kopra. In August 2002, Loria was removed from his position as STS-113 pilot three months before launch, following an accident at home which uncovered a career-ending back problem. And the latter broke his hip during a bicycle accident in January 2011, shortly before he was due to serve as chief spacewalker on STS-133. NASA had no alternative but to draft in another astronaut in his stead, producing the shortest training flow – only five weeks – in Shuttle history. A few days before Kopra's accident, Shuttle veteran Rick Sturckow was assigned to shadow the training of STS-134 commander Mark Kelly, whose wife, congresswoman Gabrielle Giffords, had been critically injured in an assassination attempt. At the time of the incident, STS-134 was only three months from launch. Kelly's wife subsequently recovered from her injuries and Sturckow was stood down.

All such events, though unanticipated and traumatic, carried the potential to impair the homogeneity of tight-knit crews and it was fortuitous that astronauts were thoroughly screened for compatibility from the point of selection by NASA. "The selection boards always had a philosophy of, every once in a while, we're going to mess up and we're going to fail to pick somebody that we really should have picked," said Gibson, who sat on several astronaut selection committees. "But we really don't want to pick somebody that we should *not* have picked. They had a very conservative viewpoint. It worked very well." Similar processes were followed when making crew assignments to specific flights. "You spend so much time working together and that's part of the process of crew selections," said Shuttle commander and former chief astronaut Dan Brandenstein. "You don't put oil and water together. You specifically look for people that are compatible. NASA looked for a good mix. They looked for people with specialties that mesh with the mission requirements."

TRAINING, TEAMWORK AND TENSIONS

Following the loss of Challenger, Shuttle commander Hank Hartsfield presented a harsh truth to the investigators of the Rogers Commission. "Had we not had the accident, we were going to be up against a wall," he testified. With only two Shuttle simulators available at JSC, two missions planned for later in 1986 would have required their crews to be fully trained in all mission tasks and emergency scenarios…with no more than 33 hours of 'sim' time. "That is ridiculous," Hartsfield said. "For the first time, somebody was going to have to stand up and say we have got to slip the launch or we are not going to have the crew trained." Training for a given mission could and did mentally and physically burn out even the most experienced astronauts. "The training is structured such that it trains to the lowest common denominator and it just takes forever," noted Shuttle commander Steve Oswald. "You're going through all the stuff again for those that haven't flown before. It got to be kind of a long, drawn-out deal. I had a great time, but afterwards I was just done."

Over the years, more than one astronaut remarked that the training regime was so laser-focused upon recovering from all manner of conceivable emergencies or failures that a totally problem-free launch (as most launches, thankfully, were) came as an anti-climax. "Ninety percent of your training is now irrelevant," joked Blaine Hammond, who piloted two flawless flights in 1991 and 1994. And although many missions were delayed past their original targeted dates, some flew sooner than planned, which imposed another burden on training workloads. One case was STS-48, whose launch moved forward from November to September 1991. Commander John Creighton and his crew, initially given low priority in the simulators, now found themselves at the head of the queue and forced to catch up quickly. "All of a sudden, we found out we were going to leapfrog a couple of other flights and be sooner, rather than later, so then we really had to scramble," Creighton recalled. "That was a tough nine or ten months of very intense training."

Dedicated training for a Shuttle flight usually required at least a year from the point of selection until launch and was designed to stretch the crew both as individuals and as a cohesive unit to respond to all manner of unexpected events. This could impose intense stress upon even the most steely-eyed astronauts. In early 1985, STS-61C commander Hoot Gibson and pilot Charlie Bolden were in the simulator, handling multiple malfunctions as they practiced a launch and ascent run. First, an engine failure was thrown at them, then a minor unrelated issue with an electrical bus. Gibson had flown before, whereas Bolden had not. Spotting the failure, Bolden worked to 'safe' the affected engine. On turning his attention to the second problem, he inadvertently selected the wrong bus and shut down a healthy engine. "When I did that, the engine lost power and it got real quiet," he reflected later. "We went from having one engine down on the orbiter – which we could have gotten out of – to having two engines down and we were in the water, dead!" If that simulation had been a real launch, Bolden would have brought death to them all. Gibson leaned across the cockpit and tapped his pilot on the shoulder.

"Charlie, let me tell you about Hoot's Law."

"What's Hoot's Law?"

"No matter how bad things get, you can always make them worse!"

Fig. 2.7 STS-60 pilot Ken Reightler practices an emergency evacuation from the full-fuselage Shuttle simulator at the Johnson Space Center (JSC) in Houston, Texas. Training was designed to be as challenging and as close an analog to the real experience as possible.

But humans are only human and even seasoned Shuttle commanders fell foul to mistakes from time to time. During one launch simulation, Gibson noticed that the SRBs had not properly separated as they ought to have done at two minutes into the ascent; a potentially catastrophic malfunction. Quickly, he reached across the cockpit, flipped a switch and hit a push button to manually command them away. But in his haste to solve the problem, Gibson hit the wrong switch and accidentally commanded the separation of the ET, rather than the boosters. The result was a rapid shutdown of the main engines, a 'fast-separation' from the tank and SRBs and certain death as the simulated 'Shuttle' plunged out of control into the ocean. "I killed my whole crew," Gibson reflected. "Why had that happened? Because I rushed it. I was in such a big hurry. Nobody had a chance to stop me because I was so quick!" Following the incident, Gibson set a rule that any action in the cockpit deemed 'irreversible' could only be performed with concurrence from another crew member. Henceforth, the commander or pilot would verbally announce their intent to throw a switch and another astronaut would have to respond "I see it and I agree" before it would be executed. It offered a stark reminder that under the conditions of Hoot's Law, the unforgiving nature of the Shuttle and the fallibility of its human pilots could in a split-second transform a survivable emergency into an unnecessary and life-threatening disaster.

Other commanders operated quite differently but no less effectively. In his memoir *Endurance*, astronaut Scott Kelly remembered training for his first mission as a pilot, STS-103 in December 1999. Leading that flight was Curt Brown, who had already flown five previous Shuttle missions and was recognized as one the astronaut office's most talented commanders. During one training run, as Brown worked malfunction after malfunction in the simulator, identifying and focusing on each problem in order of criticality to mission success, Kelly spotted a computer failure. "This would normally be his responsibility too, but because I wasn't as busy and could reach his keyboard myself, I fixed it for him," wrote Kelly. "I typed in the command while Curt's head was still buried in cooling system problems." A few minutes later, Brown looked up from his work to notice that the computer failure alert had gone.

"What happened to the port failure on FF One?"

"I port-moded it for you," replied Kelly.

"You did what?"

Brown was furious. Despite being encased in a bulky pressure suit, strapped to his seat in the simulator of the Shuttle's tiny flight deck, Brown reached across the cockpit and punched Kelly sharply in the arm, ordering him never to do it again. "He'd made his point and though I didn't agree with his method, I appreciated his directness," Kelly wrote. "I never touched any buttons or switches on his side of the cockpit again without his explicit approval."

The episode demonstrated not only that astronauts were perfectly human with very 'human' personalities, but also that training for a Shuttle mission was designed to be as stressful as the real thing. Overseeing this high-stakes training was a team of simulation supervisors, who earned a fearsome reputation for themselves. It was their task to introduce all manner of systems failures and unexpected gremlins into training exercises to stretch (and stress) the crew in conditions approximating a real mission as closely as possible. "In real life," quipped former flight director Matt Abbott, "things aren't breaking as much as they do in simulations." And whilst it was never the simulation supervisors' intent to give the crew so many emergencies that they could not possibly recover from them all, the exercise certainly helped to forge crews and Mission Control teams into tight units which grew to understand the very tone of each other's voices. One simulation supervisor who later became an astronaut could not resist bringing a little of his previous career into his new one. During one simulator run, with chief astronaut John Young in the commander's seat, he mischievously pushed the SSME shutdown switches. The cabin went ominously quiet and Young was forced to declare an emergency and execute a Transoceanic Abort Landing (TAL). Retelling the anecdote in his memoir, Tom Jones wondered what the new astronaut was playing at. "Why would you kill an engine on the most senior astronaut?" he wondered with incredulity. Finally, the culprit burst out laughing and confessed. The former instructor had just been keeping his hand in, feeding Young one last malfunction. To everyone's surprise, Young loved the practical joke.

Nor was it just the astronauts who were being tested by the simulation supervisors; so too were the flight controllers in Mission Control, who were responsible for overseeing every aspect of a Shuttle mission and whose training enabled them to sniff out the tiniest nuances in systems and the most complex software problems. "This was about the most challenging job you could ask for," remembered flight director Jay Greene. "The simulations were mental simulations that were as challenging as anything NASA has to offer. There were two things going on. One was the goal to train the crew to work with the control center and, at the same time, train maybe a dozen different operators to the max extent possible. Instead of having one failure – which is about the most you'd expect during a launch – they'd try and give everybody something to play with and the flight director would have to co-ordinate everybody's problems and come out with a solution that got the crew safely to orbit or resulted in a successful abort and recovery. During the course of a day, we'd run maybe eight launch sims and every sim had maybe ten different faults that the [simulator supervisors] would put in. By the end of the day, you had somewhere between 80–100 problems that you dealt with." And the result of those challenging days of training was a level of camaraderie unseen in almost any other walk of life. "When you get through beating each other up, you've got to 'fess up," said Neil Hutchinson, "because there's no hiding when you fouled something up."

No matter how closely the simulators could mimic the conditions of a real Shuttle launch, their value vanished at the second the SSMEs and SRBs ignited for real on the pad and the shackles of Earth were broken. "They try to fake you out a

Fig. 2.8 Flight engineer Roberto Vittori (foreground) and pilot Greg Johnson (left) practice ascent procedures in the Shuttle simulator during training for STS-134, the final voyage of Endeavour in early 2011.

little bit by tipping the Shuttle simulator," said Shuttle commander Paul Weitz, "but it doesn't compare with…three main engines and two solids going. You know you're on your way and you're going somewhere and you hope they keep you pointed in the right direction, because it's an awesome feeling." Sally Ride added that the sensations in the simulator produced a surprisingly close analogue for the real thing. "It shakes about right and the sound level is about right and the sensation of being on your back is right," she said. "It can't simulate the G-forces that you feel, but that's not too dramatic on a Shuttle launch. The physical sensations are pretty close and, of course, the details of what you see in the cockpit are very realistic. The simulator is the same as the Shuttle cockpit and what you see on the computer screens is what you'd see in flight." But the visceral sense of climbing into space atop a huge amount of highly explosive rocket fuel was a point at which all simulators unequivocally fell short.

MORALE AND FEAR

Type-A personalities though most astronauts were, the sense of fear and the sapping of morale as the Shuttle program passed through two human tragedies – the loss of Challenger during launch in January 1986 and the destruction of Columbia during re-entry in February 2003 – was palpable. Both incidents underlined two key points: a growing realization that the risk of a disaster were far greater than previously calculated, once cited as one-in-350, and the immense bravery of the Shuttle astronauts. "I don't think I'd want to go on an airliner with a one-in-350 chance that the thing is going to crash or have a terrible catastrophe," said former NASA Johnson Space Center (JSC) director Jefferson Howell. "The astronauts realized there was a risk. I don't think they realized how big of a risk. They're very brave men and women, because that reality came forth and we realized that there wasn't much else we could do to stymie that risk, given the way that thing is engineered and built."

In the months following Challenger, the astronaut corps hemorrhaged members with no less than five resignations and reassignments. For Kathy Sullivan, the media's unwavering focus upon the death of schoolteacher Christa McAuliffe was distasteful. "Excuse me, there were a few other people on that vehicle, some of whom had only sworn their life to service of their country," she told the NASA oral historian. "They didn't win the prize and get the ride. They were actually committed to this. That and a thousand other angry things, of course, are swirling through my head." For more than a handful of astronauts' spouses, there was a sense that their wives or husbands had flown and returned safely. What was the purpose in sticking around to do it again, they wondered? "Let's go home," some astronaut wives told their husbands. "Let's go someplace. I don't want to see this happen to you."

And the families, indeed, were the oft-forgotten element in the equation. George 'Pinky' Nelson, who flew the last mission before Challenger and the first flight after the accident, remembered his work to develop a family support plan. It took the form of a formally signed document which outlined how NASA would care for the astronauts' families during missions and after any contingencies. Previously, astronauts were personally responsible for getting their families out to Florida, paying their way and finding motel rooms. Spouses would be flown out to the launch and landing site, but not children, which placed an additional burden on friends and neighbors. Jeff Hoffman, who flew five Shuttle missions, was vocal about the problem. "Nowadays, NASA gets the condominiums and all the crew stays together, but back then you were on your own," he recalled. "They would take you up to the roof of the Launch Control Center on launch day, but that was basically it." Prior to the many-times-delayed STS-61C mission in January 1986, Nelson's wife, Susie, spent three weeks in a Cape Canaveral condo and their children missed a great deal of schooling. "Had the accident occurred on that flight, instead of the flight afterwards," Nelson said, "it would have been just a nightmare, because the families were scattered all over the place." During preparations for one mission in late 1990, Hoffman's children had boarded an aircraft in Houston for the flight to Florida. Just as the aircraft's engines roared to life, the launch was canceled and the disappointed children were forced to disembark and return home.

Arriving in Florida for launch was always difficult. "The worst thing is saying goodbye to your kids," said Shuttle commander Andy Allen in a Smithsonian interview. "My process in the quarantine period before launch was to get my will all squared away and write notes and letters to my kids." Before each of his three Shuttle missions, Allen created a recording of songs loved by his family. "Having been in the Marine Corps and been on aircraft carriers and had gazillions of close calls as a fighter pilot, nothing is as stretched-out as getting ready for a spaceflight," he added. "Part of it was because my kids…were at an age where they understood what was going on: Daddy might blow up and he might not come back."

In the aftermath of Challenger, 'family escorts' – fellow astronauts who knew the families as close, personal friends and knew and could explain the risks – were assigned to each mission to support wives, husbands, children and extended relatives should the unthinkable happen. And Casualty Assistance Call Officers (CACOs) were appointed to create a liaison point between the family and NASA. It would be their responsibility to deal with all manner of tough questions, ranging from the location of the dead astronauts' wills to whether loved ones wanted to be buried in military or local cemeteries. After the loss of Columbia, astronaut Mike Massimino recalled that he and others did whatever was necessary to help, but the shock affected even these steely-eyed missile-men (and women) as much as it did the families. Standing in a supermarket queue one day, he unexpectedly zoned out. Massimino had his wallet ready for the cashier, he wrote in his memoir *Spaceman*, but suddenly he completely forgot why he was in the shop. Fortunately, one of Massimino's neighbors was behind him in the queue. She paid the cashier.

"Be kind to this man," she said. "He's had a rough day."

Fig. 2.9 Commander Dick Scobee leads crewmates Judy Resnik, Ron McNair, Mike Smith, Christa McAuliffe, Ellison Onizuka and Greg Jarvis out to the launch pad for Challenger's catastrophic final launch on 28 January 1986. As a civilian schoolteacher, McAuliffe won the media spotlight, at the expense of her fellow astronauts.

Military officers among the astronaut corps handled the situation differently, to a degree. Mike Lounge acquiesced that former fighter pilots would never admit to fear, although it remained ever-present. In the final minutes before launch on STS-61B in November 1985, pilot Bryan O'Connor – about to make his first Shuttle flight – glanced across the cockpit towards his commander, veteran astronaut Brewster Shaw. He watched as Shaw calmly removed his gloves and wiped sweat from his palms. "Oh, my God," O'Connor thought. "My commander, who's been through this before…his hands are sweating! Why aren't mine sweating? I need to be nervous now, if he's nervous!"

Several years later, a six-man crew was in training for STS-44. Four of the astronauts were first-timers, with only commander Fred Gregory and mission specialist Story Musgrave having flown before. In the final minutes before liftoff, as the first-timers exchanged humorous stories and banter, they suddenly noticed that Musgrave was deathly silent.

"Hey, Story, why are you so quiet?"

"Because I'm scared to death."

At that moment, recalled STS-44 rookie Jim Voss, the rest of the cockpit grew equally silent. Musgrave had served in the Korean War. He had logged thousands of hours in high-performance jet aircraft. He had seen fellow astronauts – friends – die aboard Challenger. And now he was about to embark on his fourth Shuttle launch. It was time to get serious. If Musgrave was frightened, then so too should they be.

PASSENGERS, POLITICIANS AND THE FALLACY OF 'ROUTINE'

In its early life, the Shuttle existed as a pawn in the hands of its political masters, sold as a commercial vehicle to be operated as routinely and reliably as an airliner. And as outlined earlier in this chapter, part of its 'package' to customers was NASA's offer of seats for non-professional astronauts (known as 'payload specialists') to accompany satellites which were being launched on behalf of their parent organizations or to conduct experiments sponsored by government, industry or academia. It was a bonus that no other launch provider could offer. But in the heady days before the loss of Challenger, as missions were delayed and payloads moved around on an increasingly unsettled flight manifest, so too did the payload specialists. In early March 1985, one crew had been training for months to retrieve a large scientific satellite from orbit and bring it back to Earth, when all at once their mission was canceled. In its stead, they were assigned three communications satellites – including one for Saudi Arabia – and a French-built experiment. Along with those new payloads came two 'new' payload specialists: a Saudi prince and a French fighter pilot. At first, the astronauts considered themselves to have more in common with the Frenchman (and the mission's commander even warned his crew not to tell any harem jokes when the prince was around), but the opposite turned out to be true. On their first meeting with the prince, he introduced himself with a self-deprecating joke: "I've left my camel outside!"

Humor aside, this new category of non-professional astronaut brought other worries to the door of Shuttle commanders, including very real threats to flight safety and questions about how the reusable spacecraft – a billion-dollar national asset of the United States – should properly be used on an international stage. On one mission, Mexico's first astronaut was slotted into the STS-61B crew only four months prior to launch to observe the deployment of his country's communications satellite. This unnerved the flight's commander, Brewster Shaw, who had no idea how the newcomer might behave in space. "I'm probably a paranoid kind of guy, but I didn't know what he was going to do on-orbit," Shaw remembered. "I remember I got this padlock and…went down to the hatch on the side of the orbiter and I padlocked the hatch control, so that you could not open the hatch." Fortunately, the Mexican astronaut proved a model crew member.

These short-notice crew assignments underlined an increasingly uncomfortable rubbing of shoulders between professional and non-career astronauts. "We thought it was not appropriate to take inexperienced people and put them on there, because there's a lot of resources in getting them ready to fly," said Hank Hartsfield. "You had to determine their personality. It was hard to do in a short period of time. Were they going to be stable? If you had a problem on-orbit, am I going to have to babysit this person or are they going to be able to respond to an emergency situation and take care of themselves like the crew has to? It could be detrimental."

Fig. 2.10 Clad in their orange partial-pressure Launch and Entry Suits (LES), the STS-40 astronauts – including pilot Sid Gutierrez (right) – participate in the Terminal Countdown Demonstration Test (TCDT) at the launch pad. The astronauts are in the process of boarding the slidewire escape baskets to simulate a Mode 1 Egress from the Shuttle.

Other interpersonal issues also came into play. When Bryan O'Connor was named to command STS-40, he expressed serious reservations of another kind about the compatibility of his payload specialists: physician Drew Gaffney and biochemist Millie Hughes-Fulford. "Millie and Drew were like oil and water," O'Connor told the NASA oral historian, "and it was a pleasant surprise for me when, seeing how they operated, or didn't operate together in the office, or after the training's done – to where they would take all that baggage, old concerns with one another's performance, disagreements about the science – and we'd get into the simulator and there was none of that. These two trained like two professionals." Yet when the pair left the simulator, O'Connor likened their relationship to the 19th-century feud between the Hatfields and McCoys. "Everybody just assumed that everybody would get along," said STS-40 astronaut Rhea Seddon. "Sometimes that's hard to do, especially when you've been training together in close quarters for a long time. Everybody has their little quirks." To ensure that they would function under the extreme physical and mental duress of a real mission and potentially handle a real in-flight emergency, O'Connor requested a Myers-Briggs personality assessment for his crew. The results revealed that Shuttle crews, with their mix of military officers and civilians from widely disparate backgrounds, featured a much broader range of personalities. Another astronaut who took up the option of a Myers-Briggs assessment was STS-45 commander Charlie Bolden, whose crew included a pair of payload specialists who had been in training for over a decade. The tests revolved around six personality types, two of which described focused, goal-oriented individuals, a trait that is representative of 15 percent of the general population and a whopping 98 percent of the astronaut corps population. Such personality characteristics for Type-A astronauts were perhaps unsurprising, but Bolden found the psychometric tests particularly enlightening in that they uncovered each crew member's strategies for handling periods of calm and extreme stress.

In later years, Shuttle commanders took their crews – both NASA astronauts and payload specialists, together with flight directors – on team-building expeditions sponsored by the National Outdoor Leadership School (NOLS). Working with instructors, the astronauts hiked into the mountains for nine or ten days. "We got to learn a lot about how each of us, as individuals, deal with the kind of situations that they put us into," said STS-107 commander Rick Husband. "It's a physical challenge with the backpacks. It's also a challenge learning how to keep track of all your equipment personally, then learning to work together, pulling together and learning more about each other, so that when you come back…you know each other's strengths and weaknesses and so you can maximize that during the rest of your training flow." Matt Abbott described the NOLS excursions as an essential means of cementing bonds of trust not only among the crew, but also with Mission Control. "It's very important, especially for the lead flight directors and some of the lead flight controllers, to develop a friendship and camaraderie with the crew," he said. "To be part of that team and this huge ground team that's supporting all that, the more of that kind of bond that you can develop that you can develop, the more successful the mission will be. You develop a 'feel' for how the other people are working." Astronaut Mike Massimino, in *Spaceman*, described one team-building expedition to Cold Lake in western Canada. "When people are put in extreme circumstances for a long period of time," he wrote, "or even just removed from their normal routine, they get angry more quickly, teams split apart, trust and communication can break down."

Nor were foreign nationals the only payload specialists under consideration in the pre-Challenger era; so too were politicians, teachers and journalists. Senator Jake Garn from Utah was the chair of the appropriations subcommittee responsible for NASA's budget, and he frequently pestered NASA administrator James Beggs for a seat on the Shuttle. In his oral history, astronaut Jeff Hoffman learned about Garn's assignment whilst finishing a training session. At first, he thought it was a joke, but Garn – an experienced jet pilot – also turned into a model crewman and shone a positive spotlight of publicity on their mission. Several months after Garn's flight, congressman Bill Nelson from Florida also flew a mission. Opinion in the astronaut corps about the wisdom of this practice was mixed. Some considered it a flagrant attempt by NASA to curry favor with government officials. Others poked fun at the politicians' training by posting sign-up sheets on astronaut office bulletin board, asking for volunteers for an equally short course to become a fully-fledged senator. Their mocking fury was not entirely misplaced, for they had paid their dues through tours of Vietnam, ferocious workloads at test pilot school and years of academic study. More frustrating was the fact that other payload specialists with justifiable experiments were being shifted onto later flights to make room for politicians. One particularly tragic example of this practice was Hughes Aircraft engineer Greg Jarvis, who was bumped from STS-61C by Bill Nelson, only to end up launching (and losing his life) aboard Challenger a few weeks later.

In July 1984, the Teacher in Space Project was initiated and NASA was directed to find a gifted educator who could communicate enthusiastically with students from orbit. Eleven thousand applications were winnowed down to 114 semi-finalists and in July 1985 Christa McAuliffe, a social studies teacher from Concord in New Hampshire, was announced as the winner. She began training with the STS-51L crew the following September. "Teachers teach the lives of every kid in this country through the school system and if you can enthuse the teachers about doing this, then you enthuse the students and impress on them that's something to expect in their lifetime," said STS-51L commander Dick Scobee. "As far as I'm concerned, it's a good insurance policy for the human race." But not all astronauts agreed with the philosophy of flying civilians on the Shuttle. "This was an unhealthy environment," admitted John Fabian. "We were taking risks that we shouldn't have been taking. We were shoving people onto the crews, late in the process, so they were never fully integrated into the operation of the Shuttle and there was a mentality that we were simply filling another 747 with people and having it take off from Chicago to Los Angeles. This was not that kind of vehicle, but that's the way it was being treated at that time."

Fig. 2.11 John Glenn practices an emergency escape from a troubled Space Shuttle during training for STS-95. This exercise simulated a Mode 5 Egress, in which the astronauts would evacuate the vehicle under their own steam by means of a Sky Genie after landing.

Even payload specialists with justifiable missions to perform found themselves infuriated by seemingly inconsequential additions to the flight plan. On
STS-51F in July 1985, a powerful payload of solar physics telescopes was
accompanied by a secondary commercial 'experiment' to test drinking dis-
pensers for Coke and Pepsi. On the day before Challenger launched, the seven-
man crew – including solar physicist Loren Acton – were working through a
briefing on the latest scientific data ahead of their mission. Suddenly, NASA's
chief counsel entered the room to brief them on the Coke and Pepsi protocols.
Acton hit the roof.

"We've been getting ready for this mission for seven years," he thundered. "It
contains a great deal of science. We have a very short time to talk about the final
operational things that we need to know. We don't have time to talk about this
stupid carbonated beverage dispenser test. Please leave!"

The chief counsel left.

Sadly, in the months before January 1986, the relentless efforts to 'com-
mercialize' the Shuttle did not leave with anything like the good grace of the
chief counsel. In fact, when McAuliffe, Scobee and the remainder of the STS-
51L crew perished aboard Challenger, other civilians were queuing up for
missions of their own. A Journalist in Space project had already been pared
down to 40 semi-finalists, including Pulitzer Prize-winners John Noble Wilford
and Peter Rinearson and veteran CBS News anchorman Walter Cronkite. At
the time of the accident, they were set to be physically and psychologically
screened at JSC in April 1986, before the primary and backup candidates
began training in May for a September launch. This practice of flying non-
essential crew members on the Shuttle ostensibly ended with Challenger, or so
it seemed, for in late 1997 John Glenn – former Project Mercury astronaut and
senator – requested a mission for himself. Aged 77 at the time of his flight on
STS-95 in October 1998, Glenn became (and remains) the oldest person ever
to have launched into space. It prompted a raft of criticism from former astronauts who had seen friends die on Challenger. "The Shuttle is not an airliner,"
wrote Mike Mullane. "Visualize a launch pad emergency, in-flight bailout or a
crash landing. Trying to escape from a Shuttle in a time-critical situation…
requires lightning-fast reactions." Mullane was particularly worried about how
the elderly Glenn, clad in a full pressure suit, survival gear and parachute,
would negotiate climbing from the middeck to the flight deck in an emergency,
clambering through an overhead window and rappelling his way down the side
of the vehicle.

Aside from the smiles, the thinly veiled frustration of the astronaut corps
towards civilians, passengers and the fallacy that the Shuttle was the spacefaring
equivalent of an airliner came home on 18 January 1986, just ten days before

Challenger was lost. Her sister Columbia had just landed at Edwards Air Force Base in California, having been diverted from Florida due to poor weather. Shortly after touching down, crewman Steve Hawley called his father.

"Well," said the elder Hawley, "this just proved to me now that you guys are really operational and that the Space Shuttle is just like an airline."

His son was perplexed by the remark. "Why do you say that?"

"Because you're in California and your luggage is in Florida!"

3

Blowing The Bolts

"WHEN YOU GOT DEBTS"

"John, we can't do more from the launch team than say, we wish you an awful lot of luck," radioed Launch Director George Page from the Launch Control Center (LCC) at the Kennedy Space Center (KSC) in Florida, a few minutes before 7 a.m. EDT on Sunday, 12 April 1981. "We are with you one thousand percent and we are awful proud to have been a part of it. Good luck, gentlemen." Out on historic Pad 39A, Columbia – bolted like an ungainly butterfly onto a pure-white External Tank (ET) and twin Solid Rocket Boosters (SRBs) – stood ready to embark on the Space Transportation System's first mission, a two-day test flight designated STS-1. Commander John Young responded with thanks. Next to him was pilot Bob Crippen. Both men had been training for STS-1 for three years and, if not for a last-minute computer malfunction, they might have launched two days earlier.

In the pre-dawn darkness of the 10th, clad in bulky pressure suits, the two men left their quarters in the Operations and Checkout Building and took a short bus ride out to Pad 39A, to be buckled into their seats aboard the orbiter. The countdown proceeded uneventfully, until its final pre-planned 'hold' point at T-9 minutes, when an issue arose with one of Columbia's five General Purpose Computers (GPCs). It was described by NASA as a "timing skew"; in essence, the backup flight software was unable to properly synchronize with the primary set. Unlike earlier manned spacecraft, the Shuttle was the ultimate 'fly-by-wire' machine, totally reliant upon its GPCs to run a multitude of different functions. "Even the cockpit where the crews have control sticks and switches…they don't go back to the engines or the aerosurfaces," said former Shuttle director Arnold Aldrich. "They go to the computers and the computers then tell these things what to do." Indeed, the computers were so critical to mission success that four 'primaries' all

© Springer Nature Switzerland AG 2021
B. Evans, *The Space Shuttle: An Experimental Flying Machine*,
Springer Praxis Books, https://doi.org/10.1007/978-3-030-70777-4_3

carried the same flight software load and 'voted' before issuing commands. Known as 'Fail-Op, Fail-Safe', if one primary disagreed with the others it was outvoted and deemed faulty. The backup GPC carried an entirely different set of flight software, so that if all four primaries became corrupted, it could assume control and guide the vehicle safely home. However, whilst it could support ascent and re-entry commands and control the Shuttle in orbit, the backup GPC did not have the capability on its own to execute a full mission. The problem on 10 April 1981 hinged upon improper communication between the four primaries. Liftoff was delayed as engineers wrestled the problem, but when a solution was not forthcoming the launch was postponed until the morning of the 12th.

The rationale behind the Shuttle's computers, remembered Jack Garman, NASA's former deputy chief of the spacecraft software division, was not to have an ultra-reliable system, but a highly redundant one, with plenty of backup capability. "The whole notion on Shuttle was to try to have 'quad redundancy', to get fail-op, fail-safe," he said. But a winged vehicle with the infinite complexity of the orbiter made it difficult to incorporate such elevated levels of redundancy into every system. "It's kind of hard to do that on wings," Garman added. "There's some pretty fundamental things you can't do it on, but in general, on systems that could fail, they had quad redundancy. You have identical software in four machines and if there's a bug in that software – as we proved many times in the Shuttle program – all four machines will obediently do exactly the same thing." Great expense had already been devoted to developing the software and astronauts and flight controllers had worked around the clock in the Shuttle Avionics Integration Laboratory (SAIL) to wring out its idiosyncratic problems.

"The software became the biggest stumbling block," remembered Shuttle commander Gordon Fullerton. "The software in these computers not only controls where you fly and the flight path, but almost every other subsystem. Getting the software wrung out and simulators writing the checklists, we didn't really have it nailed down by STS-1. You cannot be 100-percent sure of everything." In fact, GPC problems cropped up repeatedly during the Shuttle's career. Launch attempts in June 1984 and August 1985 were scrubbed partly on account of problems with the backup GPC. In-flight maintenance was necessary on several missions when Cathode Ray Tube (CRT) monitors went blank and entire GPCs failed and necessitated their replacement in space by the astronauts (as happened on STS-30 in May 1989). Still others exhibited software glitches, whilst others failed prior to re-entry. The upgraded AP-101S GPC – with a memory 2.5 times larger and a threefold increase in processor speed over the original AP-101B – first flew on STS-37 in April 1991 and had been installed in all orbiters by the end of 1993. And from STS-101 in May 2000 the $80.5 million Multifunction Electronic Display Subsystem (MEDS) of 11 full-color, flat-panel displays, based upon the Boeing 777's 'glass-cockpit' architecture, entered service. This was fully installed across the entire surviving Shuttle fleet by 2005.

Fig. 3.1 STS-1 commander John Young participates in an exercise in the Shuttle Mission Simulator (SMS) at NASA's Johnson Space Center (JSC) in Houston, Texas.

But much of this still lay far in the future as Columbia prepared for her maiden launch. Early on 12 April 1981, after technicians had installed a software 'patch' to correct the GPC timing issue, Young and Crippen were back aboard the orbiter for their second attempt at getting into space. And on this occasion, the final minutes ran like clockwork. At T-31 seconds, control of the countdown was handed over from the human-tended Ground Launch Sequencer (GLS) to the autonomy of the GPCs. With ground computers now assuming a backup 'monitoring' role, it was down to the Shuttle's electronic brain to monitor hundreds of different functions and ensure that the most complex experimental flying machine ever built was ready to go.

"T-20 seconds and counting," intoned NASA public affairs announcer Hugh Harris. "T-10, nine, eight, seven, six, five, four…we've gone for Main Engine Start…"

Six seconds before 7 a.m. EDT, anything still sleeping in Florida was jarred instantly and abruptly awake by a low, growl-like rumble which intensified into a thunderous crescendo. Beneath the dark bells of the three Space Shuttle Main Engines (SSMEs), a shower of glowing sparklers suddenly materialized, generated by the launch pad's six hydrogen burn-off igniters. These removed excess quantities of hydrogen gas lingering beneath the engines to eliminate the risk of it igniting and damaging either the Shuttle or the launch pad with the resultant blast wave at liftoff. As the SSMEs roared to life at 120-millisecond intervals, a sheet of translucent orange flame gave way to three dancing Mach-diamonds, as supersonic exhaust gases surged perfectly from the engine bells. A vast cloud of steam rapidly obscured the Shuttle and billowed into the clear Sunday morning air. Suddenly, Young and Crippen realized they were about to ride a wild animal into space.

"We have Main Engine Start," continued Harris, his voice notching up an octave in pent-up excitement. His next words were drowned out by the staccato crackle of the twin SRBs, which ignited at T-0, precisely on the hour. Columbia punched her way off Pad 39A and quickly cleared the launch tower to the cheers, tears and fears of 3,500 gathered spectators and hundreds of thousands more watching on television. "We have liftoff of America's first Space Shuttle," Harris gushed, "and the Shuttle has cleared the tower!"

From their seats on the flight deck, Young and Crippen felt the vehicle rock perceptibly, back and forth, as the momentum of the engines produced a noticeable 'twang' effect, then a correspondingly sharp increase of noise in the cabin. And although Crippen later remembered that the racket of Main Engine Start certainly got their attention, it was the kick-in-the-butt ignition of the SRBs which convinced them that they were really heading out of town in a hurry. Astronaut Jerry Ross, who would fly the Shuttle seven times, described the sensation as like someone had taken a baseball bat and given the back of his seat a smart whack. For the first few seconds of the climb, Young and Crippen's instruments were blurred, though not unreadable, by the incessant shuddering. At ten seconds, as the GPCs commanded the Shuttle 'stack' to roll over onto its back and establish itself onto the proper heading for a 40.3-degree-inclination orbit, the astronauts noticed that the vibrations had lessened sufficiently for them to read their displays.

"Roll Program," radioed Young, as Columbia executed an axial roll to orient the vehicle onto the correct flight azimuth before hitting a period of maximum aerodynamic turbulence (colloquially named 'Max Q') a minute into the mission. "When you get the vehicle going uphill and you're still in the sensible atmosphere, there are tremendous aerodynamic pressures on it," reflected flight director Neil Hutchinson, who was sitting in Mission Control that morning. "You have to get the angle at which it is going through the airstream exactly correct. It has a very narrow performance corridor. You have to get that roll out of the way and get that whole thing set up long before you get the maximum [dynamic pressure], when the amount of atmosphere combined with the direction the vehicle's going and the velocity is the worst." Seated near Hutchinson was astronaut Dan Brandenstein, serving as Capsule Communicator (or 'Capcom'), the only person to talk directly with the crew. Upon receiving confirmation from Young that the roll program was underway, Brandenstein replied with a brisk, up-tempo: "Roger, Roll!"

Climbing through the low atmosphere, the wind noise outside Columbia's heavily reinforced cockpit intensified into something not dissimilar to a scream-like trill. A minute after departing Pad 39A and leaving the coastline of Florida behind, the stack approached 15 kilometers and passed smoothly through Max Q. To avoid exceeding the structural limitations on the airframe during this highly dynamic phase, the GPCs commanded the SSMEs to throttle back to 67 percent, about two-thirds of their rated performance. Shortly afterwards, with Max Q behind them, they were throttled back up to full power. One veteran Shuttle commander likened it to riding a roller coaster at the crest of a hill, the only difference being that – just before the start of the anticipated big drop – there was no drop and the adrenaline kept going.

"Columbia, you're Go at throttle up," called Brandenstein.

"Roger, Go at throttle up," replied Young.

"I still go back and listen to the tapes," Brandenstein told a NASA oral historian. "The first call or two, I was almost yelling at them. I was kind of pumped up, I guess."

The incessant guttural noise of the SRBs became more sporadic and had decreased to virtually nothing as the time approached, two minutes into ascent, for their separation. The astronauts witnessed an orange-yellow flash of light which streamed across Columbia's nose and forward windows as the boosters' separation motors fired and pushed them away. Their departure was accompanied by a harsh grating sound and a fair amount of gunk deposited on the Shuttle's windows, although both SRBs performed nominally, parachuting into the Atlantic Ocean about 250 kilometers downrange of the launch site for refurbishment and re-use.

By this point, Young and Crippen felt the odd sensation that they had ceased accelerating, as if they were riding a smooth electric engine. "It all goes quiet and peaceful," said astronaut Bill Lenoir. "The sensation you have is that you're losing out, that you're falling back into the water," added fellow astronaut Terry Hart. "You don't think you're accelerating as much as you should be going. I'd worked on the main engine program anyway, so I was very familiar with what the engines could do or not do. I think in the next minute, I must have checked the main engines to make sure they were running, because I'd swear we only had two working. It just didn't feel like we had enough thrust to make it to orbit. Then, gradually, the External Tank gets lighter and as it does, of course, with the same thrust on the engines, you begin to accelerate faster and faster. After a couple of minutes, I felt like – yes – I guess they're all working." Other crews noticed the same phenomenon. On STS-65, the sudden and unearthly silence after the departure of the SRBs was so unsettling that pilot Jim Halsell instinctively leaned forward in his seat to check that the SSME data-tapes were still registering 'green'.

Fig. 3.2 Columbia rises from Pad 39A for the first time at 7 a.m. EDT on 12 April 1981.

By this stage of flight, the Shuttle/ET combo was already well above much of the sensible atmosphere. And with the SRBs gone, Young and Crippen found it noticeably easier to flip switches. Columbia pushed on for another six minutes, reaching Mach 19 – almost 23,340 kilometers per hour – before the SSMEs were throttled back to maintain three times the force of terrestrial gravity and avoid over-exerting the airframe. Throughout ascent, Young's heart rate peaked at 90 beats per minute and that of Crippen a little more at 130. After the mission, Young quipped that his old heart simply refused to beat any faster. But Neil Hutchinson had another explanation. Perhaps, he wondered, with a twinkle in his eye, the unflappable Young had been asleep the whole time.

Eight minutes and 30 seconds after rocketing off the face of the Earth, the astronauts felt themselves getting lighter and lighter. Then, all at once, the GPCs commanded the SSMEs to shut down and the equivalent of three times the force of terrestrial gravity – like having a gorilla sitting on their chests – was instantly replaced by weightlessness. Nineteen seconds later, the ET, its liquid oxygen and hydrogen propellants depleted, was jettisoned to burn up safely in the atmosphere over a sparsely inhabited stretch of the Indian Ocean. Young and Crippen grinned as the first vestiges of weightlessness manifested themselves in the form of washers, filings, screws and bits of wire which began to liberate themselves from every nook and cranny in the cabin and float, comically, in mid-air. And less than an hour later, a 'burn' of Columbia's two Orbital Maneuvering System (OMS) engines circularized their orbit at 145 kilometers. The maiden voyage of the Shuttle was formally underway.

But the success of getting Columbia into orbit masked the reality that STS-1 was arguably the most dangerous experimental test flight in history. All other U.S. manned spacecraft, from Mercury, Gemini and Apollo of yesteryear to Crew Dragon, Starliner and Orion of today and tomorrow, were (and will be) extensively tested in an unmanned capacity before ever being entrusted with humans. On STS-1, by stark contrast, astronauts rode this volatile beast on its very first launch. The sheer gutsiness of Young and Crippen that day was profound. Yet there were technically sound reasons behind the decision. Although an 'autoland' capability was devised, many astronauts considered it extremely difficult to fly and land the vehicle without a crew. "You've got to have a pilot there to land it," insisted Shuttle commander Joe Engle. "Fortunately, the certification of the autopilot all the way down to landing would have required a whole lot more cost and development time…and I think the rationale that we put forward to discourage the idea of developing the autoland was that you can leave it engaged down to a certain altitude, but you always have to be ready to assume that you're going to have an anomaly and the pilot has to take over and land. The pilot is in a much better position to effect the final landing, having become familiar and acclimated to the responses of the vehicle."

"It would have been very difficult to have devised a scheme, in my view, to have flown that program unmanned," added fellow astronaut Fred Haise. "I guess you could've used an RF link and really had a pilot on a stick on the ground, like they have flown some other programs. But to totally mechanically program it to do that (and inherent within the vehicle) would've been very difficult." Shuttle commander Hank Hartsfield added that the relative feebleness of the Shuttle's computers also precluded the reliable deployment of autoland functionality. "We had separate software modules that we had to load once on orbit [and] into the entry," he said. "All that had to be loaded off storage devices or carried in core, because the Shuttle computer was only a 65 K machine. When you think about the complexity of the vehicle, with full computers voting in tight synch and flying orbital dynamic flight-phase and giving the crew displays…and doing that with 65 K of memory, it's mind-boggling. Take the little computer you carry in your purse and it's got more memory than that!"

Whatever one's opinion, Young and Crippen were under no illusions that they might not come home from STS-1. One day, shortly before they entered quarantine for the final time before launch, astronaut Joe Allen bought Young his lunch in the cafeteria. Later that day, Young promptly paid him back. Allen laughed and told him to forget it.

"No," retorted Young. "You don't go fly these things when you got debts! All my debts are paid."

SEATS, SUITS AND SURVIVAL

Shortly before the SRBs burned out and separated, Brandenstein radioed an unusual call to Young and Crippen, telling them "Negative Seats". It was an advisory that they had now reached an altitude too high for the Shuttle's ejection seats to be of any further use. Privately, many astronauts doubted the usefulness of these rocket-propelled seats, but both they and the high-altitude pressure suits worn by Columbia's first four crews – STS-1 to STS-4, the Orbital Flight Tests (OFTs), conducted between April 1981 and July 1982 – were intended to provide as much full-body protection as possible on this untamed vehicle. The OFT missions carried only two crewmen (a commander and pilot), which made it relatively straightforward to accommodate the seats and bulky suits in the Shuttle's tiny cockpit. But as it became 'operational', flew more regularly and carried as many as seven astronauts on each mission, practicality steadily and stealthily eroded safety and good sense.

Ejection seats had been an integral component of the Shuttle's crew-survival philosophy from the outset. During the Phase B design process, prime contractor North American Rockwell (later Rockwell International) worked on multiple systems: one derived from the seats of the SR-71 Blackbird reconnaissance aircraft, another featuring an 'encapsulated' design drawn from the XB-70 Valkyrie experimental bomber and another utilizing an entirely separate crew cabin, not dissimilar to those of the F-111 Aardvark and B-1A Lancer. In April 1971, NASA announced that "provisions shall be made for rapid emergency egress of the crew during development test flights", rationalizing this position with an expectation that Shuttle missions would become 'routine' over time. Unknowns would gradually become known, so the logic went, to permit a smooth transition from experimental flying machine to full, airliner-style operability. Of course, the Shuttle was far too immature and too complex to ever reach this level of routine operation. And by deleting a meaningful method of crew escape in the form of the seats and suits after only four OFT test flights, NASA unintentionally turned the Shuttle into a death-trap for future crews; a trap which claimed 14 lives.

Fig. 3.3 Only Enterprise and Columbia were ever fitted with the capability for ejection seats, during the Atmospheric Flight Test (ALT) and Orbital Flight Test (OFT) missions. In Columbia's case, the seats were disabled after STS-4 and the massive units and their rails were removed after STS-9.

"When you look at the envelope the Shuttle flies, the ejection seats cover a very small percentage in some highly selected areas," said former Space Shuttle program office manager Bob Thompson. But it was never NASA's intention to maintain such a capability as the vehicle matured into operational service. "Once we put those people in there, they were going to stay in the orbiter until the orbiter landed," he continued. "That's the way we designed it. We took the calculated risk. You have to make those judgement calls and that judgement call was made. If you want to get to orbit with any kind of payload, you won't put an escape mechanism other than the basic vehicle you've got them packaged in."

Rockwell International ultimately selected modified SR-71 seats for Columbia during the OFT missions. The ejection process would be commanded by the astronauts themselves and the seats could be used during uncontrolled flight, an on-board fire and an impending impact with unprepared surfaces or water. They were satisfactorily tested at the high-speed test track at Holloman Air Force Base in Alamogordo, New Mexico, in January 1977. The escape sequence required 15 seconds from the first recognition by the crew of an emergency to initiating the ejection process to getting themselves a safe distance away from the vehicle. In such an eventuality, the commander and pilot would be boosted through overhead hatches and parachuted to safety. However, although intended for use during first-stage ascent or 'gliding flight' below an altitude of 30 kilometers, it was readily apparent that (if they were forced to eject shortly after launch) the astronauts would be catastrophically exposed to SRB and SSME exhaust. Nor could they safely eject in the event of an emergency on the launch pad, for they would likely hit the ground before their parachutes had time to open.

In any case, as Columbia would horrifically demonstrate on STS-107 in February 2003, a high-altitude structural break-up of the orbiter at hypersonic speed was simply not survivable, regardless of what crew-escape provisions were in place. And whereas earlier Mercury, Gemini and Apollo spacecraft took the form of ballistic capsules, with built-in mechanisms capable of lifting the crew away from an exploding rocket at any point in the flight, the Shuttle enjoyed no such luxury. The twin SRBs, once ignited, could not be throttled back or shut down early; they ran to exhaustion, before being discarded two minutes into each flight. Those two minutes presented a survivability 'black zone' during which crews had no chance of escape should something go wrong. "To have jettisoned the SRBs while they were still burning would have created questionable structural loads on the ET," noted one former NASA contractor, "not to mention adding two unguided surface-to-air missiles to the astronauts' worry-list."

Another problem was that, unlike the two-man OFT missions, operational Shuttle flights would carry much larger crews. This made it impractical to equip everyone with a rocket-propelled ejection seat. And the seats themselves were huge, allowing barely enough room at the rear of Columbia's flight deck for one additional collapsible seat. As such, ejection seats were removed for operational missions. In marked contrast to Columbia, her sisters Challenger, Discovery, Atlantis and (much later) Endeavour were built from the outset with lightweight seats for commanders and pilots, which enabled two extra seats on the aft flight deck and room downstairs on the middeck for up to four others. As for Columbia herself, a 'Micro-Modification' period after STS-4 saw the pyrotechnics disabled and the massive seats and their rails were removed during a lengthy maintenance phase in 1984. This allowed collapsible seats to be added, which brought Columbia up to the same specification as her sisters.

Equipping each crew member with an ejection seat would clearly have limited the number of astronauts on a mission. Even as few as four seats would have necessitated an extensive redesign of the Shuttle's wiring network and imposed a significant weight penalty, which itself translated into a reduced payload capability. But on Columbia's STS-5 mission in November 1982 – the first 'operational' Shuttle flight, with a four-man crew – the ejection seats (though nominally inactive) remained in place. And this created an uncomfortable scenario in which commander Vance Brand and pilot Bob Overmyer theoretically retained ejection capability, whilst mission specialists Joe Allen and Bill Lenoir did not. For Brand, this situation was intolerable. He insisted that the ejection seats be 'pinned': if an escape mechanism was not available for all of them, then none of them should have it. "He stated it categorically in a flight techniques meeting," recalled Allen, "and NASA officials did not argue with him."

Opinions differed as to the wisdom of Brand's decision. "I got into a debate with Vance about the ejection seats," said Lenoir. "My take was, why the hell should you lose four people, if you could only lose two? If I'm one of the two that you're going to lose, I'm going anyway. Why shouldn't two of you get out? Vance was never comfortable with that, so we flew with 'hot' seats, but there were no procedures or plans to use them." Brand's rationale did not hinge upon selfless personal sacrifice for his crewmates; in fact, he considered it selfish. He knew that no Shuttle commander could live out the remainder of his days with the knowledge that he had survived and his crewmates had not. "He had been a test pilot in England for a while and some of the English bomber airplanes enabled the pilots to get out, but not the gunners," said Allen. "There was a small body of data that involved psychological studies done on individuals [who] had escaped, but by escaping had left their shipmates to certain death. They had certainly been tormented, in terms of what the data showed, for the rest of their lives. Vance didn't want to be another bit of statistic. He said it was selfish. I didn't think it was selfish at all."

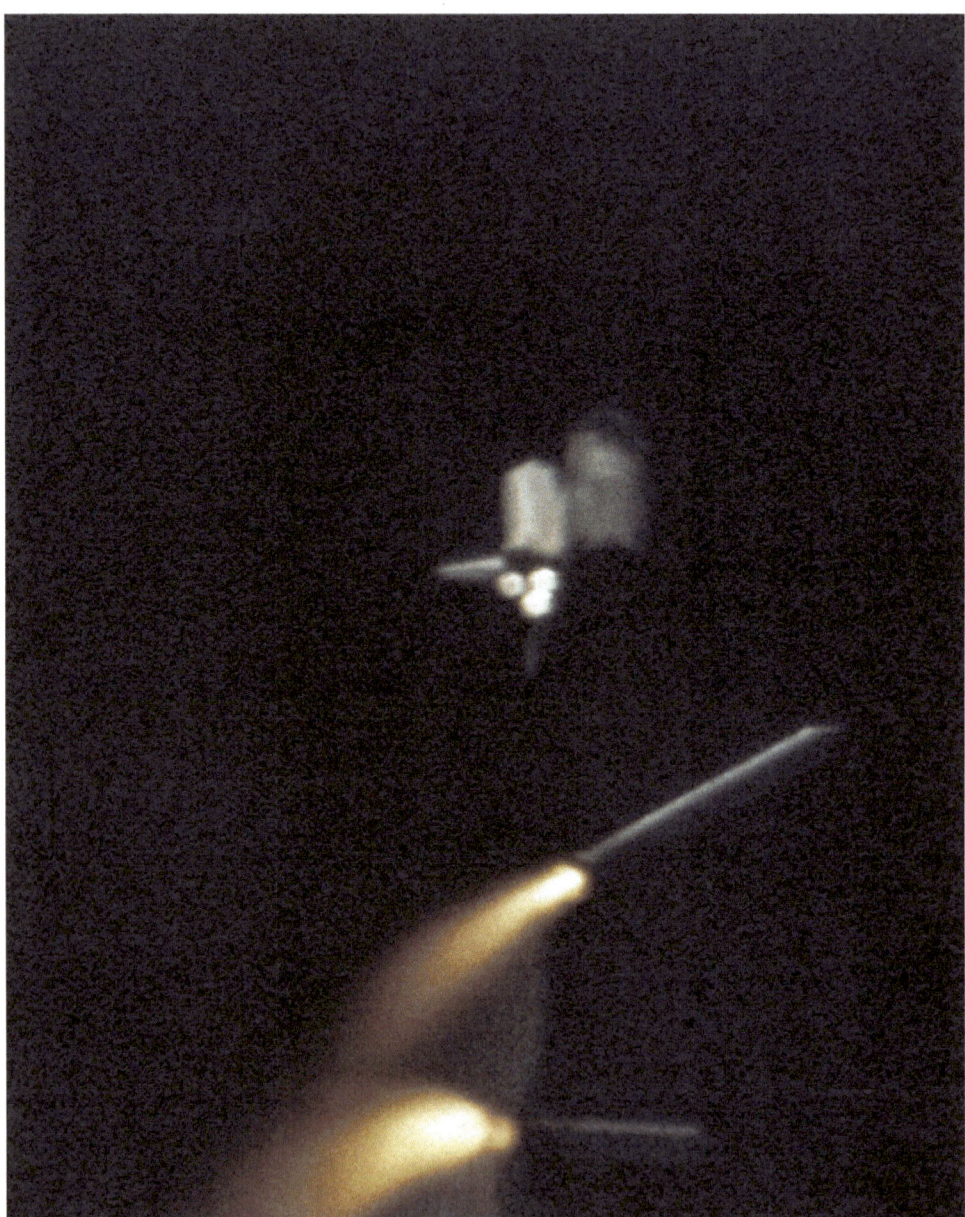

Fig. 3.4 Until the point of Solid Rocket Booster (SRB) separation, two minutes into each mission, no practical escape capability existed for the astronauts in the event of an in-flight emergency. It created an unconscionable 'black zone' for crew survivability.

Other astronauts had similarly mixed views about a reduced crew escape capability after only four OFT missions. Bryan O'Connor remembered a conversation that he once had with STS-4 commander Ken Mattingly. "I told him I just didn't feel comfortable with how we could possibly get to a confidence level after such a short test program." Mattingly just grinned and told him to pay no heed to the politically charged rhetoric from NASA Headquarters. "You and I both know that it will take a hundred flights before this thing will be operational!" In fact, even at the end of STS-135, the 135th and final mission in July 2011, the Shuttle remained a temperamental machine. However, the presence of additional crew members (one of whom sat behind the commander and pilot as a 'flight engineer' from STS-5 onwards) offered additional sets of trained eyes on the instruments. "The orbiter itself represents more of a workload than should be put onto a crew of two," said Joe Engle. "It's just too demanding as far as configuring all of the systems and switches and circuit breakers. There are over 1,500 switches and circuit breakers that potentially have to be configured during flight and some of those are in fairly time-critical times."

Unsurprisingly, when Challenger was lost shortly after liftoff, the media roundly and properly criticized its lack of an effective escape system. But even Bob Crippen – who went on to command three Shuttle missions after his stint on STS-1 – doubted that any system could adequately save astronauts' lives in a major disaster. Ejection seats would have contributed little to protect Challenger's STS-51L crew in a first-stage ascent accident or Columbia's STS-107 crew in a re-entry calamity. The only mechanism which might have made full survivability dimly possible was a totally separate crew compartment, although that would have added $300 million to the Shuttle development costs and 6,300 kilograms to its empty weight. Even if such concepts had been entertained, they demanded a robust early-warning system to give the crew timely notification of an evolving problem. Insofar as we can ever truly know, the heroes of Challenger and Columbia had not the slightest indication until their very last moments of the catastrophes which would soon engulf them.

As for full-body pressure suits, NASA's original intent was for only the four OFT crews to wear these bulky garments. Known as the Model S1030A Ejection Escape Suit (EES) and built by David Clark Company, they kept astronauts alive at altitudes as high as 24 kilometers and speeds of up to Mach 2.7. And like the ejection seats, they drew heavily upon SR-71 heritage, with the added Shuttle requirement of an anti-G suit. But from STS-5 in November 1982 until the loss of Challenger in January 1986, astronauts transitioned to wearing lightweight coveralls for ascent and re-entry, together with clamshell-like helmets which provided supplemental oxygen in the event of a cabin depressurization below an altitude of 15 kilometers. But the protective envelope that these flimsy suits provided was limited. "We had a harness, which had a life-preserver on it," said crew equipment specialist Troy Stewart. "It had no parachute and that was just about it." Astronaut John Fabian had worn similar flight garments during his military career in Southeast Asia and referred to them somewhat disparagingly as 'party suits'. He even darkly wondered if a sign reading 'Enter If You Dare' ought to be placed on the door to the launch pad gantry.

Full-body pressure suits were restored in the aftermath of STS-51L, with David Clark Company selected to fast-track an updated evolution of the S1030A for improved hypobaric and cold-water immersion protection. In its initial incarnation, this took the form of the partially-pressurized S1032 Launch and Entry Suit (LES), first used on Discovery's STS-26 flight – the first post-Challenger mission – in September 1988. This later evolved into the fully pressurized S1035 Advanced Crew Escape Suit (ACES), which first flew on STS-64 in September 1994 and was first worn by an entire crew on STS-73 in October 1995. These flame-retardant ensembles would guard their wearers against fire, cold atmospheric conditions and frigid water temperatures if the Shuttle were to lose cabin pressure or oxygen at altitudes of up to 30.5 kilometers or had to ditch in the ocean.

Some astronauts felt that the new suits aided their chances of survival in an emergency, whilst others such as STS-26 astronaut Mike Lounge regarded them as 'political eyewash'. "It's an extra weight in the crew cabin that takes away from the payload-carrying capability of the Space Shuttle and it is just no value-added; it's value-subtracted," Lounge told a NASA oral historian. "What little you could do in the event something went wrong you could do less of it when you're burdened by these suits." The capability to land in the water safely was also highly problematic, according to astronaut Rhea Seddon. "I always kind of felt like a lot of the escape modes in flight, with the suits on, were urban myths," she said. "It gave us – what was it the pilots say – something to do so you didn't stress or you didn't tighten up. It gave you things to do while you were diving at the ground!" Or as another astronaut darkly put it: the suits would keep them occupied whilst they were dying…

In his memoir *Riding Rockets*, astronaut Mike Mullane was also not immune to the irony of the suits. "I was strapped into a fortress that would keep me alive long enough to watch Death's approach," he wrote. "Even a cockpit depressurization would no longer mercifully grant us unconsciousness. We now wore pressure suits that would keep us alive and conscious through any cockpit rupture." But fellow astronaut Mary Cleave felt the suits at least contributed a new sense of 'seriousness' to the conduct of missions. On her first flight, STS-61B in November 1985, she described the attitude as "really loose", but on her second, STS-30 in May 1989, the enhanced sense of safety was readily apparent.

An intriguing footnote in the LES/ACES story is that it was STS-26 pilot Dick Covey who contributed to the choice of bright pumpkin-orange for the color of the suits. In their earliest incarnation, they were an intense deep blue, identical to the

Fig. 3.5 The Model 1030A Ejection Escape Suit (EES) was worn by Columbia's first four crews between April 1981 and July 1982.

life-rafts which the astronauts would use if the Shuttle ended up in the water. Crew members argued that after escaping a crippled orbiter, hundreds of kilometers offshore, they would never be seen by recovery forces, so a transition was made to orange life-rafts. "Well, if they're going to do that, why are we going to have blue suits?" Covey rationalized. "Why don't we have orange suits?" Aside from their color, the suits proved difficult to don and doff, particularly for smaller astronauts, who carried the same weight of equipment (including parachute and survival gear), regardless of their physical size or body weight. The problem, as Seddon explained, was about more than just the 'bulk' of the suit. Should an emergency occur after landing, the astronauts would be obliged to clamber out of the Shuttle's overhead windows and rappel down the side walls. She found herself having to do lots of 'chin-up' exercises before launch to physically condition herself for the task. "Those suits were built to be worn by high-altitude pilots; regular-size guys," said Seddon. "If they had a problem, they ejected. They didn't have to crawl out and run away. They didn't have to rappel down the side of their vehicle."

By 1994, the LES was superseded by the S1035 ACES. It differed fundamentally from its predecessor in that it functioned at full (rather than partial) pressurization. "It was always realized that the LES was not the optimum suit," wrote Dennis Jenkins in *Space Shuttle: The History of the National Space Transportation System*, "and in 1990 a new development effort was initiated that resulted in the Model S1035 ACES...designed to be simplified, lightweight, low-bulk, full-pressure...that facilitated self-donning/doffing and provided enhanced overall performance." The suits had gloves on disconnecting wrist-rings, better liquid cooling and ventilation capabilities and added insulation. Oxygen was fed via a connector at the left thigh and transmitted to the helmet through the base of the neck-ring. Textured gloves afforded better dexterity and flexion, and heavy zippered, paratrooper-style boots and survival backpack with life-raft and emergency lighting sticks completed the ensemble.

Yet the suits were complex and far from infallible. On STS-40 in June 1991, astronaut Drew Gaffney wore a medical catheter in his arm to measure the blood pressure in the great veins, close to his heart, throughout ascent. But threading the device into his suit was no mean feat. "Once you put it under a pressure suit," pointed out Seddon, "where you can't get to the catheter that's threaded up at the crook of the elbow, you had to go through another round of safety assessments." If Gaffney had needed to escape the Shuttle in an emergency, with the catheter in place, he could bleed to death if it accidentally pulled loose and specialized disconnects were engineered as an added safeguard.

A more comical incident occurred in September 1991, when STS-48 pilot Ken Reightler, anxious to perform at his peak on his first mission, found that the suit had a tendency to 'pull' on his body and hamper his ability to reach switches during ascent. "To bolster my confidence, right after liftoff, I started systematically reaching around the cockpit as the G-forces started to build," he recalled.

But his approach was not appreciated by commander John Creighton, seated next to him on Discovery's flight deck. For a few moments, he watched Reightler's antics with a measure of understanding. Then he leaned across the cockpit and politely but pointedly asked: "Would you please knock that off?"

Ejection seats and pressure suits, therefore, could only protect their occupants up to a certain point. Other escape capabilities were similarly limited. By the mid-1980s the structure of the Shuttle was too mature in design to be easily retrofitted with an 'escape pod' or other feature to realistically pluck the crew to safety in a major disaster. But in the years after Challenger, two concepts – a tractor rocket and a curved, telescoping pole – were considered as a limited means of improving their chances. Ultimately, the pole was selected for what became known as 'Bailout' and it flew on every subsequent mission from STS-26 until STS-135. Formally described as a Mode 8 Egress, in an emergency the Shuttle commander would issue the bailout call if the range to landing at the point of an unrecoverable emergency exceeded 100 kilometers at an altitude of 15 kilometers. The commander would place the vehicle into a minimum-sink-rate attitude and engage the autopilot, after which all crew members would lower their visors and activate their suits' oxygen supplies. One astronaut on the middeck would detach the 2.8-meter pole from its mounting on the middeck ceiling, vent the cabin atmosphere, then pyrotechnically jettison the Shuttle's meter-wide, 136-kilogram side hatch. The pole (which Apollo launch pad leader Guenter Wendt disparagingly nicknamed 'the beanpole') would be extended out through the hatch and each astronaut would affix a lanyard onto his or her suit and slide out. This could occur during any abort scenario, or an emergency de-orbit situation or failure which precluded a 'normal' runway landing.

"It was very challenging, though, to get seven people out of two decks on an aircraft or, in this case, a spacecraft," said engineer Kevin Templin. "There are some aircraft that have gotten as many as four people out of two decks, so a lot of work with the military…to find out what they had built in aircraft." But the reality of the situation was that it would be too expensive and impractical to retrofit anything more capable into the Shuttle.

The pole served to guide the escaping astronauts outward to a few meters and propel them onto a trajectory which eliminated the risk of them hitting the orbiter's left-hand wing. Tested in early 1988 by U.S. Navy parachutists aboard a Lockheed C-141 Starlifter aircraft, it was predicted that a full crew would require 90 seconds to vacate the Shuttle. During one training session in May 1995, STS-69 commander Dave Walker advised his crew to begin bailing out immediately after the side hatch was jettisoned, to allow enough time for all of them to escape. "There are a lot of circumstances you can get into…where you're not under control and then the pole has no function there," said Shuttle commander Steve Nagel, who closely monitored the pole trials. "However, I think you're better off having a pressure suit and a parachute and some survival gear. Even if it's out of control, somebody

Fig. 3.6 STS-51 commander Frank Culbertson (foreground) and pilot Bill Readdy (behind) are pictured in the Model S1032 Launch and Entry Suit (LES) in August 1993.

might have a chance of climbing out the hatch. There's plenty of evidence in World War Two of crew members getting out of bombers with wings off if they happen to be close to an opening or a hatch."

But the loss of Columbia during the STS-107 re-entry on 1 February 2003 brought the questionable usefulness of all such concepts into sharp focus. Neither the ACES suits, nor the pole, nor even ejection seats, could have guarded against conditions far more severe than those for which they had ever been designed. As the Shuttle broke apart that day, suits and astronauts were instantaneously overwhelmed by the effects of near-total vacuum, low ambient atmospheric temperatures at high altitude and the extreme thermal and wind-speed conditions associated with rapid deceleration from hypersonic velocities. By bitter irony, the suits themselves may have caused additional injuries to Columbia's crew. In particular, the non-conformal shape of the ACES helmet (intentionally designed to allow the wearer to freely move his or her head) not only induced blunt-force trauma, but also, via its neck-ring, caused fractures to cervical vertebrae and jaws as skulls whipped backwards and forwards. A tractor rocket or a pole would have served no useful purpose in such a situation. Both were only effective below speeds of 350 kilometers per hour, at altitudes below six kilometers and required the Shuttle to be in controlled, gliding flight; conditions which STS-107, traveling at 23,400 kilometers per hour at an altitude of over 60 kilometers and yawing uncontrollably to the left and right, did not have a hope of ever achieving.

LOADED GUNS

Given that so many things had to go exactly right to launch the Space Shuttle, many astronauts expressed quiet wonderment that any of its 135 missions managed to leave the ground at all. First and foremost, Florida's highly dynamic weather had to play ball, not just at the launch pad, but also at the nearby Shuttle Landing Facility (SLF) where the crew might need to return if an emergency arose shortly after lift-off. 'Shortly', that is, because any emergency in the first phase of flight would require the astronauts to sit through a two-minute-long survivability 'black zone' before the SRBs were jettisoned and the first abort opportunity became viable.

This was the Return to Launch Site (RTLS), in which the orbiter/ET stack would initially continue flying downrange. Although it was possible to immediately shut down the SSMEs and discard the ET, this wisdom was tempered by the fact that the tank still held a large amount of fuel at this early point of ascent. Firing the explosive bolts to separate the tank could induce uncontrollable sloshing and perhaps a collision with the orbiter itself; the safer option was to continue burning fuel, then dump residual propellant to achieve more suitable center-of-gravity conditions for a safe glide back to the runway. A Powered Pitch Around (PPA) maneuver at an altitude of 120 kilometers and a speed of Mach 5 would 'flip' the orbiter/ET at a rate of ten degrees per second to face the engines into the direction of travel. Its eastward progress would correspondingly diminish and it would rapidly shed 60 kilometers in altitude as it headed westward. Next, a Powered Pitch Down (PPD) maneuver would slightly tilt the Shuttle's nose 'downward' to afford the safest situation for jettisoning the ET. Twenty seconds later, the SSMEs would shut down and after ejecting of the tank the crew would guide their ship to a landing about 25 minutes after launch.

On paper, this might sound straightforward; in reality, it would have been quite a challenge. "The RTLS is a daunting prospect for the crew," wrote astronaut Tom Jones. "We would have to fly the orbiter and attached ET through half an outside loop, then ride backward through our exhaust plume at Mach 5. Ditching the empty tank, we would then try to make it back to the Kennedy runway. No Shuttle crew had ever flown such an emergency approach. None wanted to be the first to try." The re-entry following an RTLS would also have been far steeper than normal, imposing higher G-loads on the astronauts and potentially extreme aerodynamic loads on the Shuttle's wings. The timing of when the RTLS would be commanded was wholly dependent upon the nature of the emergency: in the event of an SSME failure at liftoff, for example, it would come immediately after SRB separation. It was perhaps the most dangerous of all Shuttle aborts and, in fact, early plans for STS-1 had envisaged Young and Crippen flying an RTLS profile to validate it under actual mission conditions.

Fig. 3.7 First demonstrated by U.S. Navy parachutists in early 1988, the pole allowed crews of seven or eight members to evacuate a crippled Shuttle in controlled, gliding flight in approximately 90 seconds.

But although Young appreciated the need to train for such an abort in the simulator, he considered it at best foolhardy – at worst, utterly insane – to unnecessarily try it on the Shuttle's very first flight. "Let's not practice Russian roulette," he told NASA senior managers, "because you might have a loaded gun there." His advice was heeded and an RTLS was never carried out other than in controlled conditions on the ground.

But in February 1996, the possibility of an RTLS must have crossed the mind of STS-75 commander Andy Allen as he received a momentary scare just four seconds after Columbia's liftoff. The data on his instrument panel indicated that one of the three SSMEs was running at only 40 percent of its rated performance. Fortunately, it was erroneous data, but had an engine really been on the brink of failure that day the crew would have had little option but to wait out the first-stage flight on the SRBs and perform an RTLS. "We had a couple of moments there that we got a little adrenaline rush," Allen recalled later. At one point, he turned to pilot Scott 'Doc' Horowitz, and remarked: "This looks like a bad simulator run!"

A decade earlier, in January 1986, another bad simulator run almost materialized for real as Columbia began STS-61C. Seconds after liftoff, commander Robert 'Hoot' Gibson and pilot Charlie Bolden noticed an indication of a helium leak in one of the SSMEs. The astronauts' attention was quickly arrested by the blaring of the master alarm. "I looked down at what I could see, with everything shaking and vibrating," Bolden said later. "Had it been true, it was going to be a bad day." Mission Control saw nothing abnormal on their displays, as Bolden and Gibson worked through their procedures. Firstly, they isolated the primary helium system, but there was no change: it still showed a leak. Isolating the secondary system equally had no effect. It turned out to be a bad sensor – nothing more than an instrumentation problem – and the astronauts reconfigured their systems back to normal. Years later, Bolden remembered that his first launch into space on STS-61C "went by really fast!"

But shortly after midnight on 23 July 1999, as Columbia began STS-93, the Shuttle came closer than it ever would to executing an RTLS under real flight conditions. With commander Eileen Collins and pilot Jeff 'Bones' Ashby at the controls, the mission turned night into day across a sleeping Florida. But the spectacular launch masked a potential catastrophe brewing in the main engines. It captured the attention of Collins and Ashby five seconds after leaving the pad, when they spotted a sharp voltage drop on one of their electrical buses. This caused one of two backup controllers on two of the three SSMEs to shut down. The third engine remained healthy and all three burned nominally to insert Columbia into orbit, albeit traveling 16.2 kilometers per hour slower than she should have been. Although this produced a negligible impact on the mission, its severity was such that managers wondered if there had been a shortfall of propellant pumped into the ET before launch.

Subsequent investigation showed that this was not the case, but imagery of ascent revealed a hydrogen leak from one of Columbia's SSMEs: a narrow, bright patch inside the nozzle, perhaps indicative of a weld-seam breach in one of more than a thousand stainless steel hydrogen recirculation tubes. The STS-93 scare was a significant one. On no other Shuttle mission did a crew come so close to having to make the RTLS call. And if the primary engine controllers had also failed that night, an SSME shutdown immediately after liftoff would have become increasingly likely and an RTLS abort – in the hours of darkness – would have turned into an inevitability. "We were prepared for that," Collins said later. "We were listening for the engine performance data calls [from Mission Control] on ascent. This crew would have been ready to do whatever was needed." But an RTLS would have been made harder by the presence of STS-93's primary payload, the huge Chandra X-ray Observatory, which would have pushed Columbia's landing weight to 113,000 kilograms, more than 590 kilograms heavier than the safety rules stated. In this case, a one-mission-only waiver was granted, and in an RTLS Collins would have had to land the Shuttle faster at about 380 kilometers per hour, very close to the maximum certification of about 395 kilometers per hour.

When one considers the white-knuckle ride that an RTLS might have been, it is understandable that flight rules demanded picture-perfect weather at the SLF before all launches. More than a third of all Shuttle flights were delayed (sometimes on multiple occasions) due to weather constraints at the launch site. Missions were postponed for a multitude of reasons, from a blanket-like pall of overcast which enshrouded half of northern Florida on the eve of STS-130's launch in February 2010 to encroaching thunderstorms and from lightning to rain showers and low cloud 'ceilings' to ground fog. Flight rules forbade launches if crosswinds at the SLF exceeded 28 kilometers per hour and even prior to the final Shuttle mission, STS-135 in July 2011 – although RTLS weather was classified as 'No-Go' on account of showers – Flight Dynamics Officer (FIDO) Mark McDonald reported that sufficient energy existed for Atlantis' crew to fly safely through the rain. Flight Director Richard Jones waived the rule and STS-135 flew safely.

As well as ensuring that the astronauts' visibility during final approach and touchdown was good, the controllability of the orbiter during the descent back into KSC was of paramount importance. The launch of Discovery on STS-51A in November 1984 was called off due to conditions of windshear which exceeded the Shuttle's structural limits, whilst other missions suffered from out-of-limits crosswinds and strong tailwinds on the runway. "History now shows we were also possibly very lucky," STS-51A astronaut Joe Allen recalled, "because both of the tragic accidents, that of the Challenger and that of Columbia, involved launching through very high wind shear conditions and there's some thinking now that high wind shears and Space Shuttles do not go safely together."

Fig. 3.8 Ominous weather at the Kennedy Space Center (KSC), as seen here in July 2009 before Endeavour's STS-127 mission, forced many flights to meet with delay.

Delaying for rain showers and thunderstorms to preserve RTLS visibility criteria sometimes generated a measure of ire among the astronauts themselves. In August 1985, Discovery's STS-51I launch was scrubbed twice due to a poor weather forecast, even though commander Joe Engle and pilot Dick Covey knew that the showers in question lay far away from the launch pad. After the second scrub, Covey vented his frustration to John Young. "I can't believe we scrubbed for those two little showers out there," he said. "Anybody with half a lick of sense would have said: Let's go! This could be a lot worse." Young told him straight: the crew could not make calls on the weather. Their knowledge was limited solely to the view from the Shuttle. They should focus on the flying, Young said, and leave other experts to focus on the weather. Ironically, on the day STS-51I finally flew, the weather looked worse than ever. (Indeed, the crew had walked out to the launch pad with yellow raincoats over their flight suits, so torrential was the downpour.) Convinced that they would scrub again, crewmen Mike Lounge and James 'Ox' van Hoften even unstrapped from their seats and took a nap. So too did Bill Fisher. As Engle and Covey started Discovery's three Auxiliary Power Units (APUs), their rhythmic humming awakened Fisher.

"What's that noise?"

"We're crankin' APUs," the pilots replied. "Let's go!"

"Yeah, sure, we're not going anywhere today. Why you startin' APUs?"

"Dammit, we're going!" came the insistent response from Discovery's flight deck. "We're going to launch. Get back in your seats and get strapped in!"

Before each flight, the chief astronaut took to the skies in the Shuttle Training Aircraft (STA) to perform practice approaches and landings on the SLF and thoroughly evaluate visibility and controllability characteristics. Notably, in June 1992, prior to Columbia's STS-50 launch, the countdown clock was held at T-9 minutes due to worries about the presence of cirrus clouds and lightning. The STA pilot's input ultimately allowed the launch to go ahead when he reported that his visibility of the runway was not impaired. Five years later, in July 1997, STS-94 launched almost an hour sooner than planned to avoid an encroaching patch of thunderstorms and rain. Before STS-121 in July 2006, SLF crosswinds were forecasted to be slightly higher than the maximum allowable limit and it was a timely evaluation from the STA pilot which allowed the launch to go ahead. And in November 2009, a cloud ceiling below 1,500 meters threatened a clear violation of the rules in the case of STS-129. Eventually, chief astronaut Steve Lindsey offered a pilot's eye perspective on the thickness of the cloud layer which he relayed to the weather officer and the launch went ahead.

Yet Florida's weather was not the Shuttle's only pre-launch concern. Over its 30 years of operational service, dozens of missions succumbed to delays, induced by Mother Nature and technical difficulties either in ground support equipment or aboard the enormously complex vehicle itself; two flights in particular – one in January 1986, a second in October 1995 – endured no less than six scrubbed launch attempts. And this posed a physical and mental toll on the astronauts. Before each flight, crews lay on their backs, legs elevated and clad in bulky pressure suits for more than two uncomfortable hours. After one launch delay, astronaut Steve Oswald could not get out of the vehicle quickly enough for a toilet break. And on 13 March 1989, the crew of STS-29 – commanded by Mike Coats – waited almost five hours, their liftoff repeatedly pushed back by the presence of ground fog and unacceptably high winds. "Mike had a very bad backache," remembered pilot John Blaha, who was seated alongside him on Discovery's flight deck that day. "He finally decided he had to unstrap, because we were being delayed for such a long time. Mike unstrapped and actually…was laying on his side and then he'd lay on the other side." Rockets awaiting their own launches from Florida also conspired against the Shuttle: delayed Atlas and Titan missions in November 1996, March 2000 and February 2001 caused delicately assembly timelines to be thrown into disarray. Such agonies continued until the very end of the program. Even on the morning of 8 July 2011, as Atlantis counted down to STS-135, the clock was held at T-31 seconds due to uncertainty over whether the gaseous oxygen vent arm on the launch pad had properly retracted. All turned out to be fine and the mission flew successfully after a delay of a few minutes, but for the astronauts the wait must have been interminable.

Fig. 3.9 STS-59 pilot Kevin Chilton, clad in his bulky Launch and Entry Suit (LES) and oriented in the vertical, legs elevated, on the launch pad, illustrated the uncomfortable conditions astronauts endured before each liftoff.

The issue of an RTLS came to the fore for a different reason on STS-87 in November 1997, in which Columbia trialed a 'heads-up' maneuver during ascent. Intended to eliminate a tracking station in Bermuda and enable the astronauts to communicate with Mission Control through the Tracking and Data Relay Satellite (TDRS) network sooner than was previously possible, the heads-up maneuver occurred six minutes into the flight, as Columbia flew at more than 16,000 kilometers per hour, and involved the GPCs commanding the orbiter/ET stack to perform a 180-degree roll from a 'heads-down' to a 'heads-up' attitude. It proved entirely nominal. "We had to do a fair amount of analysis to ensure that we weren't doing something dumb," remembered ascent flight director Wayne Hale. "One thing we didn't want to perturb was the RTLS abort mode." As a result, the heads-up maneuver was scheduled after the closure of the RTLS 'window' and simulations showed it could be done safely even with an electrical system failure and only one SSME running.

With the closure of the RTLS abort window, there was often an interval of several seconds before the opening of the next option – the Transoceanic Abort Landing (TAL) – which would have provided the conditions needed for the Shuttle to land about 45 minutes after liftoff on the far side of the Atlantic Ocean. Should an SSME fail or another significant anomaly arise (such as a cabin pressure leak or cooling system malfunction), crews would be directed toward a range of sites which over the years included Banjul in The Gambia, the Moron or Zaragoza air bases in Spain, Ben Guerir in Morocco and, from STS-114 in July 2005, Istres-Le Tubé Air Base in France. Additionally, early in the Shuttle program, Dakar in Senegal served as a TAL site, together with a group of Emergency Landing Sites (ELS) and East Coast Abort Landing (ECAL) runways worldwide, ranging from Australia to Cape Verde, the Bahamas to South Africa and Canada to Liberia. But despite Banjul's exceptional suitability for Shuttle missions launched into 28.5-degree inclinations, it was not so well-suited for International Space Station (ISS) flights at 51.6 degrees and it was phased out in 2001. So too was Ben Guerir, deactivated in 2005 after supporting 83 Shuttle launches.

Under normal circumstances, three TAL sites were considered for each Shuttle mission and conditions had to be near-perfect in at least one of them for a launch to proceed. In many cases, flights were postponed or scrubbed when this requirement was not met. Unacceptably high crosswinds and tail winds, desert haze and dust storms, low cloud ceilings and fog, rain showers and thunderstorms and overcast skies all conspired to work against the TAL sites at one time or another. Several countdowns, including that of STS-27 in December 1988, were halted only seconds before liftoff as flight controllers held out for TAL weather to improve. Just before STS-66 in November 1994, all three TAL sites declared themselves 'No-Go' on account of weather, with Ben Guerir eventually selected when its initially poor crosswind situation began trending 'downwards' in severity and became marginally more acceptable than the others. The launch of STS-34 in October 1989 was subjected to a slight delay when rain showers at one TAL site forced a rapid reconfiguration of assets to another TAL site to replace it. And STS-100 in April 2001 saw its TAL sites declare themselves as 'No-Go' until only minutes before launch, when head winds dropped and the weather outlook suddenly brightened. One of the most significant TAL weather events was the "challenging" condition encountered at Zaragoza, Moron and Istres before STS-130 in February 2010. An ominous picture of low cloud and showers at Moron had already taken it out of the running, whilst cloud ceilings at Istres changed its weather position from 'Go' to 'No-Go', late in the countdown. Eventually, weather pilots at Zaragoza reported that although there was "moisture" in the air, there were no rain droplets visible on their windshields and this TAL site was selected and STS-130 launched without incident.

Although extensively trained for by every Shuttle crew, none of the TAL sites were ever called upon in a real emergency. But a close call occurred on the afternoon of 29 July 1985, shortly after Challenger launched on STS-51F. Five minutes and 45 seconds after liftoff, with the RTLS window having long since closed and the TAL window approaching its own closure, the Shuttle suffered a premature SSME shutdown whilst traveling at over 15,000 kilometers per hour at an altitude of 108 kilometers. "The ground made the call 'Limits to Inhibit', which is, for us, an extremely serious omen," remembered astronaut Story Musgrave, the flight engineer aboard Challenger that day. "It means the ground is seeing problems that are going to shut you down. I'm looking through the procedures book and thinking we're going to land at our transoceanic abort site in Spain. I'm rehearsing all the steps and my hands are moving through the book, and I'm thinking 'We're going to Spain. Things are bad!'"

Fig. 3.10 STS-51F commander Gordon Fullerton's finger on the push button to select the Abort to Orbit (ATO) contingency option. Note that the rotary switch indicates the respective settings for the RTLS, TAL and AOA aborts.

The emergency which afflicted STS-51F arose when temperature readings for the left-hand engine's High Pressure Fuel Turbopump (HPFT) indicated 'above' its maximum operating redline and the GPCs commanded it to shut down. By this stage, the Shuttle was too high, had insufficient propellant and was flying too fast to achieve an RTLS landing, which left either a TAL or the next available option, an Abort to Orbit (ATO). Until Mission Control made the definitive call, Musgrave's focus was upon the page of his checklist which dealt with the requirements to land at Zaragoza Air Base, a joint-use military and civilian installation in the autonomous region of Aragon in north-eastern Spain. It had been selected for STS-51F because the mission's targeted orbital inclination of 49.5 degrees put it close to the nominal ascent ground-track and permitted better use of SSME propellant and cross-range steering capability. As Musgrave worked through the procedures that he was convinced he would soon have to read to commander Gordon Fullerton and pilot Roy Bridges, fellow astronaut Karl Henize, also seated in Challenger's cockpit, was getting nervous. He noticed the word 'SPAIN' at the top of Musgrave's checklist and all at once he spoke.

"Where we going, Story?"

"Spain, Karl," replied Musgrave. Then he retracted it. "We're close, but not yet."

Not yet, indeed. Challenger would not perform a TAL that day, but her unlucky crew would be forced to execute the only in-flight abort ever performed in the Shuttle program. At length, Mission Control radioed the command: "Abort ATO, Abort ATO". The Shuttle had missed the closure of the TAL window by only 33 seconds, which brought the next abort option (the ATO) online. Six minutes and six seconds after launch, Fullerton fired the twin OMS engines for 106 seconds, consuming 1,875 kilograms of much-needed propellant. This maneuver permitted Challenger to achieve a lower-than-planned but satisfactory orbit. Two minutes later, the right-hand SSME also indicated excessively high temperatures, but the Limits to Inhibit command correctly prevented the GPCs from shutting it down. With only two main engines thus firing for the final stage of ascent, STS-51F and her seven-man crew limped into a 230-kilometer-high orbit, a far cry from the 390 kilometers planned, but safe enough for her to complete a successful eight-day mission. "We even made up for the fuel we'd had to dump on the way up because of the engine failure and eked out an extra day on it," Fullerton reflected, years later. "We were scheduled for seven and made it eight!"

The incident represented the only in-flight SSME shutdown ever experienced by the Shuttle and was a surprise since all engine parameters had registered nominal during the countdown, ignition and the first several minutes of the flight. But at approximately two minutes into the ascent, at about the same time as the SRBs were jettisoned, data from Channel A – one of two measurements of the HPFT discharge temperature of the uppermost engine – displayed characteristics indicative of the beginning of failure. Its measurement began drifting and, at three minutes and 41 seconds, the Channel B sensor failed. However, its sibling continued to drift, approaching and exceeding its own redline limit five minutes and 43 seconds into the flight, which triggered the shutdown. The HPFT discharge temperature data from Channel B on the right-hand engine, meanwhile, began to climb and passed its own redline just over eight minutes after liftoff. Measurements from Channel A remained within prescribed limits and, said NASA's post-mission report, all other operating parameters relating to the left-hand and right-hand engines were normal. Post-mission analysis suggested that the problem was not with the uppermost engine itself, but with faulty sensors that incorrectly indicated an overheating situation. The sensors consisted of extremely thin wires whose electrical resistance changed whenever they were heated. They had experienced failures in the past and upgraded versions flew on the next mission, STS-51I in August 1985.

The SSMEs of several other missions also provided a reminder that the Shuttle was an experimental flying machine and its operations were far from ever becoming routine. During Columbia's STS-3 ascent in March 1982, an Auxiliary Power Unit (APU) overheated at 4.5 minutes after liftoff, triggering a caution-and-warning alarm in the cockpit and forcing crewmen Jack Lousma and Gordon Fullerton to shut it down early. This left one of the SSMEs running at only 82 percent of its rated thrust during the latter stages of the climb to orbit. Yet although mission after mission flew successfully, concerns about the long-term health of the main engines lingered. Even on the morning of 28 January 1986, when Challenger exploded in the skies above Florida, many observers aimed the finger of suspicion – at first – not upon the true culprit (the SRBs), but rather at the SSMEs, whose long and difficult evolution had made them the likeliest critical component to fail.

ABORT!

In addition to RTLS and TAL weather, several very 'human' factors conspired against the Shuttle in the form of unauthorized boats and aircraft intruding into the launch danger area. In October 1998, shortly before the launch of STS-95 (whose crew included Senator John Glenn), countdown clocks were held at T-5 minutes to shoo away a pair of intruder aircraft. Such worries took on greater significance with Endeavour's STS-108 flight in December 2001, the first mission after the terrorist attacks on New York and Washington, D.C. Small aircraft were banned from flying within 56 kilometers of the launch pad, the U.S. Coast Guard strictly enforced an impenetrable coastal buffer around KSC and boats were barred with a draconian launch danger zone extending 120 kilometers out to sea. As late as November 2008, during the final minutes before the STS-126 launch, NASA was informed of an inbound threat to the Shuttle about 3.5 kilometers offshore. It turned out to be nothing more than a hoax, but the perpetrator was arrested and later sentenced to a lengthy prison term.

Nevertheless, long before 9/11, uninvited intruders were an unwanted source of irritation for NASA. Several missions were delayed due to the antics of ships straying inside the targeted SRB splashdown recovery zone, including the *Ocean Mama* freighter, which in April 1985 had to be shooed away by the Coast Guard to enable Discovery's STS-51D launch to go ahead. Aircraft, too, managed to inadvertently sneak their way into KSC's closed airspace. One such incident occurred in the final minutes before Discovery's maiden voyage, STS-41D in

August 1984, when a pair of private pilots accidentally strayed inside the launch danger area. The countdown clock was halted as they were escorted out, but for the furious astronauts – who had endured two previous launch scrubs and having lain uncomfortably on their backs for over two hours – the mood was black. "Shoot the fucker down," one of them angrily muttered. "After nearly a seven-minute delay," wrote STS-41D mission specialist Mike Mullane in *Riding Rockets*, "its pilot pulled his head out of his ass and flew off. We all wished him engine failure."

Fig. 3.11 Smoke drifts away from Pad 39A on 26 June 1984, after Discovery's three main engines shut down in the final seconds before liftoff.

The astronauts of STS-41D had good reason for their ill-disguised frustration. Two months earlier, they had sat through one of the most terrifying incidents ever experienced in the Shuttle program; an incident which would afflict no fewer than five missions by August 1994. Their countdown to launch on 26 June 1984 had gone exceptionally smoothly and at T-31 seconds Discovery's GPCs assumed primary control of critical vehicle functions. The SRBs underwent their final nozzle steering checks, the firing chain was armed and at ten seconds the now-familiar flurry of sparklers from the hydrogen burn-off igniters gave way to the familiar, low-pitched rumble of Main Engine Start.

Inside the Shuttle's cockpit, the astronauts felt immense vibrations as turbopumps awoke and liquid oxygen and hydrogen flowed into the engines' combustion chambers. Then, shockingly, their noise died and the silence was arrested by the blaring of the master alarm. Something had gone seriously wrong. "Then there's this grinding," remembered STS-41D payload specialist Charlie Walker. "I cannot describe it. It sounded like…imagine in your mind the hand of God comes out of the sky, reaches down and twists the launch tower and structure outside the vehicle. It sounds like the place is being ripped apart." Discovery's left-hand and right-hand engines had blazed perfectly to life, but the uppermost engine had failed to ignite. "The vibrations were gone," wrote Mullane. "The cockpit was as quiet as a crypt. Shadows waved across our seats as Discovery rocked back and forth on her hold-down bolts." Pilot Mike Coats could hear nothing but the screeching of seagulls outside his window. Two glaring red lights on his instrument panel indicated that the left-hand and right-hand engines had indeed shut down, but the indicator for the uppermost engine remained dark. "Surely it couldn't still be running?" they all wondered. Coats jabbed his finger onto the shutdown button in order to make sure, but the status indicator remained dark.

Downstairs, on Discovery's middeck, the eyes of Walker were focused intently on the procedures for a 'Mode 1 Egress': step-by-step instructions for opening the side hatch and evacuating the vehicle in a hair-raising emergency described by NASA as "too critical to dispatch fire/rescue personnel to the pad". All told, there were eight egress modes, covering emergency escape procedures for the prelaunch phase (Modes 1–4) and post-landing phase (Modes 5–8), with Modes 1, 5 and 8 allowing the astronauts to exit the Shuttle unassisted, Modes 2, 3 and 4 requiring aid from the launch pad 'closeout' crew and Modes 6 and 7 requiring support from fire crews, helicopter-borne Search and Rescue (SAR) forces or the pre-positioned convoy crew on the runway. Several kilometers away from Pad 39A, the crew's families were watching this drama unfold, and they were as perplexed by what they could see as by what they could not. "A thick summer haze had obscured the launch pad," Mullane wrote. "When the engines had ignited, a bright flash had momentarily penetrated that haze, strongly suggesting an explosion. As that fear had been rising in the minds of the families, the engine-start sound had finally hit…a brief roar." The sound echoed off the walls of the huge Vehicle Assembly Building (VAB) and was gone.

Over the intercom, the astronauts heard the words "RSLS Abort", denoting a Redundant Set Launch Sequence. Something had gone wrong and liftoff had been aborted in the final seconds. The vehicle continued rocking, back and forth. Intuitively, the crew knew there were built-in safeguards to prevent the SRBs from igniting, for if that happened it would kill them all. But they also knew that only a few seconds existed on the countdown clock. "A couple of seconds in the world of electronics is a lifetime and I'm sure that all the safety devices had rotated to prevent [the solids] from igniting," wrote Mullane, "but in the back of your mind, you're thinking *What happens if those ignite?*" Although the safe-and-arm devices of the SRBs had been removed, as planned, at T-5 minutes, neither booster could be commanded to ignite until several criteria had been satisfied. Two of those requirements were that all three SSMEs had attained at least 90 percent of their rated thrust (which they had not) and that no 'SSME Fail' indicators had been tripped (which they had). But still the situation was far from coming under control. As if the indication that the uppermost SSME might still be burning was not enough, LCC now told the crew that there was a fire on the launch pad and the suppression equipment had been activated to deal with it.

The decision over whether to unstrap and evacuate the orbiter was now in the hands of STS-41D commander Hank Hartsfield. On the middeck, astronaut Judy Resnik unstrapped and peered through the window in Discovery's side hatch. She could see no evidence of a fire and asked Hartsfield if she should begin opening the hatch. Had a Mode 1 Egress been needed that day, it would have been ordered either by Hartsfield or the NASA Test Director (NTD). The Orbiter Access Arm (OAA) would have moved back into position alongside the Shuttle's side hatch within 30 seconds. The astronauts would have unstrapped and left their seats, opened Discovery's side hatch and proceeded in a 'buddy-system', two by two, to a set of emergency slidewire baskets at the launch pad's Fixed Service Structure (FSS). The 'floor' of their route to the baskets was painted yellow with black chevrons guiding the way, lest heavy smoke or water from the fire suppression system should hinder visibility. The baskets would transport them to a reinforced concrete bunker at ground level, at which point the NTD would assume command of operations. The astronauts would either remain in the bunker, with its supplies of breathing air, water, telephone communications and medical stocks, or would be instructed to evacuate the area in a tank-like M-113 armored personnel carrier.

Listening to the chatter over the air-to-ground communications loop, however, Hartsfield elected to sit tight. It was a gutsy decision which probably saved their lives. Burning hydrogen is invisible to the human eye and had already begun to ignite combustible materials on the pad surface. Subsequent inspections would reveal scorched paint all the way up the gantry, as far as the OAA, some 45 meters above the pad surface. "The flame may have been as high as the cockpit, but…we would not have seen it," Mullane wrote. "We could have thrown open the hatch and run into a fire." Years later, Walker would praise Hartsfield for not ordering a Mode 1 Egress that day. In his conversations with the launch director after the abort, Hartsfield realized that a great deal of doubt also existed over the reliability of the slidewire baskets and that concern also informed the judgement to keep the crew aboard the orbiter.

Then, with nerves as tight as elastic, came a spark of humor. "Gee," said STS-41D flight engineer Steve Hawley. "I thought we'd be a lot higher at MECO!"

There had indeed been a Main Engine Cutoff (MECO), but not at the edge of space as should have happened on a nominal mission, for Discovery remained

Fig. 3.12 Visible scorching on Discovery's aft body flap in the wake of the STS-41D pad abort. One of the Shuttle's engines can be seen at left, with the orange hue of the External Tank (ET) and one of the white Solid Rocket Boosters (SRBs) in the background.

firmly shackled to the ground. T-zero had not been reached, the SSMEs had not reached a minimum of 90 percent of their rated thrust and the SRBs had not been commanded to fire. Hawley's joke broke the ice and got the crew laughing. When they finally opened the hatch and disembarked from their ship, they did so into a torrential downpour from the waters of Pad 39A's fire suppression system. The entire gantry was sodden with droplets falling from every pipe and platform, and for the shivering astronauts it was like walking through a waterfall. "After a launch abort, you could take a gun and point it right at somebody's forehead," Mullane wrote. "And they're not even going to blink, because they don't have any adrenaline left in them. It's all been used up."

Five times between that harrowing day in June 1984 and 18 August 1994, astronauts would be plucked from the Shuttle after enduring an RSLS abort. The next such event came on 12 July 1985, when Challenger's SSMEs roared to life and her STS-51F crew instinctively braced themselves for the jolt of liftoff. But it never came. "At T-7 seconds, the main engines start with a rumble from far below," pilot Roy Bridges remembered. "As the person in charge of the engines, I watch the chamber pressure indicators come to life and surge towards 100 percent." But Bridges knew that something was not quite right. "The left engine indicator seems to be lagging behind," he continued. "Before I can say a word, it falls to zero, followed by the other engines. With less than three seconds before our planned liftoff, we have an abort. The groans from the rest of the crew are now audible." For a moment, Bridges noticed commander Gordon Fullerton looking quizzically over to his side of the cockpit. Bridges held out his hands, palms up. "Gordo, I didn't touch a thing!" It was an automatic shutdown, triggered by the GPCs in response to a slow-to-close coolant valve on the left-hand engine.

It would be almost a decade before the curse of RSLS again visited the Shuttle and, when it did, it occurred on three occasions in an 18-month period. On the morning of 22 March 1993, Columbia's engines came abruptly to life and just as abruptly – three seconds before liftoff – fell silent. An incomplete start-up of the right-hand SSME had been triggered by a leak in a liquid oxygen pre-burner check valve, which caused the purge system to be pressurized above its maximum redline. The culprit was a tiny shard of rubber, trapped inside a valve. Within seconds, the OAA was automatically moved into position alongside the Shuttle's side hatch so that the seven-man STS-55 crew could evacuate. However, as with STS-41D before it, the possibility of running into an invisible hydrogen inferno prompted commander Steve Nagel to keep his crew aboard. Ground personnel deactivated electronic systems, paramedics were rushed the pad as a precautionary measure and the bitterly disappointed astronauts ran through their post-shutdown procedures. "There's a couple of moments wondering what's happened, because all you see on board are red lights, indicating an engine shutdown," Nagel said later. "You know the computers shut down the engines, but you don't know why or exactly what went wrong with them." Matters were no less tense in the LCC, where ashen-faced launch director Bob Sieck oversaw the proceedings. "Your initial reaction is to make sure there are no fuel leaks or that there's nothing that's broken that's causing a hazardous situation," he recalled in the aftermath of the abort. "Really, it was one of those nice, boring countdowns...until the last few seconds. What *did* work, and worked very well, were the safety systems on-board. As a result, the crew is safe and the vehicle is on the pad and safe as well."

Still, the astronauts emerged from Columbia looking visibly shaken by the experience. So too were the five astronauts of STS-51, for whom the initial roar of Main Engine Start also fell quiet on the morning of 12 August 1993. As commander Frank Culbertson, pilot Bill Readdy and flight engineer Dan Bursch were faced with a blaze of red lights on Discovery's instrument panel, a flurry of messages washed over the LCC communications loop. All three SSMEs were safe and in "post-shutdown standby" as Readdy worked to power down the APUs. Luckily, no fire detectors on the launch pad had been tripped during the incident, which was later traced to a faulty fuel-flow sensor on the left-hand engine. The 'major component failure' raised by the sensor glitch, about 0.6 seconds after ignition, "caused a 'miscompare', which violated the Launch Commit Criteria", noted

Fig. 3.13 Discovery's three main engines in the process of being replaced following the STS-51 pad abort.

NASA's STS-51 Mission Report, published after the flight. "As a result of the failure, the engines were shut down and safing activities were initiated."

A year later, Bursch became the only astronaut unfortunate enough to experience two RSLS aborts in his career. Having finally flown STS-51 in September 1993, he was rapidly reassigned to serve again as flight engineer on STS-68, targeted to launch just before sunrise on 18 August 1994. The countdown that morning was smooth and uneventful, to such a degree that astronauts Jeff Wisoff and Tom Jones killed time playing rock, scissors, paper, as they lay uncomfortably on their backs in Endeavour's middeck. At 6:54 a.m. EDT, right on time, the Shuttle's main engines roared to life. From the roof of the LCC, Jones' wife, Liz, and their two children, together with the other crew families, watched as a huge pall of steam obscured the launch pad. "In the growing light of dawn," Jones wrote *Skywalking*, "she saw the gout of orange exhaust flare beneath the orbiter and saw the steam billow from the flame trench as the engines spooled up to full power." But full power, alas, was not on Endeavour's side that morning. "…Three, two, one…and…we have Main Engine Cutoff," came the call from the launch announcer. "GLS safing is in progress."

As the Ground Launch Sequencer overrode the GPCs and began safing the Shuttle's systems, the three blazing engines suddenly fell dark and silent. And whereas previous RSLS aborts occurred at around T-3 seconds, STS-68 had gotten down to just T-1.9 seconds, closer than any other mission to lifting off, without actually lifting off. Almost instantaneously, cooling water was sprayed onto the still-hot SSMEs and the attention of everyone in the LCC was riveted on the fire detectors of the Main Propulsion System (MPS). For the third time in a little over a year, an acronym-laden flood of urgent communications passed between controllers, managers and astronauts: "We have Main Engine Cutoff…RSLS safing is in progress…All three main engines are in post-shutdown standby…GLS is Go for orbiter APU shutdown…"

The famous countdown clock at KSC starkly read T-00:00:00, yet no Shuttle ascended toward the heavens and only an ominous smudge of grey cloud rose above Pad 39A. Then came the call which brought a measure of calm: "No MPS fire detectors tripped." There was no evidence of a hydrogen blaze, meaning a Mode 1 Egress by the crew was unnecessary. Through the tiny window in Endeavour's side hatch, Jones could see the launch pad's gantry swaying backwards and forwards as the Shuttle continued to rock from the initial 'twang' of the SSMEs. STS-68 had been stalled, it later became apparent, by an issue with the right-hand engine's High Pressure Oxidizer Turbopump (HPOT). A sensor detected a dangerously high discharge temperature that exceeded the flight rules, and Endeavour's computers commanded a shutdown after the Engine Start Command (ESC) had been issued. "From ESC+2.3 seconds through ESC+5.8 seconds, the HPOT discharge temperature must not exceed 1,560 degrees Rankine," noted NASA's STS-68 Mission Report, using the scale for engineering systems in which heat computations are made with degrees Fahrenheit. "The SSME HPOT discharged temperature Channel A attained 1,576 degrees Rankine.

The Channel B measurement attained 1,530 degrees Rankine and that was also higher than predicted." A 'normal' HPOT discharge temperature was about 1,400 degrees Rankine. As a result, the Shuttle's right-hand engine was shut down at 4.72 seconds after ignition, followed by the others.

"The turbopump was an amazing yet delicate piece of machinery," wrote Jones. "The size of a V-8 engine, it produced 310 times the horsepower. If the HPOT, spinning at 28,000 revolutions per minute, had come apart in the aft engine bay, it would have been a very bad day for Endeavour…and her astronauts." And for poor Dan Bursch, his crewmates teased him mercilessly for having jinxed their mission with 'his' RSLS bad luck. When the crew arrived in Florida for their second (successful) launch attempt in late September, they devised a dastardly and clever plan to convince Endeavour that the luckless Bursch was not aboard: they kitted out the beleaguered flight engineer in a Groucho Marx disguise, complete with spectacles and moustache.

And the unfortunate STS-93 (whose troubled launch was described earlier in this chapter) also suffered a dramatic abort a mere seven seconds before liftoff on the evening of 20 July 1999 and missed an RSLS by a hair's breadth. Its launch attempt was abruptly curtailed due to high concentrations of gas in Columbia's aft compartment. This was attributed to a hydrogen 'spike', which a sharp-eyed launch controller spotted briefly peaking at 640 parts per million, more than double the maximum allowable safe level. During the crisis, the mood in the LCC at KSC was tense. Sixteen seconds before liftoff, one of two gas detection systems indicated unacceptably high concentrations of hydrogen, although its counterpart showed nothing out of the ordinary: a more normal level of 110–115 parts per million. Was it an instrumentation error or a real leak? Peering intently at his data, hazardous gas systems engineer Ozzie Fish would not take the chance and urgently radioed colleague Barbara Kennedy at the GLS console to stop the countdown.

But to the assembled spectators at KSC that night, all seemed normal at first.

"T-15 seconds," came the call from NASA announcer Bruce Buckingham. "T-12…ten…nine…"

The flurry of hydrogen burn-off igniters came alive, ahead of Main Engine Start.

Inside the LCC, Fish urgently called: "GLS, give cutoff."

"…eight, seven…" continued Buckingham.

"Cutoff," interjected NTD Doug Lyons. "Give cutoff!"

"Cutoff is given," replied Kennedy.

"We have hydrogen in the aft," barked Fish, "at 640 ppm."

By now well past what would have been a 'normal' ignition of the SSMEs, Buckingham communicated the disappointing news of a launch scrub to his audience. On data screens in the LCC, the hydrogen concentration gradually crept back downward to normal levels. Lyons polled his team, asking if any emergency safing procedures were necessary, but was told that a Mode 1 Egress of the crew was not required. It later became clear that faulty instrumentation and flawed telemetry was to blame. The hydrogen burn-off igniters were duly replaced and Columbia launched safely (though not uneventfully) three nights later.

ENGINE IMPROVEMENTS

Before the loss of Challenger, most informed observers believed that if anything were to go catastrophically wrong on a Shuttle flight – resulting in a Loss of Crew and Vehicle (LOCV) – the most likely culprit would be a failure of the SSMEs. During ascent, the primary responsibility for monitoring the behavior of the engines lay with the pilot. And on STS-8 in August 1983, that seat was occupied by Dan Brandenstein, the very man who served as Capcom for the first Shuttle launch. But every so often as Challenger roared into space, his attention on the gauges was distracted by fellow astronaut Dale Gardner, sitting behind him.

"Dan, how do the engines look?"

"Yes, look fine," Brandenstein replied. Thirty seconds later, Gardner asked again.

"Fine."

Years later, Brandenstein could not remember how many times his crewmate pestered him with the same question during ascent. But when Challenger settled into orbit, he pulled Gardner to one side to ask him about it. "What was going on?"

Gardner explained that he was watching the behavior of the engine exhaust using a hand-held mirror, which gave him a spectacular rearward view through the overhead windows of the aft flight deck. But there was something about the pattern of the flame which worried Gardner. Before STS-1, the young Navy engineer had observed many SSME test-firings, in which the exhaust pattern initially looked 'solid', then suddenly began to flutter just before the engine exploded. As Challenger climbed higher and higher in altitude, and external air pressures diminished, the atmospheric factors acting on the exhaust plume caused it to expand and flutter. Gardner was convinced that he was seeing an engine on the brink of blowing up.

The STS-8 anecdote underscored a palpable sense of dread that many astronauts felt about the reliability of the SSMEs. Indeed, during plans to fly missions out of Vandenberg Air Force Base in California, it was recognized that hydrogen concentrations could become trapped in the flame trench beneath the engines, which carried an added potential for disaster. The Air Force responded with plans to install 54 outward-firing igniters inside each of the SSMEs to remove the excessive gas, but studies ultimately erred on the side of solving the problem with steam: storing hot water in pipes, recirculating it through boilers to stabilize its temperature and then spraying it into the engine ducts just before ignition.

Fig. 3.14 Impressive view looking along the length of the Orbiter Access Arm (OAA) towards Endeavour's side hatch before STS-134 in May 2011.

Certainly, the problems with the engines during their development, during actual flight conditions on STS-51F and STS-93, and during five on-the-pad RSLS aborts between June 1984 and August 1994 seemed to point in this direction. But long before STS-1, consideration was being given to potential upgrades (most notably a 'large-throat' combustion chamber for additional thrust) and this work returned to the fore in August 1983 when a formal improvement program got underway. It aimed to reduce the maintenance required by the engines between flights, as well as to provide and maintain the requisite margins for an enhanced SSME thrust up to 109 percent to support high-energy or high inclination launches. (Two missions intended to fly at this higher power setting in the pre-Challenger period were the Galileo and Ulysses probes, both scheduled in May 1986.) Over the next several years, changes to the high-pressure turbopumps, hydraulic actuators, high-pressure turbine discharge temperature sensors and the main combustion chamber itself enhanced the durability of the system. Following the loss of Challenger in January 1986, plans for 109-percent thrust as a 'nominal' mission requirement were shelved as an increasingly cold-footed engineering community opted not to push the system beyond its safe limits. Nevertheless, plans to upgrade the SSMEs continued unabated, with particular emphasis on the HPOT and HPFT, engine powerhead, heat exchanger and main combustion chamber.

In July 1995, Discovery roared into space on STS-70 with two 'original' engines and Rocketdyne's all-new Block I engine. Designated 'SSME No. 2036', it sat in the uppermost engine position and included the 23,700-revolution-per-minute Alternate High Pressure Oxidizer Turbopump (AHPOT). Three months later, Columbia flew STS-73 with a pair of Block Is and an old-style engine and in May 1996 Endeavour lifted off to begin STS-77, the first Shuttle mission with a full suite of three Block Is. The new turbopump, developed by Pratt & Whitney, had half as many moving parts as its predecessors and was produced through a casting process which eliminated all but six of 300 welds from the original design. The complete Block I engine also included a two-duct powerhead which improved fluid flow through the engine to decrease pressures and loads and a single-coil heat exchanger. The new powerhead replaced three smaller fuel ducts in the original SSME with two enlarged ducts to improve engine performance. As part of planning to reduce maintenance between missions, the Block I AHPOT did not need to be subjected to detailed inspections until it had flown ten times. It also included a new ball bearing material of silicon nitride, which proved far superior to the steel bearings previously in use, being 30 percent harder and with an ultra-smooth finish which yielded less friction wear during operational service.

Block I was part of a billion-dollar effort to overhaul the Shuttle's entire MPS architecture, with expectations that by the end of the 1990s a further upgraded Block II would bring Pratt & Whitney's Alternate High Pressure Fuel Turbopump (AHPFT) online. The main desired aim, of course, was to demonstrably improve flight safety and reliability, but the new engine also helped to support proposed high-energy, high-inclination missions under consideration to the Russian Mir space station, which operated in a 51.6-degree orbit. Already, an Advanced Solid Rocket Motor (ASRM) program to upgrade the SRBs, enhance their safety features and assist with transporting heavier payloads aloft had been canceled in October 1993 and the Block I and II engine campaign itself had stalled in December 1991 when it ran $260 million over-budget. Work resumed in May 1994 and the remaining Block I tests were concluded in nine months, with an expectation that the new engine reduced the computed LOCV probability from one in 262 flights in the earlier design to one in 335 flights for the new one.

Fig. 3.15 A full suite of three Block II Space Shuttle Main Engines (SSMEs) ignite for the first time on 8 April 2002 to launch Atlantis on STS-110.

And despite difficulties with Pratt & Whitney's AHPFT – including a test stand failure in January 1996 – that repeatedly delayed the timeline for bringing the complete Block II SSME into service, the ground tests of the new Large Throat Main Combustion Chamber (LTMCC) successfully demonstrated reduced operating pressures and a redesigned hot gas manifold. As a result, in January 1998, Endeavour's STS-89 mission to Mir saw the first use of the interim 'Block IIA' engine using all the Block II enhancements except for the AHPFT. Nine months later, Discovery flew STS-95 for the first time with a full suite of three Block IIAs. Certified to function at 104.5-percent thrust, the Block IIA allowed the Shuttle to lift an additional 225 kilograms of payload to orbit. For a while there was renewed talk of pushing it as high as 109 percent using the all-up Block II. It was hoped that this might shorten the available 'window' for the risky RTLS abort by 30 seconds and extend the availability of the more benign TAL. "We're out there on the edge," admitted engineer Chris Singer at NASA's Marshall Space Flight Center in Huntsville, Alabama, "so we run our test program…to demonstrate we've got margin against that edge, but we don't like to walk along it."

In July 2001, the maiden flight of a Block II engine formally got underway when Atlantis lifted off for STS-104 to deliver the Quest airlock to the ISS. Weighing just 3,500 kilograms, the upgraded engine was significantly lighter than the

4,200-kilogram first-generation SSME. It was expected that the Block II reduced the chance of an LOCV during ascent from one-in-438 to one-in-483. Nine months later, on Atlantis' very next mission, STS-110, she carried a full suite of three Block IIs for the first time. "We've basically eliminated the welding inside the pump that we could not inspect," Shuttle program manager Ron Dittemore commented at the time. "We have made the pump beefier, more robust." The consequence, admittedly, was a weight increase for the 70,000-horsepower AHPFT, but this sacrifice in performance was more than compensated by the safety enhancements. Stronger turbopumps and more robust turbine blades meant that issues which blighted the Shuttle's earlier career (and might have caused a catastrophic explosion) could now be managed safely. In the weeks preceding STS-55 in early 1993, concerns about the presence of obsolete tip-seal retainers on the blades of Columbia's original HPOT had forced a spate of launch delays as documentation was scrutinized and her SSMEs were replaced for extensive checks. The following June, STS-57's launch was stalled by an incorrectly fitted inspection stamp on one of two springs in the HPOT. And in April 1994, concerns about deformed nickel-alloy liquid oxygen guide vanes in the pre-burners of the turbopumps delayed the launch of STS-59. "Some of the failures that we have seen on this new pump during testing would have taken out a pump – and an engine – in the past," Dittemore remarked after the introduction of the Block II turbopump architecture. "With this new pump, it just keeps on chugging."

But at the end of the day, hardware is only hardware and problems cropped up from time to time, including leaking liquid hydrogen transducers, higher than allowable gas concentrations in the Shuttle's aft fuselage, suspected cracks in coolant manifolds and issues with HPFT tip-seals. Late in 1999, the flight of Discovery on STS-103 was halted firstly by a 1.3-millimeter drill bit found to be stuck in one of her SSMEs, a failed leak check and finally frayed wiring and a crushed hydrogen conditioning line on an engine. Then in June 2002, several cracks (each measuring 2.5 millimeters in length) were found on metal flow liners inside one of Atlantis' Block II SSMEs, present to help cryogenic propellants to flow past a set of accordion-shaped 'bellows'. Since the liners were not intended to hold pressure, NASA noted that the problem did not constitute a leak, but rather a concern that debris from the crack might work its way into an engine and trigger an explosion. Inspections and repairs of all four orbiters were ordered – with Discovery and Columbia each turning up three cracks and Endeavour just one – and the fleet was grounded for several months. The cracks were welded and rough edges polished to prevent further damage.

Yet the Block I, Block IIA and Block II incarnations contributed to a relative paucity of in-flight problems with the SSMEs for the remainder of their lifetime. Moreover, their impeccable safety and reliability contributed in 2011 to their selection to power the first-stage core of the Space Launch System (SLS), NASA's new super-heavylift rocket. As a result, descendants of the same engines which exploded on the test stand in the 1970s, which failed the Shuttle five times on the launch pad and which almost ended the missions of STS-51F and STS-93 will, in a few years' time, propel humans back to the Moon and eventually onward to Mars.

4

The External Tank

WOODPECKERS, WEIGHT AND TALES OF WOE

Throughout its career, the Space Shuttle was remorselessly attacked from many angles: from politicians eager to slash NASA's budget to anti-nuclear protesters keen to prevent the launch of plutonium-powered missions like the Galileo and Ulysses probes and from poor weather conditions to dropped tools and deadly technical maladies. That is not to forget, of course, two appalling tragedies in January 1986 and February 2003 which claimed not only Challenger and Columbia and the lives of their astronauts, together with several other missions which came within a hair's breadth of disaster. But of all the lethal incidents which bedeviled the fleet over those three decades, none was stranger than that which hit Discovery in the summer of 1995. In fact, when STS-70 mission specialist Don Thomas walked out of a training simulation to be told that a woodpecker had just scrubbed his flight, he was certain it was a prank.

They could not be serious? Could they?

Across the vast expanse of Florida's Merritt Island, nearly 570 square kilometers play host to a dedicated National Wildlife Refuge, home to over a thousand types of plants, 330 breeds of birds and dozens of varieties of fish, amphibians, reptiles and mammals. Twenty-one of them are acknowledged as endangered species and among their number is a medium-sized migratory woodpecker, the grey-capped, beige-faced Northern Flicker, whose penchant for noisily drumming on trees or large metallic objects as a means of communication earned it the ire of many at the Kennedy Space Center (KSC) that summer. For over Memorial Day weekend in late May 1995, right in the high heat – literally – of the woodpeckers' mating season, one of their number happened to spot the biggest tree it had ever

© Springer Nature Switzerland AG 2021
B. Evans, *The Space Shuttle: An Experimental Flying Machine*,
Springer Praxis Books, https://doi.org/10.1007/978-3-030-70777-4_4

seen. And as woodpeckers do, it proceeded to start tapping holes into it. But this was no ordinary tree. Bright orange, covered with Spray-On Foam Insulation (SOFI) and an underlying metallic skin, the perplexed woodpecker might have wondered what it had gotten itself into. For the 'tree' upon which it labored so fruitlessly was not a tree at all, but the External Tank (ET) of Space Shuttle Discovery, primed and ready for launch.

Two types of insulating foam were used on these gigantic tanks: the low-density, closed-cell SOFI which coated the majority 'acreage' of their surfaces and a much denser composite known as 'ablator', made from silicone resins and corks and employed on areas which endured the most extreme thermal duress of up to 1,650 degrees Celsius, including the ET's aft dome, situated closest to the Shuttle's three main engines. More than 90 percent of the foam was applied using automated systems, with the remainder smoothed manually into place by hand. In most areas of the tank, the foam was only 2.5 centimeters deep, although it was thicker (between 3.8 and 7.6 centimeters) in regions subjected to more intense heating. As well as keeping the cryogenic propellants at their requisite temperatures – the liquid oxygen at minus 182.8 degrees Celsius and the liquid hydrogen at minus 252.8 degrees Celsius – in the hot, humid Florida air, as well as preventing the accumulation of ice, this insulation could withstand up to 180 days sitting on the launch pad, daytime extremes of 46 degrees Celsius, humidity of 100 percent and could resist the ravages of sand, salt, fog, rain, solar radiation and even fungus.

But not woodpeckers.

The STS-70 stack had been transported from the Vehicle Assembly Building (VAB) to Pad 39B on 11 May, tracking an opening launch attempt in early June to deploy an important NASA communications satellite. On Memorial Day, 29 May, Discovery's crew were wrapping up one of their final training sessions in the simulator at the Johnson Space Center (JSC) in Houston, Texas, before flying down to the Cape for launch. Even as they did so, the outermost layer of SOFI on the ET had already succumbed to dozens of deep 'excavations' and multiple beak and claw marks, with over 200 holes now in need of repair. Some were quite small at about 1.2 centimeters across, others as broad as 10 centimeters. And several had punched straight through the foam to the underlying metallic skin of the ET. Technicians managed to scare away further attacks using air horns and Predator Eye balloons during the working week, but with little more than a skeleton staff at Pad 39B over the weekend, keeping the pesky avians at bay proved more problematic. A solution was to ask the KSC secretaries to stand at various levels on the launch pad and blow air horns if woodpeckers came close. They were even issued with pale blue T-shirts, emblazoned with the tongue-in-cheek legend, 'Pecker Patrol'.

Fig. 4.1 Repairs to the uppermost portion of Discovery's External Tank (ET) before STS-96 in May 1999 are clearly visible, following a battering by hailstones.

Two other events, of equivalent strangeness, occurred several years later. In April 1998, as Columbia was being readied for STS-90, a bat attached itself to the ET. Engineers took his body temperature with an infrared camera, revealing him to be about 20 degrees Celsius, and it was assumed that he was trying to cool off on the 16-degree-Celsius foam surface. According to launch director Dave King, it was theorized that the bat heard the chirping of hundreds of crickets being flown aboard the Shuttle and it remains unknown if he took flight before the three Space Shuttle Main Engines (SSMEs) roared to life as his sensitive ears would not have survived the acoustic shock. And in March 2009, during the STS-119 countdown, a bat with a broken left wing and perhaps a shoulder or wrist injury was found clinging to Discovery's ET. Like his STS-90 predecessor, the bat was probably trying to keep cool. Launch imagery revealed that he clung on to the tank for at least the first few seconds of ascent and was almost certainly incinerated shortly afterwards.

Birds, of course, come with the territory in the Merritt Island National Wildlife Refuge. During the launch of STS-114 in July 2005, a vulture hit Discovery's ET and plunged to its death in the fiery exhaust. Weighing at least twice as much as the piece of foam insulation which had hit and doomed Columbia on STS-107 (as will be described in Chapter 6), bird strikes were a significant threat. Efforts to mitigate the risk included a novel 'bird abatement program', involving the rapid removal of roadkill as a readily available food source from KSC as a means of discouraging them from turning up at these least opportune times. Acoustic deterrents were also established, together with trap-and-release protocols and an Aircraft Bird Strike Avoidance Radar to track avian movements and afford real-time data up to a minute before liftoff. The system was first tested on Discovery's STS-121 mission in July 2006.

Prime contractor Martin Marietta's zeppelin-like ET was by far the largest part of the Space Shuttle stack, standing 47 meters tall and comprising a pair of large tanks to feed liquid oxygen and hydrogen propellants through a pair of 43-centimeter-wide disconnect valves into the combustion chambers of the SSMEs. Due to the offset position of the orbiter (which appeared to 'hang' off the side of the tank), it was an engineering and aerodynamic necessity for the thrust vector of the SSMEs to pass directly through the center-of-gravity of the ET after Solid Rocket Booster (SRB) separation. This structural consideration dictated the internal layout of the ET. The smaller of the two tanks, at 15 meters tall, held almost 651,700 liters of liquid oxygen and occupied the upper third of the ET, whilst its much larger sibling, at 29.5 meters tall, held over 1.75 million liters of liquid hydrogen and resided in the lower two-thirds of the ET. Separating the tanks was the 6.8-meter-tall 'inter-tank' which housed instrumentation and electronics. In March 1977, structural articles for the two propellant tanks were extensively tested at NASA's Marshall Space Flight Center (MSFC) in Huntsville, Alabama, after which they were disassembled, inspected and mated with a new inter-tank for delivery to the National Space Technology Laboratory (NSTL) in Mississippi the following August. Here the first real 'tanking' of an ET with propellants was performed. Elsewhere, more than a dozen acoustic tests were conducted on miniature versions of the tank to evaluate the SOFI and ensure that it would not crack under intensities as high as 170 decibels during a Shuttle launch.

But even as the 35,000-kilogram Standard Weight Tanks (SWTs) were being built for the first Shuttle missions, NASA was looking at ways of reducing their weight still further. Admittedly, small incremental savings of between 500 kilograms and 1,100 kilograms were achieved in the first production batch of SWTs, but Martin Marietta promptly set to work to reduce the weight of later tanks by at least 2,700 kilograms and up to 3,400 kilograms. It was expected that shaving as much unnecessary weight as possible from the ET would aid high-energy launches, particularly those planned into polar orbit from Vandenberg Air Force Base in California, as well as high-inclination missions out of KSC. Many of the savings from the early SWTs were accomplished by slightly shaving the tolerances for metallic and Thermal Protection System (TPS) elements to the minimum allowable thickness for their specification. Even paint carried a penalty. On the first two Shuttle missions in April and November 1981, Columbia's SWTs were painted with a topcoat of Fire Retardant Latex (FRL), which alone amounted to over 360 kilograms in weight. But the SOFI was considered more than adequate for the ET's thermal protection needs during ascent and the white-colored FRL was deleted from STS-3 in March 1982 onwards. As such, STS-1 and STS-2 remain the only missions to have flown with white-colored ETs; all others went unpainted and carried the natural rusty-orange hue of the SOFI. However, during Discovery's May 1995 battle with the woodpeckers, in addition to air horns and fearsome-looking Predator Eye balloons, a helpful member of the public suggested repainting the ET in blue. Apparently, it was a color that woodpeckers hated. NASA declined the offer.

Right from the start, it was recognized that additional weight savings would demand an extensive redesign of the entire tank. "Because the ET is the structural backbone of the integrated vehicle," wrote Dennis Jenkins, "load paths are complex, making weight reduction a difficult task." As part of this redesign, an anti-geyser line for liquid oxygen circulation during the fueling process was deleted, the orbiter/ET aft crossbeam was reduced in thickness, electrical modifications were implemented and portions of longitudinal structural stiffeners (known as 'stringers') for the liquid hydrogen tank were removed. Heavy SRB attachment points were replaced with lighter, stronger and cheaper titanium alloy ones. And whilst the earliest SWTs were built primarily from aluminum, their second-generation Lightweight Tank (LWT) successors employed aluminum-lithium alloy, with walls up to 0.12 millimeters thinner than their predecessors. When Challenger flew the first LWT on STS-6 in April 1983, its total

Fig. 4.2 Pre-launch frosting is visible on the lowermost hydrogen tank of STS-99's External Tank (ET) before flight in February 2000. The 'ribbed' texture of the instrument-laden inter-tank is clearly visible, with the liquid oxygen tank above. Note the gaseous oxygen venting hood (known colloquially as the 'beanie cap') at the apex.

weight of around 30,300 kilograms represented an approximately 12-percent reduction over the SWTs. And in June 1998, Discovery's STS-91 mission flew the first Super Lightweight Tank (SLWT), which brought this down still further to 26,500 kilograms.

The development of this final incremental phase in the tank's design began in 1991 as part of efforts to empower the Shuttle to deliver larger, heavier payloads into high-inclination orbits for the construction of the International Space Station (ISS). "Each pound we can take from the External Tank is one more pound we can take to orbit," said ET Project Manager Parker Counts of MSFC. "This becomes especially important when launching the International Space Station into its proper orbit." Under a $172.5 million contract, Martin Marietta developed an aluminum-lithium alloy for the SLWT, some 40 percent stronger and 10 percent less dense than the LWT. Its walls were machined in an orthogonal, waffle-like pattern, which afforded added strength and durability. As STS-91 commander Charlie Precourt remarked, "It would be akin to taking your car as it sits today, removing all four doors and the engine and still have something that drives down the street."

However, the SLWT endured a troubled evolution. Prime contractor Lockheed Martin – the result of a 1995 merger between Martin Marietta and the Lockheed Corporation – experienced difficulties with cracking at the junctions between circumferential welds and vertical ones. The cracks appeared when the welds were reheated during the process of joining sections. "We learned a lot about aluminum-lithium," said Counts. "It's definitely the material of the future, but it's definitely not as easy to work with as the old stuff." Certification testing ended in July 1996 and the first production SLWT emerged from NASA's Michoud Assembly Facility (MAF) in New Orleans, Louisiana, in January 1998. It arrived at KSC a few weeks later and was mated to the STS-91 stack in May. After rollout to the launch pad, it underwent a full 'tanking test' to assess its responses to the fueling process. In its last evolved incarnation as the SLWT, the tank which flew the final Shuttle mission in July 2011 was 25 percent lighter than the one which had helped power Columbia uphill for the first time in April 1981.

ONE-WAY TICKET

Near the base of the ET were a pair of mechanical 'disconnects', through which propellants flowed from the great tank via a pair of 43-centimeter-wide valves in the Shuttle's belly and into the combustion chambers of the three SSMEs. Under conditions of maximum engine thrust during ascent, liquid oxygen was fed at a rate of 80,000 liters per minute and liquid hydrogen at 215,300 liters per minute. Both 'sides' of the vehicle – the orbiter and the ET – possessed mechanical disconnects and shortly before the separation of the tank, about 18 seconds after

Main Engine Cutoff (MECO), a pair of flapper valves were commanded shut by pneumatic helium pressure to prevent further propellant discharge and contamination. The criticality of these disconnects cannot be underestimated, for any inadvertent closure whilst the SSMEs were still firing would have precluded propellant flow from the ET and precipitated a catastrophic failure. In the early years of the Shuttle program, the system operated without incident, despite a liquid hydrogen primary seal leak in the disconnect on the ground which resulted in a launch delay for the (ultimately canceled) STS-51E mission in February 1985.

But five years later, the disconnects brought the Shuttle fleet almost to its knees. Before each mission, the Flight Readiness Review (FRR) would scrutinize every scrap of pertinent documentation, before reaching formal consensus on a target launch date. Early in May 1990, Columbia was provisionally scheduled to launch on STS-35 on the 16th, but a problem was noticed with a proportioning valve on her Freon coolant loop and this required replacement. Launch was moved to the 30th. In the pre-dawn darkness of that morning, as engineers at Pad 39A worked to load liquid hydrogen into the ET, a tiny leak was detected near the Tail Service Mast (TSM) on the Mobile Launch Platform (MLP). Further investigation revealed a more troubling leak, which apparently had its point of origin somewhere in the disconnect hardware in Columbia's belly. The launch attempt was canceled and the ET was emptied of propellants and 'inerted'. Since the STS-35 crew would be running a 24-hour, dual-shift operation on their mission, they had already begun 'sleep-shifting' to prepare for flight. Astronaut Jeff Hoffman was awake to hear the news of the scrub and woke his crewmates to deliver the disappointing news. Several days later, a miniature tanking test attempted to pinpoint the leak. But with the vehicle sitting vertically on the launch pad, conditions were far from ideal for exploratory work and on 12 June the Shuttle stack was rolled back to the VAB. Columbia was detached from her ET and SRBs and returned to the Orbiter Processing Facility (OPF) for repairs. The Shuttle 'side' of the disconnect hardware was replaced with a set borrowed from another orbiter and brand-new hardware for the ET 'side' was delivered a few days later from MAF.

Then, as if matters could not get worse, another mission was struck by a similar malady. As work progressed on Columbia, her sister Atlantis was advancing towards the launch of her own STS-38 mission in July 1990. As a precautionary measure, after rollout to the pad on 18 June, NASA performed a tanking test to ensure that she too had not fallen foul of the problem. Two modes of propellant loading were followed during the test. The first, called 'slow-fill', acted to chill down the pipework and structure of the ET so that when liquid hydrogen was later pumped at a higher rate, known as 'fast-fill', it would not boil-off and generate excessive amounts of gas. On 29 June, as liquid hydrogen was pumped into STS-38's ET, engineers gasped in dismay as the same problem arose: concentrations of gas were detected right in the vicinity of Atlantis' disconnect hardware after the fueling process moved from its slow-fill to fast-fill regimes. The leak was infinitesimally small (being described by NASA as "both temperature and flow-rate

Fig. 4.3 The 43-centimeter disconnect hardware transported propellants from the External Tank (ET) at left into the Shuttle's belly and from thence to the combustion chambers of the three main engines.

dependent"), but a lingering sense of unease hinted that the leaks in both Columbia and Atlantis were more than sheer coincidence. The hardware was telling them something.

To identify the source, more instrumentation was fitted around the disconnect and another tanking test was performed on 13 July. Once again, concentrations of leaking gas were found. Sealants were added to halt the leak, but a third failed test on the 25th underlined the reality that this would be a tough nut to crack. Two weeks later, the STS-38 stack was rolled back to the VAB, passing the STS-35 stack as it returned, newly repaired, to the launch pad. Yet NASA managers were convinced that the two leaks were distinct from each other. "Incorrect torqueing of bolts around the flange interface between the tank and the orbiter caused the Atlantis mishap," *Flight International* reported in August. "The Columbia leak was caused by a faulty seal in the drive mechanism used to close the flapper valve in the disconnect." At length, STS-35 was rescheduled for another launch attempt on 1 September. That was delayed a few days due to payload problems, but on the evening of the 5th, as engineers were working to load the ET, they detected leaking hydrogen gas yet again. The maximum allowable rate of leakage was 660 parts per million, but their instrumentation told them that the actual rate from Columbia was 6,500 parts per million.

It soon became apparent that two leaks had struck STS-35. The data revealed that the initial disconnect leak had been resolved, but a new one had arisen, somewhere in the Shuttle's aft compartment. Diagnostics put it near Columbia's recirculation pump package inlet or manifold. Three hydrogen recirculation pumps were promptly replaced, as was a damaged Teflon seal on one of the SSMEs, to no avail. In the hours preceding another launch attempt on 18 September, more leaking hydrogen gas was detected, this time at a rate of 6,700 parts per million. NASA was out of options. STS-35 was indefinitely postponed until the problem could be resolved. Former astronaut Bob Crippen, now the Shuttle program manager, assembled a 'tiger team' to investigate the problem and retorque Columbia's entire liquid hydrogen propulsion system. He assigned veteran engineer Bob Schwinghamer from MSFC to lead the investigation. Years later, Schwinghamer remembered NASA deputy administrator James 'J.R.' Thompson telling him, without a hint of humor, that he had a one-way ticket to Florida and should not return to MSFC until the leak was fixed.

Schwinghamer's team spent three months at KSC, from September until December, setting up an intricate fault tree and co-ordinating a huge number of personnel, spread across several NASA centers. When they wrapped up the final tanking test on 30 October, Schwinghamer could confidently declare that Columbia was now "the soundest leak-free orbiter at that time in the fleet". The very nature of hydrogen meant that a leak of any sort could not be tolerated – even though the Shuttle's Main Propulsion System (MPS) was designed to overpower leaks with a nitrogen purge – and it certainly surprised STS-35 commander Vance Brand that an orbiter was being grounded for such a long period of time. But it also represented a clear and encouraging mindset change compared to the workings of NASA management prior to the Challenger accident four years earlier.

Fixing the hydrogen leaks cost the agency $3.8 million and it was suspected that the problem (at least in Columbia's case) originated from a complete disassembly of the SSMEs following her previous mission, as part of efforts to remove polishing grit from her propellant lines. When the engines were reassembled, their seals had been improperly fitted and minute glass beads introduced contaminants into the disconnect hardware. From the start of June, the attention of engineers had been drawn solely to the disconnect leak, which posed a more serious problem, and the seals were overlooked. "As a result," explained *Flight International* in mid-November, "NASA has introduced a new processing program in which key engine components will be checked for leaks before the engines are finally assembled." In any case, the STS-38 and STS-35 missions both flew safely in November and December 1990. But the ET umbilical hardware would go on to wreak further havoc a few months later, as Discovery was being prepared for her STS-39 mission. In late February 1991, during final checks at Pad 39A, technicians found cracks on all four lug hinges on the two ET umbilical door drive mechanisms, which forced a rollback of the Shuttle to the VAB for repairs. "The cracks are not in the door hinges," explained NASA Headquarters in a news release, "but rather in metal that supports the mounts for electric mechanisms that open and close the doors." The safety-of-flight implication was that the doors had to close properly after the jettison of the ET to ensure that the Shuttle could endure the furnace of

Fig. 4.4 Technicians work to replace the disconnect hardware on Columbia during the summer of 1990.

re-entry. Although documented events were suspected of overly stressing the doors and perhaps initiating the cracks, no conclusive evidence as to their origin could be found and NASA conservatively opted to investigate further. Replacement hinges from another orbiter were removed, strengthened and fitted to Discovery.

In the meantime, NASA announced that a new hydrogen dispersion apparatus would be added to the MLP for future missions, beginning with STS-39. The system would provide a nitrogen-rich air flow around the disconnect hardware and help to disperse concentrations of hydrogen. As for the disconnects themselves, NASA was already working on plans with Shuttle prime contractor Rockwell International to develop an upgraded system, narrower at 35.5 centimeters in diameter, and contracts worth $27.6 million were awarded in February 1991. It was expected that the new disconnect hardware would prevent inadvertent closure of the flapper doors during ascent. Unfortunately, the new disconnect project was canceled by NASA in 1993, but much of its technology was employed to improve the safety and reliability of the existing 43-centimeter hardware. Furthermore, NASA instituted more rigorous rules around the issue of hydrogen leakage, embedding them more firmly within its Launch Commit Criteria (LCC) framework. Nor was this an issue which only affected the Shuttle in its early career. As late as April 2002, during preparations to launch Atlantis on STS-110, the ET umbilical drive door motors came under scrutiny. Testing of similar mechanisms revealed Endeavour's motors were slow to close. Fortunately, Atlantis' own motors turned out to be healthy and the mission flew safely.

BATTLING ICE, INSULATION AND OTHER ILLS

The unpredictable Florida weather, with its frequent storms and rains, labored incessantly over the years to damage the fragile insulation of several ETs before they ever left the ground. In May 1999, Discovery was a week away from launching STS-96 when her tank's SOFI took a battering in a violent weekend downpour of hail. With the Shuttle stack standing vertical on Pad 39B, technicians could implement some repairs, but the limited accessibility to reach more serious areas of damage required a rollback to the VAB. Admittedly, some of the damage was little more than pinpricks, but there were dings and dents measuring up to 5 centimeters long. Inspections in the VAB revealed 648 divots in the SOFI, of which 189 were small enough to pose an insignificant risk. More than 200 others, however, had to be fixed by blending and sanding work on the tank, whilst an additional 250 were 'patched' with injections of new foam. A similar episode occurred after a February 2007 storm, during preparations for STS-117, when hailstones as big as golf balls pummeled the ET and left nearly 2,000 holes in the SOFI and caused minor damage to 26 heat-resistant tiles on Atlantis' left-hand wing. To effect repairs, a 45-kilogram tool (nicknamed the 'pencil sharpener') was constructed to sand down patches of replacement foam at the top of the tank. One end of the sharpener fitted over the 'spike' at the top of the ET, whilst the other end housed a 60-centimeter-long cylinder of sandpaper which technicians used to grind the foam flat and smooth. But repairs of this complexity took time. STS-96 suffered only a week of delay, whereas STS-117 was put back three months. And for a system which, by the time of STS-117, had already shown that it was more than capable of causing enough damage to take human lives, such repairs were critical.

Right from the start of the Shuttle program clear links were drawn between the vulnerability of the ET and the vulnerability of the orbiter's TPS – that delicate combination of thermal blankets, a patchwork of thousands of individually-designed and hand-applied silica tiles and the Reinforced Carbon Carbon (RCC) panels lining the nose cap and leading edges of its wings – to damage from falling ice or foam. With liquid oxygen and hydrogen chilled to cryogenic temperatures, a corresponding build-up of ice on the ET's external surfaces before launch was an accepted inevitability.

To mitigate the danger, in the last few hours before each launch, the Final Inspection Team (better known as the 'Ice Team') were dispatched to the pad, heavily armed with cameras, binoculars, infrared scanners and a wealth of knowledge and expertise. And from Discovery's STS-120 launch in October 2007, they were also able to transmit real-time digital imagery of their inspection back to mission managers via fiber-optic cables. For a couple of hours ahead of each mission, their keen eyes and high-tech tools scoured the vehicle for unusual accumulations of ice or abnormal temperatures which might jeopardize a safe launch. Starting their work high up the launch pad gantry, they inspected the gaseous oxygen vent hood (otherwise known as the 'beanie cap'), atop the ET, then worked their way downwards to check the Shuttle's flight surfaces, main engines and the SRBs. Excessive ice on the ET's main surface 'acreage' had already scrubbed one attempt to fly STS-51C in January 1985. And the December 1992 launch of STS-53 was itself delayed by more than an hour as the Final Inspection Team advised waiting for sunrise to melt hazardous accumulations of ice. Moreover, in the final hours before Challenger's tragic STS-51L in January 1986, conditions were so cold that the team used broom-handles to physically dislodge icicles from the ET. But ice formation was not just a function of temperature; it was also highly dependent upon localized humidity, wind speed and wind direction, which resulted in a whole range of other variables which complicated the work of the Final Inspection Team.

The team also proved crucial in the hours preceding other missions. In October 2000, one eagle-eyed member of the team with binoculars observed a 226-gram 'pip-pin', which had somehow gotten itself lodged into the liquid oxygen pipeline leading from the ET to the orbiter. The pin was little more than a shard of debris from a dismantled worker access platform, but concerns swirled around the possibility that if it fell away during launch it could damage the ET's aft dome or the Shuttle's main engines. The location of the pin was about 12 meters above the pad surface and ideas of dislodging it with high-pressure hoses were not entertained on safety grounds. Ultimately, Discovery's launch was delayed and the pin was removed.

Fig. 4.5 Before each launch, the Final Inspection Team scoured every inch of the flight surfaces of the Shuttle, its External Tank (ET) and twin Solid Rocket Boosters (SRBs) to search for unsafe accumulations of ice or frost which might hamper a successful mission.

Incidents of falling SOFI debris during ascent were far from uncommon and occurred on virtually every Shuttle mission. Even as Columbia speared towards orbit for the first time in April 1981, crewmen John Young and Bob Crippen noticed white material (most likely foam liberated from the ET) streaming past their cockpit windows. After the mission, more than 300 of the Shuttle's TPS tiles had incurred sufficient damage as to warrant replacement. Forty more tiles were damaged on STS-4 in June 1982. And during STS-7 in June 1983, the first instance of foam shedding from the 'Bipod Ramp' area of the tank was recorded. Each ET carried a pair of bipod fittings which linked the top of the tank to the orbiter via forward attachment points. To prevent excessive accumulations of ice, each bipod was covered with a wedge-like, hand-sprayed, hand-carved 'ramp' of SOFI, measuring 76 centimeters long, 35 centimeters wide and 30 centimeters tall. Although intended to protect this critical area against ice build-up, the ramps also turned into a source of debris themselves.

During the application of the foam, wire brushes were used to create air holes to permit the escape of trapped gases and reduce the risk of void formation. In June 1992, during the STS-50 ascent, a chunk of SOFI measuring 60 centimeters by 25 centimeters fell from the left-side bipod ramp and left a sizable dent of 20 centimeters by 10 centimeters in one of the Shuttle's tiles. "Instead of hitting the wing, it went about a foot below the leading edge of the wing," remembered STS-50 commander Dick Richards. "We launched through an overcast, so we never knew that it even happened. We took photographs of the tank, but the photographs weren't downlinked to anybody." It was only when Richards and pilot Ken Bowersox performed their customary walk-around inspection of the orbiter on the runway after landing that they spotted the damage. Further instances of bipod ramp shedding were recorded on several missions, with dozens of tiles sustaining damage or being lost altogether. In October 2002, during Atlantis' STS-112 launch, a chunk of foam from the left-side bipod ramp detached, fell away and dented the attachment ring linking the ET to the SRBs. Following that incident, efforts to redesign the bipod ramp got underway, but – tragically – not in enough time to prevent tragedy on STS-107. Nor was the Shuttle fleet grounded in the interim period. In circumstances for which NASA would be harshly criticized, the very next mission, STS-113, went ahead in November 2002, to be followed three months later by the loss of Columbia and all hands during the STS-107 re-entry. As will be discussed in Chapter 6, Columbia sustained fatal damage to an RCC panel on the leading edge of her left wing when foam detached from a bipod during the ascent to orbit.

In the August 2003 report of the Columbia Accident Investigation Board (CAIB), the presidential inquiry into the STS-107 disaster, chaired by former U.S. Navy admiral Harold Gehman, it became clear that the sheer regularity of ET foam-loss occurrences led them to be regarded not as a safety-of-flight (and safety-of-life) issue, but rather a post-flight maintenance irritation. In July 1985, during a sweep-down inspection of KSC beaches following the STS-51F launch, fragments of fallen SOFI from Challenger's tank were found lying in the sand. Photographs taken in orbit in January 1992 and October 1993 by two Shuttle crews after ET separation revealed large dents in the inter-tank section, likely caused by falling insulation. And in November 1997, Columbia (through what can only be described as unalloyed good fortune) returned safely to Earth following STS-87, having survived the worst TPS mauling ever sustained as a direct consequence of ET foam loss. As will be described in Chapter 6, it was a minor miracle that the STS-87 crew made it home at all.

In fact, to commandeer a phrase from Richard Feynman of the Rogers Commission which investigated the loss of Challenger, the foam-shedding phenomenon was another case of NASA and its contractors luring themselves into the false sense of security that since earlier missions had dodged bullets, so too would future ones. Following the loss of Columbia and renewed emphasis upon the vulnerability of the bipod ramps, the CAIB found that hand-spraying SOFI over the "complex geometry" of the fittings generated a proneness to internal voids and defects. This was a contributory factor in the liberation of foam. An extensive redesign of the hardware after STS-107 enabled the bipods to fly mainly exposed, without insulating ramps. To prevent the formation of ice, four cartridge-like 'rod heaters' were positioned beneath each bipod in a new copper plate.

In addition to tending to the principal cause of the tragedy, NASA also completed a top-down assessment of the whole ET, which prompted a further round of modifications. The 21-meter-long propellant feedline, which ran along the exterior of the liquid hydrogen tank, up and into the inter-tank, and from thence into the base of the liquid oxygen tank, also revealed itself to be a source of debris. Within the liquid oxygen portion of the feedline were a set of five 'bellows' – two internal to the inter-tank, three external to it – which catered for differences in thermal expansion and relative motions in flight. Analysis revealed that the three external bellows were a potential source of ice accumulation and copper-nickel-alloy strip-heaters were added for the first post-Columbia mission, STS-114. A small

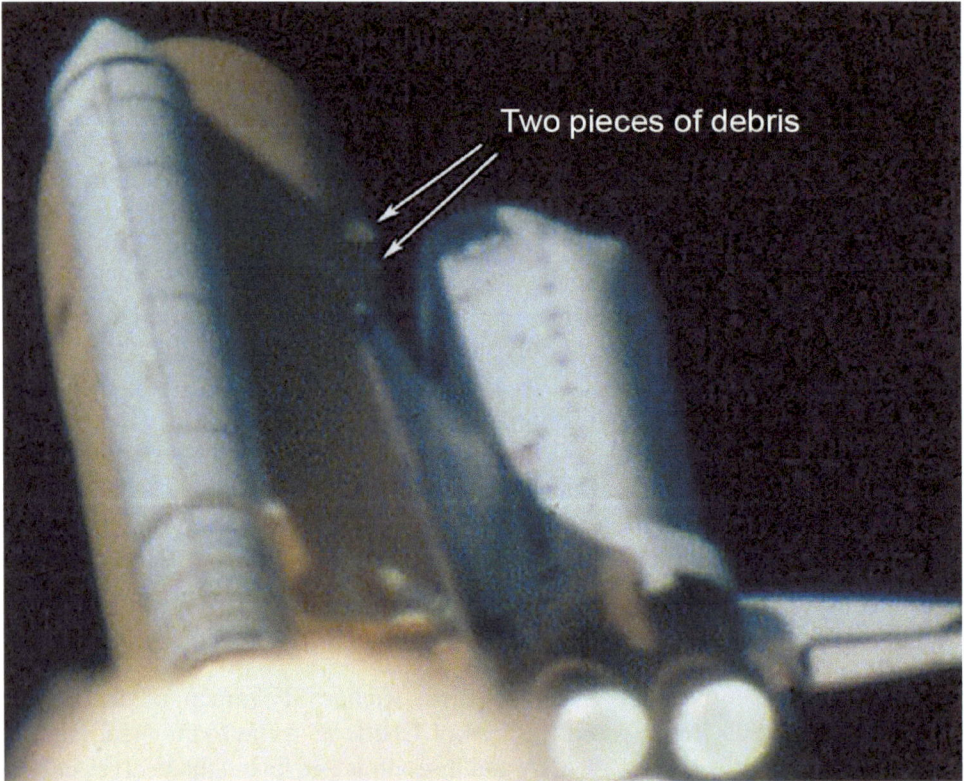

Two pieces of debris

Fig. 4.6 During the ascent of tragic STS-107 in January 2003, foam debris liberated from the Bipod Ramp on the External Tank (ET) caused crippling damage to one of Columbia's wings, which led to the destruction of the Shuttle during re-entry, 16 days later.

Sony XC-999 'lipstick' video camera had flown on STS-112's ET in October 2002 as a technology demonstration, but after the loss of Columbia it was repositioned onto the liquid oxygen feedline to provide clear, real-time views of foam liberation during ascent. And seam-like flange joint bolts were reversed to preclude the risk of gaseous nitrogen leakage from the inter-tank causing further foam loss.

But even when STS-114 finally flew in July 2005, the foam-loss phenomenon was by no means done with the Shuttle. Two minutes after Discovery's launch, a large fragment of foam fell off a Protuberance Air Load (PAL) ramp on the ET. There were two ramps, one near the top of the ET and the other just below the level of the inter-tank. They were designed to prevent unsteady air flow beneath the cable trays and pressurization lines. Covered with thick, manually sprayed layers of foam, they exposed yet another debris source. The fragment only weighed around 450 grams (about half as much as the briefcase-sized chunk which crippled Columbia) and fortunately it did not impact Discovery. Shortly afterwards, a much smaller piece of foam hit the right-hand wing with negligible effect.

However, worrisome images taken by the astronauts as the tank fell away after separation revealed multiple areas of foam liberation. The fleet was grounded for another year as NASA investigated this seemingly intractable problem. At one stage, as recriminations flew, it was hinted that MAF engineers had incorrectly handled and installed the ET during processing, but the cause was eventually traced to thermal expansion and contraction during the application of the foam which caused areas of the insulation to crack. This led the Shuttle program manager, Wayne Hale, to formally apologize to the MAF workforce, who had already borne the brunt of criticism following STS-107. Inspections and engineering analysis revealed the PAL ramps to be unnecessary and they were removed from all tanks, starting with the second post-Columbia flight, STS-121 in July 2006. Described by NASA as "the biggest aerodynamic change to the tank in the Shuttle's history", the PAL ramps were replaced with small foam 'extension' ramps to make the geometry of the area consistent with the remainder of the ET. On STS-124 in May 2008, an 'all-new' tank was flown, marking the first flight of an ET built from scratch with all the post-Columbia safety modifications as recommended by the CAIB. "This essentially is the completed return-to-flight tank," said John Shannon, who had succeeded Hale as the Shuttle program manager in February 2008.

A key player in the insulation story was the Environmental Protection Agency (EPA), which in 1995 directed the gradual withdrawal of Chloroflurocarbon-11 (CFC-11) from large-area, machine-sprayed foams under the requirements of the Clean Air Act. It was replaced by a hydrochloroflurocarbon known as 'HCFC-141b', first used as part of the ET foam on STS-79 in September 1996. But some 'detail-pieces' of foam still required a hand-spraying method of application and continued to use CFC-11 until the end of the Shuttle program. Several areas in which CFC-11 continued to be used were the bipod and PAL ramps. The CAIB also noted that STS-107's ET included a type of foam whose 'blowing agent' was CFC-11.

With the modifications to the bipod and PAL ramps satisfactorily tested by STS-114 and STS-121, subsequent missions flew with what NASA called "acceptable" levels of foam-loss risk. Notable exceptions included STS-118 in August 2007, when a 10-centimeter chunk of foam or ice from an ET feedline attachment bracket ricocheted off one of the aft struts and hit the underside of Endeavour's wing, damaging two TPS tiles. Foam loss from the hydrogen tank was also noted on STS-117 in June 2007, STS-120 in October 2007, STS-122 in February 2008 and STS-124 in May 2008. And when Endeavour launched STS-127 in July 2009, foam was lost from several points close to the inter-tank, together with part of the Ice Frost Ramp (IFR) of the liquid oxygen tank and another location close to the bipod. Corrective actions and improvements continued throughout the final few missions. From STS-120 onwards, the IFR on the liquid oxygen tank was filled using a manually applied foam to reduce the risk of de-bonding and cracking, as well as minimize the formation of voids. And from STS-129 in November 2009, the IFR of the liquid hydrogen tank was replaced in several spots with a new type of foam and less thermally conductive titanium brackets in place of the aluminum ones to reduce ice formation in less well-insulated areas. Nevertheless, as late as STS-132 in May 2010, only a year before the end of the Shuttle program, foam loss from the inter-tank was still being observed.

A HAZARD TO THE VERY END

As well as falling insulation and ice and the risks posed by the Floridian weather, ETs suffered numerous other issues during their three decades of service. On the very first Shuttle mission, STS-1 in April 1981, excessive shockwaves caused by the SSMEs and SRBs produced an 'over-pressure' event which buckled a strut linking Columbia to the tank. Had the strut failed, it would have resulted in a Loss of Crew and Vehicle (LOCV), only seconds after liftoff. "Those struts were critical," said NASA engineer Charles 'Tom' Hyle. "We had to know whether they could sustain loads that occurred during the ascent, like at high dynamic pressure when you get this big shockwave. And if the trajectory was not behaving as you'd expect, could the wings stay on, would the struts break and the orbiter flop back onto the tank?" Any such eventuality would be catastrophic. Strengthened struts were flown on STS-2 and improvements to the launch pad's sound-suppression water system kept pressures to less than a tenth of those on STS-1. And from an attach-point liner repair conducted prior to Endeavour's maiden flight to 'flaking' insulation on the inter-tank during processing for STS-89 to problems with oxygen-line welds on STS-103, these tanks more than proved their susceptibility to failure.

Fig. 4.7 The spindly attachment struts linking the Shuttle to her External Tank (ET) and Solid Rocket Boosters (SRBs) are clearly visible in this launch pad perspective.

Late in 2000, as Discovery was being readied for her STS-92 launch, an issue arose concerning bolts that were meant to retract during the ET separation sequence. Two such bolts (each measuring 35 centimeters long and

6.3 centimeters wide) extended from the tank into the Shuttle's belly and were secured with explosive nuts. The nuts were detonated at the instant of ET separation, a little over eight minutes after launch, after which the bolts were 'flown' down a housing into the tank as it was discarded. However, during Atlantis' STS-106 launch in September 2000, the explosive nut operated correctly, but the right-hand bolt remained extended and continued to stand proud by about 5.7 centimeters. Engineers feared that such a 'hung' bolt could impart undesirable forces on the ET after separation and even, in a worst-case scenario, potentially cause it to impact the orbiter. Analysis revealed that previous missions had weathered similar 'hung-bolt' situations and Discovery flew safely. "They have seen a number of times that this has occurred in the past, on the order of a half-dozen or so," said NASA Test Director (NTD) Steve Altemus. "There's never been any [damage] associated with that at all." But it was yet another indicator that past precedent was being used as a predictor of future success.

Following the return to flight after STS-107, several missions also experienced problems with their Engine Cutoff (ECO) sensors. The Shuttle was equipped with four sensors for liquid oxygen (housed in the orbiter's oxidizer feedline manifold) and four for liquid hydrogen (on a shock-isolated carrier plate at the base of the hydrogen tank of the ET). Their purpose was to safeguard the SSMEs by triggering a shutdown in the unlikely event that the propellant flow from the ET happened to 'cavitate', or unexpectedly run dry. The tank carried an additional quantity of liquid hydrogen to guarantee that at the point of MECO the shutdown process was 'fuel-rich' rather than 'oxidizer-rich', because the latter could significantly harm the engines. Under nominal flight conditions, at a pre-determined velocity, the Shuttle's General Purpose Computers (GPCs) would 'poll' the ECOs just prior to MECO and if two of the four oxidizer or fuel sensors indicated 'dry', this was taken to be an indication that the ET was almost empty and the SSMEs were commanded to shut down.

But during efforts to prepare Discovery for STS-114, one of the ECOs registered abnormal readings. After a detailed engineering analysis, a safety-of-flight rationale was implemented, whereby three out of the four sensors reading 'normal' would be acceptable for a launch to go ahead. Improper readings on an ECO sensor during pre-flight testing, however, conspired to delay STS-121 from March to July 2006. Another mission, STS-115 in September 2006, was similarly postponed. In both cases, the liquid hydrogen ECO sensors declared themselves 'wet' when they were 'dry' during tests. The concern was that in actual flight conditions (and if the ET really was almost empty) the sensors might not correctly recognize the problem and fail to command a timely SSME shutdown, with an LOCV being the likely outcome. From STS-118 onwards, a new system was tested to monitor the ECOs' circuit voltage to fuel sensors. This enabled flight controllers on the ground to recommend that the crew perform a manual SSME shutdown in the event of an ECO failure.

An additional, temporary modification to flight rules in this area was made after another incident in December 2007. During tests before STS-122, two of Atlantis' liquid hydrogen ECO sensors reported themselves 'wet' when they were 'dry'. The safety-of-flight rules were modified to allow Atlantis to fly with only two of the four ECO sensors working and augment this 'reduced-redundancy capability' with new procedures which gave flight controllers on the ground additional insight into the system during ascent. However, shortly after fueling got underway that December, a third ECO sensor failed 'wet', forcing a lengthier delay. Suspicion centered upon the possibility that part of the ET's wiring architecture was at fault. The sensors were thoroughly tested using time-domain reflectometry equipment over the following weeks, and this traced the issue to open circuits in the ET's liquid hydrogen feed-through connector. This was modified by soldering the connector's pins and sockets to address the false-reading glitch and STS-122 flew safely in February 2008.

Fig. 4.8 Impressive view of Atlantis on the pad, prior to the final Shuttle mission, STS-135 in July 2011.

Three other missions were afflicted by problems with the Ground Umbilical Carrier Plate (GUCP), an interface between the Shuttle and a vent line meant to regulate hydrogen pressures during the fueling process and route excess gas overboard to be harmlessly burned off by a flare-stack near the launch pad. However, in March 2009, as Discovery's ET reached the 98-percent-full point before the STS-119 launch, a significant leak was detected at the GUCP interface. Suspect seals and quick-disconnect fittings were duly replaced, the hardware was retorqued and Discovery flew safely into orbit a few days later. But similar GUCP leakage issues would go on to haunt and extensively delay two more missions, STS-127 in July 2009 and STS-133 in late 2010.

Other issues late in the ET's career included a stuck fill-and-drain valve which delayed STS-128 in August 2009 and a cracked flange between the liquid oxygen tank and inter-tank found in November 2010 during pre-launch preparations for Discovery's final voyage, STS-133. In this latter case, it was the first (and last) instance in which cracks in an ET had been detected whilst on the launch pad. The mission was postponed as investigations discovered a pair of additional cracks, each measuring 23 centimeters long, on opposite sides of one of the structural stringers. Further exploratory work uncovered more cracks, leading to suspicions that the use of lightweight aluminum-lithium alloys in the SLWT were a causal factor. Structural 'radius blocks' were installed to afford these stringers additional strength and Discovery flew her last mission successfully in February 2011. So too did the last pair of Shuttle missions, Endeavour's STS-134 in May 2011 and Atlantis' STS-135 in July 2011. But this episode in the dwindling twilight of the ET's career was a sobering reminder that there was nothing truly 'operational' about this experimental flying machine and that nothing could ever be taken for granted.

5

Big Dumb Boosters

THREE-TENTHS OF A SECOND

For many years, an unusual piece of artwork hung in Don Lind's house. Painted by the astronaut himself, it depicted the day of his launch into space and he gifted copies of it to his children, in the fervent hope that they and their own children would come to understand its message. Through Lind's artistic eye, Space Shuttle Challenger rises from Pad 39A at the Kennedy Space Center (KSC) in Florida at high noon on 29 April 1985 to begin STS-51B. Propelling the reusable spacecraft on her way are three Space Shuttle Main Engines (SSMEs) and a component which, nine months later, would be responsible for the taking of seven human lives: a pair of enormous Solid Rocket Boosters (SRBs). In the painting, Challenger is seemingly cradled by two great celestial hands, as if God himself were somehow willing Lind and his crewmates onward to a safe journey. Even more perplexingly, the title Lind chose for his work was 'Three-Tenths of a Second'. When Challenger was destroyed during launch in January 1986, the tragedy hit Lind particularly hard. For when he and the STS-51B crew flew that very same Shuttle, a few months earlier, they came close to losing their own lives.

Strapped to either side of the External Tank (ET) for each Shuttle launch, the SRBs provided 80 percent of the power – some 2.5 million kilograms of thrust – to get the spacecraft off the ground and to a sufficient altitude and speed to reach the edge of space. Each booster stood 45.5 meters tall and comprised four stainless steel 'segments', each weighing around 170,000 kilograms, together with a forward and an aft 'skirt' and the Solid Rocket Motor (SRM). This 'segmentation' of the SRBs was a practical means not only of assembling and testing the boosters at

© Springer Nature Switzerland AG 2021

B. Evans, *The Space Shuttle: An Experimental Flying Machine*,
Springer Praxis Books, https://doi.org/10.1007/978-3-030-70777-4_5

prime contractor Morton-Thiokol's facility in Utah, but also transporting them overland by rail to KSC in Florida. A 'tang-and-clevis' mechanism secured each segment to the next and a mixture of zinc chromate putty and a pair of rubberized O-ring seals were meant to prevent any leaks (known as 'blow-by') of hot gases from within.

Fueled by a powdery substance of ammonium perchlorate composite propellant, the SRBs fired for two minutes at the start of each mission and, when combined with the 'parallel-burn' of the SSMEs, pushed the Shuttle to an altitude of 45.7 kilometers and a velocity of nearly 5,000 kilometers per hour. But these solid-fueled behemoths could neither be throttled back nor turned off during those critical two minutes, which opened a yawning survivability chasm, known as a 'black zone', for the crew. Should anything go catastrophically wrong whilst the SRBs were still attached and firing, the astronauts could do nothing but sit tight and wait for them to be jettisoned. Only then would realistic abort options become available. "The Shuttle was unique in that…there was no escape tower and there were no launch abort rockets, so once they lit that thing off the ground, the crew was locked onto it for around two minutes or so," recalled NASA engineer Charles 'Tom' Hyle. "The aerodynamics pretty much precluded the Shuttle from getting away safely because of all the aerodynamics and the wings were fairly sensitive." Former Shuttle director Arnold Aldrich agreed. "If you try to separate them, there was no way to shut them down, and if you try to separate them while they're burning, the momentum of the thrust holds them in their places," he recalled grimly. "Even if you blow the bolts to separate them, they won't separate, because the thrust keeps them where they are. Everybody knew that. They'd already accommodated their thinking to the fact that this was going to be the way the Shuttle ought to be put together and that we would simply rely on the very high reliability of that system and press on in that manner."

Following the jettison of the boosters, they would descend to a parachute-aided splashdown in the Atlantic Ocean, after which they would be picked up by the NASA-owned recovery vessels *Freedom Star* and *Liberty Star*, towed back to port for refurbishment and used on another mission. The separation process was achieved by means of pyrotechnic devices and eight solid-fueled motors – four at the top and four at the bottom of each booster – which fired to move them away from the orbiter/ET stack. Four minutes later, having descended to 4.7 kilometers, a baroswitch triggered thrusters to discard the nose caps and permit deployment firstly of their pilot parachutes, then their stabilizing drogue parachutes and eventually three main parachutes for a controlled splashdown some 225 kilometers off Florida's east coast. It was anticipated individual SRB segments could fly 25 times.

Fig. 5.1 Engineers work at the base of one of Discovery's Solid Rocket Boosters (SRBs) in July 1993.

During pre-launch preparations, the boosters were paired in matching sets and filled with propellant ingredients from identical 'batches' to minimize the risk of thrust imbalances during ascent. Early plans featured a 'thrust termination system' – which would blow out the top ends of both SRBs and allow their thrust to diminish to zero in an emergency – but this concept was ultimately dropped, primarily because it might impart harmful structural loads on the Shuttle, as well as triggering unacceptable thrust imbalances. And redesigning and strengthening the orbiter to handle such loads threatened to impose a highly undesirable 3,500-kilogram weight penalty. The thrust termination system was canceled in April 1973. And as the SRBs flew successfully on mission after mission from April 1981 onwards, they were viewed as 'big and dumb', but totally reliable and impeccably safe. As such, none of the astronauts were ever assigned to oversee issues with the boosters. "We had lots of people paying attention to the main engines, because everybody knew about turbine cracks and potential pump problems," remembered Jeff Hoffman. "But the general feeling about the solid boosters [was] that solids don't fail and there's not much you can do about it." It was a fatal error in judgement.

Problems with the boosters did not take long to rear their heads and some became more readily visible than others. During Columbia's STS-4 launch in June 1982, the parachutes for both SRBs failed during descent and both sank in the Atlantic. Underwater remote cameras photographed the wreckage, but it proved too expensive to recover them. The cause of the incident was traced to a new feature intended to separate the parachutes from the boosters at the instant of splashdown. It was meant to stop them from being dragged through the water by their deflated canopies. The system had been 'active' on the first three Shuttle missions but was partially disabled on STS-4 and frangible nuts holding one of two 'risers' for each chute were replaced by a pair of solid nuts which would not separate the riser.

In the aftermath of the loss, all frangible nuts were replaced with solid ones from STS-5 onwards and when Columbia flew next in November 1982 both parachutes remained attached to the risers until they could be removed by recovery forces. But the STS-4 incident was neither the first nor the last such problem, with failures occurring on four more missions between November 1981 and April 1984. Even in the twilight of the Shuttle's career, SRB parachute anomalies cropped up periodically. In February 2008, a booster returned to Earth with a severely damaged canopy, whilst another in August 2009 exhibited an extensive vertical tear in its main parachute. Yet the greatest irony on STS-5 was that most of the ascent photography centered not upon Columbia and the safety of the astronauts as they headed for space, but upon whether the parachute modifications worked and the boosters returned safely. "They diverted all the photography to the SRBs, and we're pressing on," said STS-5 astronaut Bill Lenoir, with typical gallows humor. "Who knows where *we're* at?"

In the early 1980s, several performance upgrades were made, including replacement of the heavy SRB/ET attachment points with lighter, stronger ones manufactured from titanium alloy. The thickness of the segment 'walls' was also reduced by almost a millimeter to achieve an overall weight reduction of 1,800 kilograms. Some of the propellant inhibitors used in earlier SRBs were removed to allow the ammonium perchlorate to burn more efficiently. And when Challenger lifted off on STS-8 in August 1983, her SRM nozzles had longer exit-cones and narrower 'throats' to increase their thrust output by 4 percent. But following the return and disassembly of the STS-8 boosters, inspections revealed excessive corrosion in the throat of one of the nozzles. A protective carbon-fiber resin lining, 8 centimeters thick, had eroded much more severely than intended. Engineers estimated that this would have left only 14 seconds of additional firing time before the nozzle ruptured, which would almost certainly have resulted in a Loss of Crew and Vehicle (LOCV).

The fault was traced to the same batch of resin which would also be used on STS-9's boosters and, as a result, that mission found itself delayed from October to November 1983 as its nozzles were replaced. "NASA now believes that excessive corrosion of the SRB nozzle throat…relates to processing of the nozzle during the cure cycle," *Flight International* reported in October 1983. "The resin used to line the throat of the nozzle is available from two manufacturers and material supplied by one of them is apparently more sensitive to the manufacturing process than the other. High pressure applied early in the cure cycle of the 'sensitive' material is said to have prevented proper escape of volatiles, resulting in a weaker lining." The consequence was that 'spalling' had occurred in the char-layer of the throat material,

Fig. 5.2 Pre-launch closeout photograph of the top of the aft segment of STS-51L's right-hand Solid Rocket Booster (SRB), with the O-ring clearly visible.

causing it to 'flake' away, rather than steadily eroding as it should. The STS-9 stack was returned from Pad 39A to the Vehicle Assembly Building (VAB), but since the nozzle resided at the base of the SRB the repair necessitated the detachment of both the Shuttle and the ET as well as the complete disassembly of the booster.

But the real Achilles heel of the system, and the singular component upon which the loss of seven lives in January 1986 would ultimately hinge, were the primary and secondary O-rings: that combination of zinc chromate putty and rubberized seals whose *raison d'être* was to seal the segment joints and ensure that 'blow-by' of hot gases did not occur. "It was intended that the O-rings be actuated and sealed by combustion gas pressure displacing the putty in the space between the motor segments," noted the Rogers Commission, the presidential inquiry into the Challenger tragedy, chaired by former Secretary of State William Rogers. "The displacement of the putty would act like a piston and compress the air ahead of the primary O-ring and force it into the gap between the tang and clevis. This process is known as 'pressure actuation' of the O-ring seal. This pressure-actuated sealing is required to occur very early during the solid rocket motor ignition transient, because the gap between tang and clevis increases as pressure loads are applied to the joint during ignition. Should pressure actuation be delayed to the extent that the gap has opened considerably, the possibility exists that the rocket's combustion gases will blow-by the O-rings and damage or destroy the seals. The principal factor influencing the size of the gap opening is motor pressure, but gap opening is also influenced by external loads and other joint dynamics."

This inherent vulnerability of the O-rings first came to NASA's attention following STS-2 in November 1981. During post-flight 'tear-down' inspections of both boosters, significant erosion of the right-hand SRB's primary O-ring caused by the impingement of hot gases was noted, but the secondary O-ring seal remained intact and the anomaly went unreported at the Flight Readiness Review (FRR) for the next flight, STS-3 in March 1982. Prime contractor Morton-Thiokol – the result of Thiokol's 1982 merger with Morton Salt – believed the erosion to have been triggered by blow holes in the zinc chromate putty and began tests to alter both the method of its application and the assembly of the booster segments. The manufacturer of the putty, Fuller-O'Brien, discontinued its use and a new putty from the Randolph Products Company was selected in May 1982; however, after more changes, it was substituted for the original putty.

During the first five Shuttle missions between April 1981 and November 1982, the O-rings were labeled 'Criticality 1R' items by NASA, meaning that, although "total element failure…could cause loss of life or vehicle", the presence of primary and secondary O-rings lent 'redundancy' to the design. The secondary O-ring (so the logic went) would seal the joint if the primary one failed. But in its Critical Items List of November 1980, NASA acquiesced that "redundancy of the secondary field joint seal cannot be verified after motor case pressure reaches approximately 40 percent of maximum expected operating pressure. It is known that joint rotation occurring at this pressure level…causes the secondary O-ring to lose compression as a seal". High-pressure tests of the O-rings by Morton-Thiokol in May 1982 found that the secondary seal did not provide sufficient redundancy and the criticality listing was changed later that year. In December 1982, the O-rings were redesignated a 'Criticality 1' item, meaning that any failure in their performance would destroy the Shuttle and kill the crew. According to NASA's associate administrator for spaceflight Michael Weeks, who signed a waiver to accept this new criticality level in March 1983, "we felt at the time that the Solid Rocket Booster was probably one of the least worrisome things we had in the program". It was a view shared by many astronauts and managers.

But they were wrong.

Following STS-2, evidence of hot gases working their way through the zinc chromate putty and eroding the O-rings was found after STS-6 in April 1983, with both boosters affected. Ten months later, on STS-41B, the left-hand SRB's forward field joint and the nozzle of its right-hand counterpart were badly degraded, to such an extent that NASA asked Morton-Thiokol to investigate how further damage could be prevented. A week before Challenger launched STS-41C in April 1984, the company concluded that blow holes in the zinc chromate putty were one "possible cause" for the malady and NASA's booster project office at the Marshall Space Flight Center (MSFC) in Huntsville, Alabama, decided that if the secondary O-ring seal could survive hot-gas impingement, missions were safe to fly. According to the Rogers Commission, this was the start of an unsettling chain of thought that "there was an early acceptance of the problem" as both NASA and Morton-Thiokol "continued to rely on the redundancy of the secondary O-ring long after NASA had officially declared that the seal was a non-redundant, single-point [Criticality 1] failure". One of the Rogers investigators, the celebrated physicist Richard Feynman, likened this cavalier attitude to Russian roulette. "We can lower our standards a little bit because we got away with it last time," he fumed. "You got away with it, but it shouldn't be done over and over again." In his memoir *Riding Rockets*, astronaut Mike Mullane – whose own flight in August 1984 suffered O-ring erosion and blow-by – scornfully called it the "normalization of deviance".

But it was not until 24 January 1985 and the launch of STS-51C that a definitive line could be traced between the vulnerability of the SRBs and the adverse effect of cold weather conditions upon O-ring performance. Launched in frigid conditions of just 11 degrees Celsius, the mission's recovered booster nozzles (designated SRM-15) exhibited clear signs of blow-by between the primary and secondary O-rings. Even worse, this was the first Shuttle mission in which the secondary seal (as well as the primary) displayed the effects of heat. "SRM-15 actually increased concern because that was the first time we had actually penetrated a primary O-ring on a field joint with hot gas," noted Morton-Thiokol structural engineer Roger Boisjoly, "and we had a witness to that event because the grease between the

Fig. 5.3 A 1,900-kilogram fragment of the aft segment of Challenger's left-hand Solid Rocket Booster (SRB), pictured on the seafloor in the Atlantic Ocean by a robotic submersible craft.

O-rings was blackened, just like coal. That was so much more significant than had ever been seen before on any blow-by on any joint." When it was analyzed, Boisjoly and his team found products of both the zinc chromate putty and the O-rings themselves within the blackened material. Several days later, Lawrence Mulloy, manager of the SRB project office at MSFC, expressed concern about the impact O-ring problems might have on the next Shuttle mission. One of Morton-Thiokol's conclusions was that whilst "low temperature enhanced probability of blow-by…the condition is not desirable but is acceptable".

It was the first time that a link between cold weather and O-ring damage had been officially acknowledged. A red flag had been raised. Tragically, it would not be heeded.

Three months later, in April 1985, Don Lind and his six crewmates launched aboard Challenger for STS-51B. Subsequent inspections of their SRBs also indicated not only the failure of the primary O-ring, but also severe erosion of the secondary seal. The problem was blamed on leak-check procedures. But the incident proved so serious that "a launch constraint" was placed on several upcoming missions. However, as the Rogers investigators found, those launch constraints were routinely imposed and waived for mission after mission in the summer and fall of

1985. And perhaps most damning of all, on the eve of Challenger's final tragic flight in January 1986, "neither the launch constraint, the reason for it or six consecutive waivers" had been communicated up the chain of command to NASA associate administrator for spaceflight Jesse Moore or STS-51L launch director Gene Thomas.

But it did not end there. Prior to the loss of STS-51L, four more missions endured similar damage. In June 1985, both of Discovery's boosters returned with evidence of O-ring erosion and blow-by, as did three back-to-back flights by Challenger, Atlantis and Columbia the following fall and winter. For his part, Boisjoly took his concerns to Thiokol's vice president of engineering, Bob Lund. "The mistakenly accepted position on the joint problem was to fly without fear of failure and to run a series of design evaluations which would ultimately lead to a solution or at least a significant reduction of the erosion problem," he wrote. "This position is now changed as a result of the [STS-51B] nozzle joint erosion, which eroded a secondary O-ring with the primary O-ring never sealing. If the same scenario should occur in a field joint – and it could – then it is a jump-ball whether as to the success or failure of the joint because the secondary O-ring…may not be capable of pressurization. The result would be a catastrophe of the highest order: loss of human life."

Boisjoly recommended a Morton-Thiokol team should investigate and resolve the problem as a matter of urgency and, on 20 August, Lund announced the formation of a team to do just that. However, only a day earlier, in a joint briefing to NASA Headquarters on the issue, managers concluded that the O-rings were a 'critical' issue, but that, so long as all joints were leak-checked and met specified stabilization pressures, were free of contamination in the seals and fulfilled O-ring 'squeeze' requirements, it was safe to continue flying. As the year wore on, and the storm-clouds precipitating disaster loomed unseen, Morton-Thiokol's O-ring team (which comprised only a handful of members) found their efforts frustrated at every turn by senior management. "Even NASA perceives that the team is being blocked in its engineering efforts to accomplish its task," Boisjoly wrote in a 4 October memo. "NASA is sending an engineering representative to stay with us, starting 14 October. We feel that this is the direct result of their feeling that we are not responding quickly enough on the seal problem."

A little over three weeks later, Challenger flew STS-61A, again experiencing nozzle O-ring erosion and blow-by at the field joints; neither of these problems were identified at the FRR for the next mission, STS-61B in November. Indeed, that flight also suffered erosion and blow-by. By early December, Morton-Thiokol recommended that their testing equipment needed to be redesigned. Only days later, the company requested closure of the O-ring critical problem issue, citing satisfactory test results and work carried out by its task force. This closure request would be harshly criticized the following year by the Rogers Commission. One panel member told the Morton-Thiokol senior leaders: "You close out items that you've been reviewing flight by flight – that have obviously critical implications – on the basis that, after you close it out, you're going to continue to try to fix it. What you're really saying is [that] you're closing it out because you don't want to be bothered."

And when Don Lind, astronaut and painter of 'Three-Tenths of a Second', traveled to Morton-Thiokol's booster testing facility in Utah, he was in for a shocking revelation about how close his own mission had come to disaster. "The first seal on our flight had been totally destroyed and the [other] seal had 24 percent of its diameter burned away," he told a NASA oral historian. "All of that destruction happened in 600 milliseconds and what was left of that last O-ring – if it had not sealed the crack and stopped that outflow of gas; if it had not done that in the next 200-300 milliseconds – it would have *gone*. You'd never have stopped it and we'd have exploded. That was thought-provoking!"

Suddenly, Lind realized that for three-tenths of a second on 29 April 1985, his life hung quite literally by the shredded fragments of a single O-ring. "Each of my children have a copy of that painting," he told the NASA historian, "because we wanted the grandchildren to know that we think the Lord really protected Grandpa."

THE TRAUMA OF CHALLENGER

The night of 27 January 1986 witnessed the coldest conditions ever recorded before the launch of a Space Shuttle. And as the STS-51C experience a year before had shown, cold weather and O-rings did not mix well. In fact, temperatures at KSC that night plummeted to an unseasonal 5.5 degrees Celsius, forcing technicians at one point to activate fire hoses at Pad 39B to prevent water pipes from freezing. On the pad sat Challenger, in the final hours of preparation for STS-51L, the 25th Shuttle mission and her own tenth journey into space. She and a crew of seven (including a civilian passenger, schoolteacher Christa McAuliffe) would spend six days in orbit, deploying an important NASA communications satellite and a free-flying instrument to observe Halley's Comet as it made its latest visitation to the inner Solar System. Launch had already been postponed several times, due to delays in landing the previous Shuttle mission, concerns about unacceptable Return to Launch Site (RTLS) weather and niggling technical issues, including a stuck handle on Challenger's side hatch.

But one of the great tragedies of STS-51L is that the weather for one of its scrubbed launch attempts, on 26 January – Super Bowl Sunday – was predicted to be unacceptable yet turned out to be good. "It's another case of fate playing tricks on you," remembered former NASA administrator James Beggs. "They decided not to launch on Super Bowl Sunday, but Super Bowl Sunday was a fine day. If they'd have launched then, they wouldn't have had any trouble, so they held it over and caught the cold spell and lost the vehicle and the crew." Years later, STS-51L flight director Jay Greene remarked that, were it not for the stuck handle and flawed weather forecasting, Challenger might have launched on another day and flown safely. "Somebody else might have had it happen to them," he said, sadly, "but not those guys. We also scrubbed on a beautifully clear day, based on a bad weather forecast."

Fig. 5.4 Black smoke (visible at lower right) pours from the aft segment of the right-hand Solid Rocket Booster (SRB) within milliseconds of Challenger's liftoff on 28 January 1986.

In the pre-dawn darkness of the 28th, with Challenger brilliantly bathed in the glare of xenon floodlights and primed for launch later that morning, the Final Inspection Team gathered at the pad to begin their lengthy sweep-down of the facility. Their task was to locate any unusual concentrations of ice, debris or damage to the stack which could hinder a successful flight. Armed with their formidable arsenal of cameras, binoculars and infrared sensors, the team checked each surface of the immense vehicle, from the gaseous oxygen vent cap at the top of the ET to the aft skirts at the base of the SRBs, and conditions were so severe that they had to use broom-handles to knock 30-centimeter icicles off the gantry. "There were huge amounts of ice on there," remembered Shuttle astronaut Don Williams, "like an inch or more; big icicles and frost." When somebody told Williams that Challenger was 'Go for Launch', he thought they were kidding. Temperatures crept a little above zero degrees Celsius as dawn broke. Challenger's launch that morning was delayed another two hours to allow the Sun to thaw out the copious quantities of ice. "You've got to depend on the tactical on-site commander at the Cape to make the call on the temperature and the ice on the pad," said former chief flight director Randy Stone. "That was the first time in the process that I'm nervous, because this is out of the ordinary. We weren't in the discussions at all about the temperature effects on the SRBs. It was all taking place behind the scenes."

STS-51L finally got underway at 11:38 a.m. EST, the thunderous roar of Challenger's engines and the dazzle of her boosters piercing the morning stillness and pushing the stack into the cold Florida sky. "Here we go," said pilot Mike Smith from the flight deck, his words recorded for posterity by the on-board intercom. "Shit hot," added flight engineer Judy Resnik, seated behind him on the flight deck.

The calamity that would engulf Challenger began almost instantly. Half a second after liftoff, as revealed by high-speed photography near the pad and later noted by the Rogers Commission, "a strong puff of grey smoke" was detected, "spurting from the vicinity of the aft field joint of the right Solid Rocket Booster". The automated camera had picked up the tell-tale signature of both the primary and secondary O-rings failing, disintegrating and streaming away only moments after launch. As if that were not bad enough, the point of failure directly faced the ET and its highly flammable 2-million-liter load of liquid oxygen and hydrogen. Any flame emanating from the compromised SRB could now play on the thin-skinned tank like a blowtorch, igniting its contents in a fireball. The result would be the destruction of Challenger, the incineration of her crew and the obliteration of the entire launch pad.

But astonishingly, it did not happen. A chunk of solid fuel serendipitously dropped into the O-ring breach and temporarily plugged it. Challenger seemed to have dodged a bullet and the first minute of ascent proceeded without incident.

However, the temporary plug was just that: temporary. It would not hold.

Several more puffs of denser, darker smoke – a further indication that the products under combustion were indeed the grease, insulation and rubberized O-ring material from the crippled joint seal – were recorded by other ground-level cameras between 0.836 and 2.5 seconds after liftoff, as the SRBs' hold-down posts were severed and the Shuttle began its climb away from Pad 39B. As each puff was left behind by Challenger's upward trajectory, the next fresh puff could be seen near the failed joint. The frequency of the emissions, as Rogers investigators would later demonstrate, was linked to the rhythmic flexure within the SRB as the gap in the joint cycled open and closed. The last incidence of smoke above the joint came at T+2.733 seconds. In the moments which followed, a combination of atmospheric factors and the exhaust of the boosters made it difficult to ascertain whether more smoke was pouring out of the joint.

Eight seconds into the mission, the vehicle cleared the Pad 39B tower and began a pre-programmed 'Roll Program' maneuver, moving onto the 'due-east' flight azimuth for a 28.45-degree-inclination orbit, as Challenger rolled over onto her back under the command of her General Purpose Computers (GPCs).

"Houston, Challenger, Roll Program," radioed STS-51L commander Dick Scobee.

"Roger Roll, Challenger," replied Capcom Dick Covey from Mission Control.

Shortly thereafter, at T+19 seconds, to prepare herself for passage through maximum aerodynamic turbulence, the SSMEs were throttled down initially to 94 percent of their rated thrust. Thirty-seven seconds into the flight, Challenger encountered the first of several high-altitude wind shears, lasting until just past a minute after launch. In its inquiry, the Rogers Commission found that the guidance, navigation and control system correctly detected and compensated for these conditions and – although STS-51L's aerodynamic loads were higher than previous missions in both the yaw and pitch planes – the SRBs also responded crisply and effectively to all commands. The three main engines continued to incrementally throttle themselves back, limiting aerodynamic stress on the spacecraft's airframe and reaching 65 percent of rated thrust at 48 seconds after launch.

"Three at 65," came the call from booster systems officer Jerry Borrer.

"Sixty-five," acknowledged flight director Jay Greene. "FIDO?"

"T-del confirms throttles," verified Flight Dynamics Officer (FIDO) Brian Perry.

"Thank you," replied Greene.

All seemed to be going well. And it is possible that the mission itself may have proceeded normally, had the plug of solid fuel remained jammed into the O-ring breach. It might even have turned out to be another 'Three-Tenths of a Second' moment of salvation, another successful effort by the Shuttle to escape Death's clutches. However, by an incredible stroke of cruel fortune, Challenger happened to pass through the most severe wind shear ever encountered by an ascending Shuttle; a wind shear which snatched away the plug somewhere around a minute into the mission. After passing through Max Q, at 51 seconds into ascent, the SSMEs were throttled back up to full power. Shortly thereafter, at 58.788 seconds, a frame of video recorded the first evidence of a flickering flame emerging from the right-hand SRB's aft joint. The temporary plug of solid fuel had gone and, although they were oblivious to anything amiss, the crew's fate was now sealed.

Fig. 5.5 A well-defined flame breaches the outer casing of the right-hand booster and begins to play on the External Tank (ET) and its load of highly volatile propellants, approximately 58 seconds into Challenger's ascent.

Scobee and Smith had no data on the performance of the boosters. The only indication they might receive on their cockpit instruments that anything was wrong would relate to a malfunction in the steering mechanism.

Unseen and unknown to the crew, the flame from the right-hand SRB rapidly established itself, growing into a well-defined plume within half a second. Exactly a minute into the mission, telemetry indicated an unusual chamber-pressure differential between the boosters; the pressure of the right-hand SRB was 11.8 psi lower than its left-hand counterpart, pointing to a developing leak in its aft joint. As the flame increased in size, aerodynamic 'slipstream' deflected it backward and circumferentially by the protruding structure of the upper ring which linked the SRB to the ET, focusing the flame directly onto the surface of the tank. Sixty-two seconds into ascent, the left-hand booster's thrust vector control moved to compensate for the yaw motion caused by the reduced thrust from its right-hand counterpart. A couple of seconds later came the first visual indication that the flame from the damaged booster had ruptured the ET: an abrupt change in the shape and color of the flame, indicating that it was now mixing with leaking liquid hydrogen.

From his seat in Mission Control, all seemed to be progressing without incident. Covey called the crew to advise them that the engines had now throttled back up to full power. "Challenger, Go at throttle up."

Scobee came back a moment or two later. "Roger," he replied. "Go at throttle up."

In the seconds that followed, an incredibly rapid sequence of events took place which ended with the destruction of the ET, the separation of both SRBs and the structural disintegration of Challenger into several large pieces. Seventy-two seconds after liftoff, the flame from the right-hand SRB finally burned through the lower of two struts holding it to the ET. Pivoting about its uppermost strut, the top of the booster hit the instrument-laden inter-tank and the base of the liquid oxygen tank, rupturing them both. At T+73.1 seconds, clouds of white vapor were spotted at the top of the tank and around its bottom dome: the former indicating leaking liquid oxygen, the latter pointing to outright structural failure.

At T+73.6 seconds, came a massive – "almost explosive", read the Rogers report – burn of both the liquid oxygen and hydrogen. At this point, STS-51L was flying 14 kilometers over the Atlantic Ocean, traveling over 2,400 kilometers per hour and Challenger was lost from view. Her Reaction Control System (RCS) ruptured, setting off a hypergolic burn of its propellants, evidenced by a reddish-brown hue at the edge of the fireball. Meanwhile, the two boosters, now released of their loads, rapidly climbed away from a catastrophe of their causing, their plumes corkscrewing across the cold morning sky. Both were remotely destroyed in commands issued by the Range Safety Officer (RSO) at 110 seconds.

As the Shuttle broke apart in full view of the astronauts' families and millions watching on live television, what should have been a spectacular mission ended in the worst and most public disaster ever witnessed at that time. The sheer lack of comprehension as to what had happened was shared at all levels. "Flight controllers here looking very carefully at the situation," a stunned Steve Nesbitt, the NASA launch commentator, told his audience. "Obviously a major malfunction."

In Mission Control in Houston, the picture was initially no clearer. Radio communications and downlinked telemetry from the Shuttle had been severed in a heartbeat as flight controllers watched their displays in bewilderment. Then Ground Control (GC) officer Norman Talbott called Greene.

"Flight, GC, we've had negative contact, loss of downlink."

"Okay," said Greene. "All operators, watch your data carefully."

The seconds dragged on. At length, Perry called Greene.

"Flight, FIDO?"

"Go ahead."

"RSO reports vehicle exploded."

Greene's stony face was etched with the indescribable agony of recognition that he had just lost seven friends. After a long pause, finally he spoke. "Copy. FIDO, can we get any reports from the recovery forces?"

"Stand by."

Fig. 5.6 Jay Greene (foreground) and fellow flight director Lee Briscoe look on in stunned disbelief as Challenger disintegrates before their eyes on 28 January 1986.

At the Capcom's console, veteran Shuttle pilots Covey and Fred Gregory sat silently, their eyes transfixed by the inexplicable data and incomprehensible image on their monitors. Having themselves flown this vehicle only months before, both men knew instinctively that Challenger and her crew were gone. "I did see Fred and Dick," remembered Greene. "I saw both their jaws drop." At one point, cameras in Mission Control caught Gregory, head in his hands, saying a brief prayer.

Picking up his commentary, Nesbitt relayed the Mission Control chatter to his audience. "We have a report from the Flight Dynamics Officer that the vehicle has exploded. The Flight Director confirms that. We are looking at checking with the recovery forces to see what can be done at this point."

But nothing could be done.

RESIGNATION, RECRIMINATION AND REDESIGN

It might have been hoped that the explosive burn of the ET's propellants would have destroyed Challenger's crew cabin or fractured her flight deck windows to induce rapid depressurization and a merciful end for STS-51L commander Dick Scobee, pilot Mike Smith, mission specialists Ellison Onizuka, Judy Resnik and Ron McNair and payload specialists Greg Jarvis and Christa McAuliffe. But the Shuttle's cockpit was built like a fortress, tested to 140 percent of its design strength to safeguard the lives of its occupants. On the morning of 28 January 1986, it became their tomb and their window-seat from a peak altitude of 18 kilometers all the way down to the Atlantic Ocean.

Dredged from the water six weeks later, about 27 kilometers northeast of the launch site, the crumpled wreckage showed no signs of having endured a quick explosive depressurization. No upward 'buckling' of the flight deck floor (as would be expected as air from the middeck rapidly expanded) was apparent and, in any case, impact damage to Challenger's windows was so extreme that it was impossible to tell if they had sustained their fractures in flight or upon hitting the ocean. "The estimated breakup forces [on the vehicle] would not in themselves have broken the windows," wrote astronaut Joe Kerwin, NASA's head of life sciences in a July 1986 letter to Dick Truly, the agency's associate administrator for spaceflight. "Impact damage was so severe that no positive evidence [either] for or against in-flight pressure loss could be found." Tumbling uncontrolled at 400 kilometers per hour, the waters of the Atlantic would have been as unyielding to Challenger as solid rock.

But it remains likely that at least some of the STS-51L astronauts were conscious following the breakup. Four Personal Egress Air Packs (PEAPs), designed to provide a few minutes' worth of breathable air in emergencies, were recovered in March 1986 and one of them, belonging to Smith, showed signs of having been intentionally activated in the moments after breakup. Mounted on the back of Smith's seat, it could only have been switched on by one of his crew-mates (either Resnik or Onizuka) in a valiant effort to save his life. Moments before the ET disintegrated, a bright sheet of white vapor flooded across Challenger's nose and was probably visible to Smith through his window. And the last words from the Shuttle that day, recorded by the on-board intercom, were from Smith himself: a brief, clipped "Uh-oh". What that exclamation meant and precisely what he saw in those final moments can never be known, but it is not impossible to suppose that Smith was an eyewitness to the right-hand SRB pivoting violently into the top of the ET.

In the days and weeks after the tragedy, the finger of blame pointed directly to an explosion in one of the SSMEs, which proved so difficult to tame during the Shuttle's lengthy genesis. After all, they had required hundreds of tests on the ground, had been responsible for two pad aborts and had almost caused an unhappy end for STS-51F in July 1985. But when their remnants were plucked from the Atlantic in late February – each still attached to its thrust structure and with two engine controllers still intact – it became apparent that they had played no part in causing Challenger's end. Their controllers were disassembled, flushed with de-ionized water, dried and vacuum-baked, before their data was extracted. The results, according to the Rogers Commission, unsurprisingly revealed burn dam-age caused "by internal over-temperature typical of oxygen-rich shutdown". The inevitable conclusion to be drawn from this was that all three SSMEs began shut-ting themselves down in the milliseconds after breakup, as hydrogen propellant spilled from the ruptured ET.

However, in general, the engines' performance was normal and they did not exhibit any unusual behavior until just before the instant of breakup, whereupon their fuel tank pressures suddenly dropped and the controllers responded by opening the fuel flow-rate valves. Turbine temperatures correspondingly increased in response to a leaner fuel mixture being fed into their combustion chambers from the compromised ET, but it was apparent from the data that the SSMEs had not contributed to the disaster. Nor, indeed, had the ET itself, of which about 20 percent was recovered from the Atlantic, mostly fragments from the central 'inter-tank' and the hydrogen tank. Thirty percent of the Shuttle her-self was recovered and inspections revealed that Challenger disintegrated due to massive aerodynamic overloads, not explosive forces. No issues with her pay-loads were apparent. As the weeks wore on, the 'smoking gun' responsible for STS-51L moved to the SRBs.

It was already well-known that low launch temperatures and the effects of water and ice build-up negatively affected the resiliency of the O-rings and impaired their ability to properly seal the boosters' field joints. Although several missions

Fig. 5.7 Fred Gregory (left) and Dick Covey sit dumbstruck at the Capcom console in Mission Control, following Challenger's destruction.

had flown successfully in freezing conditions (notably STS-51C), Challenger launched in the coldest weather ever experienced by the Shuttle. Moreover, as investigators would discover, she had sat on Pad 39B for 38 days before launch, thereby increasing the potential for the vehicle to endure the damaging effects of significant rainfall and perhaps even the 'unseating' of her O-rings by frozen water. But although the SRBs were the technical root of the tragedy, the Rogers Commission also leveled blame on a flawed decision-making process by senior leaders. Safety concerns from engineers were not being effectively communicated up the chain of command, 'critical' issues were left unresolved, and schedule pressure – as NASA sought to fly the Shuttle 'routinely' – received greater priority than flight safety. And if the space agency bore culpability for the seven STS-51L lives, then so too did Morton-Thiokol.

Nowhere was this more shockingly demonstrated than at a teleconference between senior leaders from NASA and Morton-Thiokol on the evening of 27 January 1986. During that teleconference, concerns were raised about the wisdom of launching Challenger under the cold weather conditions forecasted later that night and the following morning. Bob Lund argued that Morton-Thiokol's 'comfort level' was not to launch SRBs below temperatures of about 12 degrees Celsius, for fear that the cold might impair the ability of the O-rings to adequately seal the field joints. But in what can only be described as a perversion of engineering logic, Lund's concerns were downplayed when his team could present no definitive case which 'proved' the boosters were unsafe to fly. In essence, they were asked not to prove that the system was safe, but rather that it was unsafe. In one exchange, Lawrence Mulloy angrily exploded: "For God's sake, Thiokol, when do you expect me to launch? Next April?"

Had Lund stood his ground, it is highly unlikely that NASA would have ignored a formal recommendation from a major contractor. In the minutes which followed, Morton-Thiokol requested a short recess from the conference to consider its data. Five minutes became half an hour. Throughout the recess, Roger Boisjoly and fellow engineer Arnold Thompson argued fiercely that it was unsafe to fly outside of the SRBs' proven temperature range, but senior leaders countered that the O-rings could still 'seat' themselves and function adequately in the cold weather. "Arnie actually got up from his position and walked up the table, put a quarter-pad down in front of the management folks and tried to sketch out once again what his concern was with the joint," Boisjoly later told Rogers investigators. "When he realized he wasn't getting through, he stopped. I grabbed the photos and tried to make the point that it was my opinion from actual observations that temperature was indeed a discriminator and we should not ignore the physical evidence that we had observed. I also stopped when it was apparent that I couldn't get anybody to listen."

At length, Morton-Thiokol's executive vice president Jerry Mason explicitly asked Lund to remove his 'engineering hat' and put on his 'management hat'. When the hook-up resumed, Lund changed his vote and so too did Morton-Thiokol's position. The new recommendation was that, although cold weather remained a problem, the data was inconclusive and the launch of STS-51L should go ahead. None of the engineers wrote the new recommendation and only the executive leadership signed it. However, when NASA managers asked for any additional comments from around the Morton-Thiokol table prior to ending the conference, no concerns were voiced. After analyzing the teleconference notes, the Rogers investigators concluded that "there was a serious flaw in the decision-making process leading up to the launch", and added that "a well-structured and managed system, emphasizing safety, would have flagged the rising doubts about the Solid Rocket Booster joint seal".

In its findings, the Rogers Commission condemned the test and certification program for the SRB joints as inadequate and noted that the mechanism by which the O-ring sealing process took place was imperfectly understood by both NASA and Morton-Thiokol. There was a sense that escalating risks were accepted, and safety standards correspondingly lowered, in what Richard Feynman aptly likened to high-stakes Russian roulette. The agency's ability to track anomalies and flag systemic issues through its pre-launch review process was fundamentally broken, with six launch constraints prior to STS-51L – all related to O-ring erosion and blow-by – having been waived. And the investigators noted that the list of O-ring problems presented by Morton-Thiokol to NASA Headquarters in August 1985 were more than sufficient to have required corrective action before the next Shuttle mission was permitted to fly.

"We were trying to make the Shuttle more operational," said former JSC director Gerry Griffin. "We were trying to make it more of a 'check-it-out-and-go'... and we weren't trying to cut corners. We were trying to do it to make the system work like a real operational system. And I think we learned out of that process that you can't. With these kinds of systems, with these kind of energies – stored energies involved and releasing it so fast – that you really can't be that kind."

Clearly, the most important recommendation made by the Rogers Commission was for a comprehensive redesign of the SRB field joint and seal, with a recertification process which approximated actual flight conditions as closely as possible. Safety-critical items had to be thoroughly reviewed, procedures for granting launch waivers had to be tightened, an office of safety, reliability and quality assurance (reporting directly to the NASA administrator) had to be established, communications between centers had to be improved, leadership structures had to be overhauled and astronauts should be put in key positions of management to lend their expertise. In July 1986, NASA announced a $680 million plan to redesign the SRB field joint's metal components, its insulation and its O-ring seals to provide "improved structural capability, seal redundancy and thermal protection". New capture latches would reduce joint movements caused by SRM pressure or structural loads and the O-rings themselves were redesigned to prevent leaks under conditions of structural deflection that were twice as great as the predicted levels. Internal insulation in the boosters was modified to be sealed with a deflection relief flap instead of putty, and new bolts and strengtheners and a third O-ring were incorporated. External heaters with integrated weather seals would prevent water intrusion and ensure that SRB joint temperatures never fell below 24 degrees Celsius. "The strength of the improved design," read NASA's response to U.S. president Ronald Reagan, "is expected to approach that of the [SRB] case walls."

The blueprint for what became known as the Redesigned Solid Rocket Motor (RSRM) evolved between the Preliminary Design Review (PDR) in October 1986 and the Critical Design Review (CDR) in February 1988, with Morton-Thiokol devising a comprehensive test regime to evaluate and verify all changes. Two full-scale simulators were built, with the Joint Environment Simulator (JES) test-fired

Fig. 5.8 Challenger vanishes from view in the rapid burn of propellants after T+73 seconds.

seven times between August 1986 and July 1988 to evaluate field joint hardware, insulation and O-ring seal performance and the Nozzle Joint Environment Simulator (NJES) test-fired nine times between February 1987 and August 1988 to evaluate case-to-nozzle joint hardware, insulation and O-ring seal performance.

As part of the effort to certify the new motor, Engineering Test Motor (ETM)-1A was fired in May 1987 to demonstrate the effectiveness of external graphite composite stiffener rings to reduce joint rotation and evaluate the new joint heaters. Two demonstration motors were tested in August and December of that year to verify the major changes and in 1988 qualification and production verification motors with deliberate manufacturing flaws were fired to evaluate the effectiveness of the redesign and the ability of the joints to hold up under unanticipated duress. Full-scale Flight Support Motors (FSMs) were tested approximately every 18 months thereafter to verify the performance and continued safety of the RSRM system, as well as to test and certify design, material or process changes.

But in many quarters in the aftermath of STS-51L, redesign was not enough; some felt that a whole new booster, built from the ground up, was necessary. In September 1986, study contracts were awarded for what was initially labeled an "alternate or Block II Space Shuttle Solid Rocket Booster". During the following year, proposals for what became known as the Advanced Solid Rocket Motor (ASRM) were submitted and in April 1990 NASA selected a Lockheed-Aerojet team for the billion-dollar effort to design, develop and certify the new booster. It was expected that the ASRM would make its first flight as early as 1996 – a decade after the loss of Challenger – and the design included substantial safety improvements. The number of segments was reduced from four to three and it eliminated the use of 'factory joints', the ET attachment 'ring' and other individual parts to produce a stronger booster casing less susceptible to flexion. Post-flight maintenance on the ASRM was expected to be much less labor-intensive than its predecessor. Tests got underway in April 1991 and continued through the following year, but as the RSRM gradually proved itself and the ASRM's projected costs climbed past $3.5 billion, it was eventually shelved by NASA in October 1993. "It waned and waned and waned," said Arnold Aldrich, "to the point where finally, in one of the budget-years, it was decided not to proceed."

Other concepts which grew after Challenger's destruction included a Liquid Rocket Booster (LRB), with study contracts awarded to General Dynamics and Martin Marietta in October 1987. Fuels under consideration included liquid hydrogen, methane, propane, monomethyl hydrazine and a highly refined form of rocket-grade kerosene, with oxidizer candidates of either liquid oxygen or nitrogen tetroxide. "The whole idea of liquid boosters presented many more options for intact-abort scenarios," wrote Dennis Jenkins. "Unlike the SRBs that cannot be shut down during ascent, the LRBs could be throttled to allow a more rapid RTLS abort or over-throttled to ensure a TAL or AOA abort could be accomplished at almost any point during ascent." However, issues of gradually expanding cost, the proven reliability of the RSRM from its first flight in September 1988 and most fundamentally the LRB's lack of reusability sounded its eventual death-knell.

THE SWINGING PENDULUM

Redesigning the boosters after Challenger also spelled the end of another kind of SRB that was originally planned for military and polar-orbiting Shuttle flights out of Vandenberg Air Force Base in California. Rather than stainless steel cases, the Vandenberg boosters would have used a 'filament-wound' design, composed of a more flexible graphite-epoxy material with a 15,000-kilogram reduction in weight. The principal rationale was that filament-wound SRBs afforded an improved weight saving, enabling the Shuttle to carry 3,600 kilograms of additional payload into low orbit. "They had the same joint as the steel cases," noted astronaut Jerry Ross, who would have flown the first Vandenberg mission. "Since the graphite ones would have been more flimsy we always were wondering what would have happened to us had we tried to launch with those, considering the Challenger accident."

The filament-wound cases were built by Hercules, Inc., subcontractors to Morton-Thiokol, and included a unique 'capture' feature to eliminate rotation in the joints. Developmental test firings were conducted between October 1984 and May 1985, with a full qualification test firing to certify the concept for operational service planned for February 1986 at the time of STS-51L. But in the aftermath of the Challenger tragedy, and to meet the Rogers Commission's recommendations, additional fixes would have been necessary to bring the filament-wound SRBs up to the necessary standard. And those fixes would have contributed extra weight and virtually canceled any advantages in terms of payload-to-orbit capability. In February 1987, the Air Force accepted that Rogers-enforced structural changes meant that the Shuttle would no longer be able to lift heavy cargoes to polar orbit. Vandenberg was finally abandoned as a Shuttle launch site in November 1989.

"We caused the whole Shuttle program to be very conservative," recalled Arnold Aldrich of the years after STS-51L. "There was this tremendous feeling of caution and concern and oversight and checks and balances, to the point where the team had a very hard time getting to each flight, because no one wanted anything that looked at all like it might have some issue with it to go untouched. We really built in a tremendous amount of conservatism that, I think, over some number of years returned to a proper balance for that." But when Discovery launched on STS-26 in September 1988 for the first post-Challenger mission, the problems with the SRBs – and the intrinsic fallibility of the system – did not go away. Astronaut Tom Henricks, who piloted two Shuttle flights and commanded two others during the 1990s, also remarked that as mission after mission flew success-fully, little by little, safety was again eroded. "The pendulum [after Challenger] had swung to as conservative as they could make it," Henricks said, "but then that pendulum started swinging back almost immediately." Nowhere was that imme-diacy more obvious than STS-27, the second post-Challenger flight in December 1988. The full details will be discussed in Chapter 6, but the fundamental cause of STS-27's brush with disaster was the SRBs; not the field joints or the O-ring seals, but a tiny shard of ablative insulation. A review of that mission's launch video revealed something breaking away from the nose of the right-hand SRB, about 85 seconds after liftoff, and hitting the Shuttle's fragile Thermal Protection System (TPS). If Atlantis' heat shield had been breached by the impactor, it threatened disaster during the fiery return to Earth. Fortunately, STS-27 dodged its own bullet

Fig. 5.9 During Atlantis' STS-27 ascent in December 1988, a fragment of insulation became detached from the top of the right-hand booster at T+85 seconds, causing damage to the orbiter.

and in January 1989 a damage review team traced the cause to a manufacturing change which was intended to improve the performance of ablative materials meant to protect the SRBs from aerodynamic heating. Members of the review team attended the FRR for the next mission in March 1989, and duly scrutinized all ET and SRB manufacturing records to identify potential sources of debris.

For STS-27 commander Robert 'Hoot' Gibson, the incident was a stark reminder of the singular principal directive of engineering: 'Better' is the enemy of 'good enough'. In other words, if a system worked, there existed little operational reason to change it. That the safety pendulum had begun to swing backwards (on only the second mission after Challenger) was rendered even more obvious when the crew's concerns about damage incurred by Atlantis went unheeded by mission controllers. Only after Gibson and his men landed did it become clear how close they had come to death. In his mind, the loss of STS-27 would have marked the definitive end of the Space Shuttle program. "We had spent all that money and all that time, rebuilding and revamping, and we launched one successful mission and we lost the very next one," he remarked. "I think the Congress would have said, okay, that's the end, guys. We just don't need to do this again."

Several years later, in July 1995, disassembly of Discovery's SRBs after STS-70 revealed a tiny air pocket (known as a 'gas path') in an internal joint at the base of the right-hand booster. The gas path was observed to extend right up to the primary O-ring. Although the O-ring itself was not breached, it did exhibit "a slight heat effect", according to NASA, in the form of singe marks and small concentrations of soot. Also present were heat-affected insulation and eroded adhesive. The incident raised perhaps the worst memories of what had happened to STS-51L and it was significant that a similar gas-path concern (with four instances of singeing) had also been identified in one of Atlantis' SRBs following her STS-71 launch in June 1995. "Gas paths or small air pockets are the result of nozzle fabrication involving backfilling of the joint with insulation material," explained NASA's anomaly report. "Similar paths had been expected or observed following previous flights, but…STS-71 and STS-70 marked the first time [that] slight heat effect was noted on the primary O-ring."

As investigators dug into the problem, the launch of the next mission was postponed from the end of July to the end of August. Although the primary O-ring had not been breached and the secondary O-ring was untouched in both the STS-71 and STS-70 boosters, the incident raised awareness that similar instances of exhaust burning through to the seals had happened on no fewer than eleven post-Challenger missions. *Flight International* reported in August that there were "no design issues" with the RSRM and that neither the STS-71 or STS-70 astronauts had been placed in any danger, as the NASA investigation team labored on a procedure for performing ultrasound inspections and minor adjustments to the injection of an insulating putty. It was hoped that this putty "may reduce" the possibility of gas paths reaching primary O-rings on future flights. "Working inside the nozzle of the boosters," *Flight International* reported in late August, "technicians are replacing the insulating material protecting the nozzle O-ring joints with a new material. Seven sets of [SRBs] will undergo this work and the new material will be introduced on future production."

Only months later, in June 1996, Columbia flew STS-78 and her recovered boosters revealed worrisome signs of blow-by of their field joints. Although the safety of the seven-member crew was reportedly never compromised, it was the first occasion than an RSRM had experienced what NASA called "combustion product penetration". A hot gas path had worked its way through the field joint, but not through the 'capture' joint, and both boosters performed within their design parameters. Nevertheless, the episode underlined growing issues with a new adhesive and cleaning fluid recently added to the SRBs to comply with Environmental Protection Agency (EPA) regulations. Even as Columbia was flying, the adhesive problem was expected to delay the next mission, STS-79, planned for late July. NASA was unable to revert directly back to the old-style adhesive and cleaner because the EPA had prohibited its methyl-based material. Years later, Gibson poured scorn on the decision. "We went to another cleaner; that didn't work," he said. "The first launch we tried with this new process, we just about killed seven astronauts. But we were protecting the snail-darters and the spotted owls and the baby seals, I guess."

Devising an entirely new adhesive would take some time, but the severity of the situation prompted Shuttle program manager Tommy Holloway to report that it was "a serious situation until we determine it's not serious". He stressed that the damage did not affect the O-rings and described the RSRM joint as "an order of magnitude more robust than the joint on Challenger". All told, engineers found six joints in which hot gas had penetrated their heat shields. STS-79's original boosters were replaced with a new set, bearing the old-style adhesive, but one of them failed a leak check at the junction between its aft-center and forward-center segments. It turned out that an applicator brush bristle had gotten lodged in the secondary O-ring. New seals were installed and the segments were restacked and leak checked.

STS-79 flew safely in September but suffered erosion of its SRB nozzles, which caused the next mission, STS-80, to be delayed. Concerns centered upon the boosters' insulating material, which had experienced higher-than-normal 'grooving' erosion in the nozzle throat. Although it was not expected to present a safety concern, NASA opted to evaluate and clear the problem before committing STS-80 to launch. Engineers concluded that the most likely reason for the 'out-of-family' issue was a 'pocketing' erosion effect, triggered by slight ply distortions in the ablative material of the nozzle's throat-ring and normal variations in other material properties. Under standard circumstances, the throat-rings were manufactured by wrapping the ablative material in a crisscross fashion and curing it at elevated temperatures and pressures. However, it was suspected that during the curing process the material close to the surface of the insulation had shifted slightly and generated the distortions. When hot gases flowed through the SRBs, the distortions significantly increased stresses in the material, which resulted in the pocketing effect and an uneven wearing-away of the ablator.

Fig. 5.10 The base of one of the Solid Rocket Boosters (SRBs), affixed to the launch pad prior to Endeavour's STS-99 mission in February 2000.

Three years later, in May 1999, a problem with an electrical connector on the left-hand booster forced the entire STS-96 stack to be rolled back from the launch pad to the VAB for repairs. This relatively minor incident forced a week-long delay and Discovery flew safely later in the month. However, a much more serious issue arose during Endeavour's STS-97 launch in November 2000, when one of two pyrotechnic devices designed to separate the base of the left-hand SRB from the ET failed to fire. Fortunately, its backup worked and both boosters were safely jettisoned at two minutes into the mission. But had both pyrotechnic devices failed, and the boosters not separated from Endeavour, the result would have been catastrophic. "This is the first time we have seen this in the life of the program," said Shuttle program manager Ron Dittemore. "It's one of two, which means we were perfectly safe. You only need one of two to fire to separate properly, but what concerns us is we don't like going into a mission knowing that perhaps we have lost one leg of our redundancy."

The STS-97 incident was traced to a cable in the External Tank Attachment (ETA) ring, which exhibited clear evidence of wear and tear. Engineers made detailed X-ray inspections of the ETA cables in the boosters for the next flight, STS-98, which revealed a faulty connector and crumbling woven outer shielding. Repairs were performed and the Shuttle was rolled out to the launch pad in early

January 2001. However, as engineers worked to test other electrical hardware in the SRB inventory, they found four cables whose electrical connectivity was poor. And in an interesting reversal of the infamous 27 January 1986 teleconference between NASA and Morton-Thiokol, the recommendation this time was to return the STS-98 stack to the VAB and postpone the mission until additional wiring tests had been satisfactorily accomplished. "In this instance, we had some statistical analysis that we had performed that indicated, based on statistics alone, that you might be comfortable proceeding with flight," said Dittemore. "And there were a number of folks who felt comfortable with just a statistical analysis. There was another camp that felt…the hardware was telling us something. We had four failures here that we discovered in our testing…and we ought to listen to what the hardware may be trying to tell us and do further inspections." The inspections revealed no further damage to the cables and STS-98 was returned to the pad and flew a safe and successful mission in February 2001.

Six months later, as Discovery was being readied for her STS-105 flight, concerns were raised about potential cracks and stress corrosion in a Hydraulic Power Unit (HPU) injector stem on her left-hand booster. Engineers inspected 39 other injectors in the Shuttle's inventory and determined that none of them exhibited cracks and the launch went ahead as intended. But perhaps the most significant close call came during Atlantis' STS-112 liftoff in October 2002, when the primary circuit to detonate the hold-down bolts to release the SRBs to allow the stack to lift off failed to fire at T-0. Although the backup circuit functioned normally and successfully severed the eight bolts – four at the base of each booster and each measuring 63 centimeters in length – the consequences of both these Criticality-1 items failing together did not bear thinking about. Had that occurred, the Shuttle would have been destroyed as it tried to tear itself off the launch pad. It was not the first time that such a brush with tragedy had occurred. In July 1993, Discovery's STS-51 launch was postponed by a flaw in a pyrotechnic initiator controller that was responsible for commanding the detonation of the bolts. And former Shuttle commander John Creighton was in no doubt about the criticality of their role. "The whole vehicle starts rumbling and shaking," he said of the seconds after Main Engine Start, "and you can't believe that these big bolts are still holding you to the ground. It feels like it's trying to rip itself off the ground."

The STS-112 episode, and the others which preceded it, were a reminder that the corrective functions built into the SRBs after Challenger did not suddenly transform them into the big, dumb, totally reliable and impeccably safe boosters that NASA had touted in years gone by. Riding these two great beasts to the edge of space, with their incessant guttural growl and harsh acceleration, remained without doubt the most violent and bone-jarring phase of every launch from the first in April 1981 to the last in July 2011. And so it will be again for astronauts in the coming decade, when the SRBs' beefed-up, five-segment descendants provide the first-stage muscle to get the huge Space Launch System (SLS) rocket off the ground. If any reminder were needed of the sheer power and naked ferocity of the boosters, it came on STS-5 in November 1982. As Columbia climbed away from Earth, pilot Bob Overmyer – a tough, crew-cutted U.S. Marine Corps fighter pilot – turned to commander Vance Brand with a frightened look on his face. The roughness of the ride, like taking a ramshackle truck down a dirt road at high speed, convinced him that the Shuttle was about to break apart.

It took flight engineer Bill Lenoir, sitting behind him on Columbia's flight deck, to chip in with a few calming words of encouragement.

"Relax, Bob," he said. "Ain't no use dying all tensed up!"

6

The Shuttle's Fragile Skin

WATCHING THE ELEVON TRIM

The astronauts of Space Shuttle Atlantis were more than certain that 6 December 1988 would be the final day of their lives. Commander Robert 'Hoot' Gibson, pilot Guy Gardner and mission specialists Mike Mullane, Jerry Ross and Bill Shepherd had spent four quiet days in space, deploying a top-secret satellite for the Department of Defense. Problems with the huge payload required them to re-rendezvous with it at one stage and effect repairs. Three decades later, the nature of those repairs remains shrouded in mystery and the National Intelligence Medals of Achievement that Gibson and his crew won for their work remain locked away in a classified vault. But what is known about STS-27 (aside from the fact that it was only the second Shuttle mission after the Challenger tragedy) is that the crew came within a hair's breadth of losing their lives during re-entry. Years later, Gibson was so firm in his conviction that Atlantis' Thermal Protection System (TPS) had been critically damaged that throughout the worst aerodynamic heating of re-entry his keen eyes were transfixed on one specific gauge on the instrument panel. Had that gauge – which provided 'trim' data for elevons at the trailing edge of Atlantis' right-hand wing – shown even the tiniest deviation from its normal position, Gibson would have known that his mission was lost and the Shuttle would soon tumble out of control and begin breaking apart in the high atmosphere. And if that happened, he knew that he and his men had only a few seconds of life left to them.

Atlantis had launched smoothly from Pad 39B at the Kennedy Space Center (KSC) in Florida on 2 December, her three Space Shuttle Main Engines (SSMEs) and twin Solid Rocket Boosters (SRBs) providing the necessary

© Springer Nature Switzerland AG 2021

B. Evans, *The Space Shuttle: An Experimental Flying Machine*,
Springer Praxis Books, https://doi.org/10.1007/978-3-030-70777-4_6

impulse to lift her into a 440-kilometer orbit, inclined at 57 degrees to the equator. Early the following morning, having successfully deployed their payload, the astronauts were awakened to disturbing news. An inspection of the launch video revealed something (probably a fragment of ablative insulation) breaking away from the nose-cap of the right-hand SRB, about 85 seconds into the ascent, and striking Atlantis' TPS. Asked about their experiences during launch, the crew reported that they had indeed seen some 'white' material on the Shuttle's cockpit windows.

Fig. 6.1 STS-27 suffered one of the most severe Thermal Protection System (TPS) maulings of any Space Shuttle mission.

If the heat-resistant tiles coating the spacecraft had been damaged or breached, it could spell disaster as they knifed their way back through the sensible atmosphere at blistering hypersonic velocities at the end of the mission. Fortunately, Atlantis was equipped with Canada's Remote Manipulator System (RMS), a 15.2-meter-long camera-tipped robotic arm, and on 3 December the astronauts were instructed to use it to make a visual inspection of the tiles. Mullane began the procedure by gingerly maneuvering the RMS across the forward section of the payload bay and then tilting it over the starboard side of the orbiter's nose cap. All looked good: the pristine 'checkerboard' of black tiles was undamaged. Delicately, he moved the RMS further along the fuselage, towards the Shuttle's belly, and as all five men watched the camera imagery on their monitors in the aft flight deck they gasped in horror. Everywhere they could see white streaks: a clear indication that some form of kinetic impact had totally stripped a multitude of High Temperature Reusable Surface Insulation (HRSI) tiles of their black outer coating to leave only the white substrate material. Those tiles were designed to protect the Shuttle from temperatures as high as 1,260 degrees Celsius during re-entry. "We could see that at least one tile had been completely blasted from the fuselage," wrote Mullane in *Riding Rockets*. "The white streaking grew thicker and faded aft beyond the view of the camera. It appeared that hundreds of tiles had been damaged and the scars extended outboard toward the carbon composite panels on the leading edge of the wing."

As shall be seen later in this chapter, those carbon composite panels – known as Reinforced Carbon Carbon (RCC) – were responsible for protecting the Shuttle's nose cap and the leading edges of the wings from temperatures of more than 1,500 degrees Celsius during re-entry. The RCC consisted of a laminate of graphite-filled rayon fabric, together with phenolic resin and layered with a unique mold, which was cured, rough-trimmed, drilled and inspected. The outer layers of the carbon substrate were covered in a 0.8-centimeter-thick layer of silicon carbide to prevent oxidation. During most of the Shuttle's flight history, experience showed that although the RCC was regularly hit by debris, it had never been totally penetrated; rather, the panels succumbed to a smattering of cracks, chips, scratches, pinholes and abnormal discoloration. In fact, there was consensus that the RCC panels were almost indestructible. But on 1 February 2003, a breach in an RCC panel on the leading edge of Columbia's left wing would lead directly to the loss of seven human lives.

From his position aboard Atlantis on that December day in 1988, Gibson knew that a single missing tile from the roughly 20,000 tiles which coated the fuselage was probably survivable, but critical damage to the RCC on the Wing Leading Edges (WLEs) could mean only one thing: the crew of STS-27 were all dead men. The length of the RMS precluded it from reaching and carrying out a full inspection of the RCC panels, but it was clear the mission was in dire peril. "Houston, we're seeing a lot of damage," radioed Mullane. "It looks as if one tile is completely missing." He knew the flight controllers could see the downlinked video and was perplexed when Mission Control replied that the damage did not look too severe. The astronauts were dumbstruck. "Are they blind?" Mullane wrote later. "Did they think the white streaks were seagull shit?" Gibson urgently keyed his microphone. "Houston, Mike is right," he emphasized, putting the weight of the mission commander into his words. "We're seeing a lot of damage." Again, he was reassured that it did not point to a catastrophic breach of either the TPS tiles or the RCC panels.

Matters were further complicated by the status of STS-27 as a secret flight. The astronauts had no way to send their imagery reliably and rapidly to the ground. Due to its classified focus, the mission used a slow, encrypted form of transmission, which meant that the images received at the Johnson Space Center (JSC) in Houston, Texas, were of poor quality. Efforts by NASA to request higher-resolution imagery were frustrated by the Department of Defense, which refused to release it owing to the nature of the classification. As such, with only the encrypted images at their disposal, Mission Control was convinced that poor lighting and grainy images had misled the crew into seeing damage which was not there. No

Fig. 6.2 Clad in their blue flight suits, STS-27 astronauts (from left) Bill Shepherd, Guy Gardner and Mike Mullane inspect the damage to Atlantis after her 6 December 1988 landing.

one accepted the opinion of Gibson, the mission's veteran commander and a future chief of NASA's astronaut corps, who knew exactly what he was seeing and could see it with the crystalline clarity afforded not by cameras, but by his own eyes. He was impotently forced to tell his crew to enjoy the rest of the flight. There was no point, he advised them, in dying all tensed up.

Early on 6 December, the astronauts donned their bulky partial-pressure suits and Gibson performed the customary firing of the twin Orbital Maneuvering System (OMS) engines to begin the hour-long descent back to a landing at Edwards Air Force Base in California. As re-entry heating levels began to increase on the flight surfaces, Mullane lingered on the flight deck and shot some impressive video of the light show out of the overhead windows. But when the first traces of gravity began to act upon the vehicle and her crew, he retreated downstairs to the middeck and strapped himself into his seat. Atlantis plunged like a meteor across the sleeping Indian Ocean, over Australia and across the vast yawning gulf of the Pacific in only 25 minutes, a clear indicator not only of her phenomenal speed, but also of her rapid rate of deceleration, as she shed her orbital speed of 28,200 kilometers per hour to prepare for a runway landing at less than 350 kilometers per hour. Air compression against her belly heated the molecules into a white-hot glow. "I wondered what was happening beneath us," Mullane would write. "I had visions of molten aluminum being smeared backwards, like rain on a windshield. None of our instruments or computer displays showed Atlantis' skin temperature. Only Houston knew that data."

After an interminable period of waiting, punctuated only by Gibson's calm voice as he relayed altitude, velocity and G-meter readings, the Shuttle – through what could only be described as providence and sheer good fortune – somehow made it through the period of maximum aerodynamic heating and the astronauts heard the blissful sound of wind noise outside the reinforced cockpit. By this point, still moving at Mach 5, Gardner deployed a set of air data probes from Atlantis' nose to provide airspeed and altitude measurements for guidance. Thirty kilometers above Earth, and still traveling at almost 5,000 kilometers per hour, Gibson spotted the dry lakebed of Runway 17 at Edwards, far off in the distance. In the minutes which followed, the spacecraft descended further, passing below Mach 1, and observers in California were momentarily startled as the Shuttle's trademark twin sonic booms rattled windows and eardrums and set off a cacophony of car alarms.

Throughout what had been a traumatic re-entry, in addition to his nominal duties, Gibson kept a close eye on the gauge which measured the level of 'deflection' of the elevons at the trailing edge of Atlantis' right wing. "If we started to burn through, we would change the drag on that wing – which is exactly what happened to Columbia," he said later. "We would start seeing 'right elevon trim', which means moving the left elevon down. I knew we would start developing a 'split' between the right and left wing elevon positions if we had excessive drag on the right side. The automatic system would try to trim it out with the elevons. That is one of the things we always watched on re-entry, anyhow, because if you had half a degree of trim, something was wrong. You had a bunch of *something* going on if you had even half a degree. Normally, you wouldn't see even a quarter of a degree of difference on that thing." Prior to achieving 'Entry Interface' – the point at which Atlantis began to feel the first effects of atmospheric friction at an altitude of about 120 kilometers – Gibson privately told himself that if he *did* see an elevon split of more than a quarter of a degree, he had a minute of life remaining in him. And he would use that time wisely, telling Mission Control exactly what he thought of their 'no-damage' analysis.

The crew of STS-27 was lucky, touching down on Runway 17 and wrapping up a successful mission. In its post-flight report, published two months later, NASA described the re-entry as "nominal in all respects", but the real horror of what Atlantis had endured and survived became abundantly clear on the runway. "There was already a knot of engineers gathered at the right forward fuselage, shaking their heads in disbelief," wrote Mullane. "The damage was much worse than any of us had expected." In total, 707 TPS tiles from a total of 20,000 on the airframe had been shredded. One of them, close to the Shuttle's RCC-shielded nose cap, was gone entirely. It had been located directly over a dense aluminum mounting-plate for the L-band antenna, which (again, through unalloyed good fortune) had prevented a burn-through to the underlying aluminum skin. All told, the damage incurred by STS-27 extended from beyond Atlantis' nose cap, right along the length of her belly and terminated just short of the WLEs. The lower surfaces of her elevons were untouched, but the Orbital Maneuvering System (OMS) 'pods' in the aft compartment had sustained 14 impacts and the starboard rudder speed brake had been repeatedly hit.

It was clear from the damage that Gibson and his men had escaped death by a hair's breadth. A Thermal Protection System Damage Review Team was immediately convened, chaired by senior NASA officials Jay Honeycutt and John Thomas and including astronaut Don McMonagle on its panel. They traced the most likely cause to a change in the SRB insulation manufacturing process, the rationale for which was to improve the performance of the ablative materials needed to protect the boosters from aerodynamic heating. For the STS-27 astronauts, it reminded them of the singular principal directive of engineering: Better is the enemy of good enough. If a system worked, and worked well, there existed little reason to change it.

But one other footnote to the near-disaster of STS-27 left Gibson frustrated and amused in equal measure. After landing, an unnamed NASA official came up to him and asked: "Why didn't you guys *tell* us about this?"

Fig. 6.3 "Did they think it was seagull shit?" Mike Mullane scornfully wondered. Humor aside, the scarring suffered by STS-27, on only the second post-Challenger mission, was critical not only to the astronauts' safety, but also to the survival of the Shuttle program.

AN ACCEPTED RISK

The damage incurred by Atlantis on only the second mission after Challenger could quite easily have gone the other way. Indeed, the dense aluminum mounting-plate for the L-band antenna had completely burned through and the superheated plasma of re-entry had begun to wreak havoc on the underlying aluminum skin when the Shuttle exited the worst period of re-entry heating. Had STS-27 been lost on 6 December 1988, Gibson is convinced that the Shuttle program would have ground to a halt; few politicians would have countenanced flying the reusable spacecraft after two Loss of Crew and Vehicle (LOCV) incidents within three years. But even before STS-27, and indeed on the eve of STS-51L, tile damage had become a reality of Shuttle operations. It was regarded as an acceptable risk and although the criticality of the tiles and RCC was well understood, as mission after mission endured damage and returned safely, a creeping insidious mix of "flight experience" and what Richard Feynman likened to Russian roulette and what Mike Mullane called "normalization of deviance" conspired to lull engineers and managers alike into a false sense of security that the heat shield was virtually impregnable.

In fact, the problems did not take long to manifest themselves. On 12 April 1981, only hours after Columbia reached orbit for the very first Shuttle mission, astronauts John Young and pilot Bob Crippen turned their cameras through the aft flight deck windows and into the cavernous payload bay. The bay itself, and its opened clamshell doors, were not their principal concern. At the far end sat the two bulbous OMS pods, which housed the powerful engines used by the Shuttle for orbital insertion and for the de-orbit 'burn' to achieve re-entry. On the right-hand (starboard) pod, several white-colored tiles were missing, evidently torn off during Columbia's violent climb to space. Known as Low Temperature Reusable Surface Insulation (LRSI), they coated not only the OMS pods and vertical stabilizer, but also parts of the main fuselage and guarded the ship against thermal stresses of up to 650 degrees Celsius during re-entry.

"Okay," Young called Mission Control. "What camera are y'all lookin' at now?"

"We're looking out the forward camera," replied the Capcom.

"Okay. We want to tell y'all here we do have a few tiles missing off the starboard pod. Basically, it's got what appears to be three tiles and smaller pieces; and off the port pod, looks like…I can see one full square and looks like a few little triangular shapes that are missing. We're trying to put that on the TV right now."

Young also noticed that no tiles appeared to be missing from the vehicle's wings, vertical stabilizer or nose, although it was impossible to determine if any others had been torn away from her belly, which would bear the full brunt of aerodynamic heating during re-entry. (The camera-tipped RMS arm which would aid the STS-27 astronauts several years later was not available on STS-1.) Back on Earth, NASA engineers watched the transmissions, but decided that none of the lost tiles were in a critical heating area of the airframe. The most that could happen, they assured Young and Crippen, was that after landing a patch of aluminum skin beneath the tiles might need replacement. Interestingly, STS-1 revealed that NASA had the capacity to tap into its links with the military and their state-of-the-art imaging equipment to photograph Columbia in space. When the missing tiles were first noted, astronaut Terry Hart, watching from the ground, was concerned. "Was there something underneath missing too?" he wondered. "We had no way to do an inspection, so we were all wringing our hands after the first shift or two." Then, oddly, they were advised that they had no further need to worry. Gene Kranz, NASA's deputy director of Mission Operations, showed up in Mission Control with a sheaf of images that clearly showed the underside of the Shuttle, completely devoid of damage. They could only have been taken with highly classified equipment on the ground.

"How did you get those?" came the astonished question.

"I can't tell you," grinned Kranz.

The images had clearly been acquired from what Hart described as "some of our national technical assets". For the flight controllers in Mission Control, it came as a sense of relief that they now definitively knew that Columbia was undamaged and she brought Young and Crippen safely home on 14 April. But over the next few years, other crews – of whom the men of STS-27 were perhaps the most visible case – would literally dice with death. During the program's first 15 flights, from April 1981 through to January 1985, missions came home with an average of 123 dings, dents or scrapes to tiles.

Fig. 6.4 Missing tiles on Columbia's Orbital Maneuvering System (OMS) pod are clearly visible in this STS-1 photograph.

As previously recorded in this book, near-perfect weather at KSC was essential for Shuttle crews to safely land the vehicle in a Return to Launch Site (RTLS) abort scenario. But good weather in Florida was critically important for other reasons, too. On 26 June 1982, the night before the launch of STS-4, a severe hailstorm damaged several of Columbia's TPS tiles and deposited water behind the covers of two Reaction Control System (RCS) thrusters. Despite the possibility that this might cause freezing during ascent, it was dismissed as an insignificant problem and astronauts Ken Mattingly and Hank Hartsfield flew safely the following day. More worrying, though, were traces of moisture behind several tiles, which prompted NASA to fly STS-4 in a 'belly-to-Sun' orbital attitude for 12 hours to evaporate the water. This did not immediately do the trick and a second period in the belly-to-Sun attitude was ordered for 23 hours, later in the mission, successfully resolving the problem. After Columbia returned to Earth, several tiles were removed and checked for moisture. Nothing was found and this 'solar inertial' conditioning was formally adopted as part of the operational 'toolkit' to handle future incidents of a similar nature.

But the TPS woes continued. Challenger returned home from her maiden voyage, STS-6 in April 1983, having sustained superficial damage to her Advanced Flexible Reusable Surface Insulation (AFRSI). This insulation took the form of layers of silica tile material sandwiched between sewn-composite quilted fabric. These lightweight 'blankets' protected lower-temperature areas where re-entry heating was not expected to exceed 400 degrees Celsius. More durable, cheaper and faster to produce than standard LRSI tiles, the AFRSI blankets were no more than 2.4 centimeters thick and they helped to reduce the overall weight of the orbiters. But observed AFRSI damage on Challenger's OMS pods in the wake of STS-6 ranged from missing outer sheets and insulation to broken stitches. In the very worst cases, this was attributed to "some type of undetermined flow phenomena" during re-entry. To further examine the problem, four AFRSI 'test blankets' were flown on Challenger's wings and the upper portions and sides of her fuselage during STS-8 in August 1983.

Nor were tiles the only damage incurred by the Shuttle, for bits of 'space junk' (properly termed 'micrometeoroid orbital debris') moving at hypersonic velocities in space also posed a very real threat. During the STS-7 mission in June 1983, Challenger suffered a 4-millimeter-wide 'pit' in the outer layer of one of her six forward flight deck windows. Commander Bob Crippen did not report the incident and it remained unknown until after landing. "His rationale, I assume, was that there wasn't anything that the ground could do to help us," remembered STS-7 astronaut John Fabian. "The event had already occurred. We were perfectly safe… and elected not to say anything. I think it was the right decision." The damaged windowpane was subjected to detailed energy-discursive X-ray analysis and titanium oxide and small quantities of aluminum, carbon and potassium were detected. The overall morphology of the impactor suggested a fleck of spacecraft paint. Although probably less than 0.2 millimeters in size, at an impact speed of 21,600 kilometers per hour it was a lethal projectile. Similar window impacts occurred on at least six other Shuttle flights.

In addition to the window damage incurred by STS-7, several tiles were lost – including a fragment torn from a landing gear door – and scarring was noticed near to the left-hand 'chine' (the junction between the leading edge of the wing and the fuselage), together with severe AFRSI discoloration. A year later, in October 1984, Challenger returned from her STS-41G mission and it became clear that almost 5,000 tiles had become 'de-bonded' from her airframe during re-entry. One tile, near the left-hand chine, had completely detached from the vehicle's fuselage. Fortunately, this flight avoided disaster, but post-flight investigation revealed that a vulcanizer (known as 'screed'), which was meant to smooth the metallic surface beneath the tile bonding material, had somehow softened and its 'holding' qualities had been impaired. Repeated injections of a waterproofing agent known as 'sylazane', together with repeated high-temperature re-entries through the atmosphere, had conspired to degrade the bonding material. Challenger's next mission, STS-51C, originally planned for December 1984, was postponed and shifted onto another orbiter and she spent more than six months on the ground undergoing repairs. By the time she flew next in April 1985, the use of sylazane had been scrapped.

As the storm-clouds of foreboding gathered in the months before January 1986, other missions suffered their own levels of damage: missing tiles from the OMS pods, large strips of AFRSI loosened or peeled away entirely, long gouges under the Shuttle's wings and – during the STS-51F ascent in July 1985 – such a battering from falling Spray-On Foam Insulation (SOFI) that more than a hundred tiles needed replacement. So severe, in fact, was the STS-51F damage that the astronauts performed an RMS camera survey in orbit, which uncovered a significant number of debris impacts and a definitive inspection after the Shuttle landed revealed 553 'hits'.

Following the loss of Challenger, the surviving fleet of orbiters returned to service in September 1988, but even the very first mission, STS-26, sustained moderate TPS damage to Discovery's right wing. The in-flight trauma endured by the crew of STS-27 has already been discussed and as the next few years would amply demonstrate, vehicles routinely returned to Earth with loose AFRSI blankets and damaged tiles. In June 1991, Columbia roared into orbit for STS-40, the Shuttle's first mission totally dedicated to medical research. But within minutes of opening the payload bay doors, cameras revealed that several AFRSI blankets had become detached from the OMS bulkhead. Additionally, a portion of the payload bay door sealant strip had become displaced, creating a risk that this might hamper their successful closure at the end of the mission. The flight director was convinced that

Fig. 6.5 Discoloration and superficial Thermal Protection System (TPS) damage are visible on Columbia in this post-landing view after STS-5 in November 1982.

there was no concern for alarm, noting that the doors' latches were strong and that even if the seal did cause an obstruction, it could be easily 'collapsed' and the doors safely sealed prior to re-entry. However, a section of door seal was shipped to JSC, where astronauts encased in bulky space suits practiced a spacewalk repair. This validated managers' belief that the doors could be safely closed, even with a loose seal, and it was noted that even if the crew did have to conduct a repair they would need only to cut it loose or push it back into its retainer. STS-40 commander Bryan O'Connor expressed concern that the seal might become snarled on a mechanical 'fork', which assisted in the closure of the doors, but was placated by a detailed explanation of the procedures. Early on the morning of landing, the doors were closed successfully (carefully watched and videotaped by the crew) and Columbia landed safely and without incident.

Six years later, however, that same vehicle returned to Earth after surviving one of the worst TPS mutilations ever recorded. In December 1997, Columbia came home from STS-87 in a condition which NASA engineers described as "not normal". Many of her tiles were severely scarred, with over a hundred deemed irreparable: over double the usual expectation in terms of damage. The impacts endured by STS-87 did not follow aerodynamic predictions and were far higher in number than they ought to have been. No fewer than 308 'hits' were recorded during post-flight inspections, more than a hundred of which were larger than 2.5 centimeters in diameter and in some cases had penetrated up to 75 percent of the total depth of the tile. The physical cause was identified as falling SOFI or ice from the External Tank (ET), due to new, environmentally friendly products, including an upgraded foam insulator and a defective coating of primer. This had literally 'pelted' Columbia's nose and fuselage during her raging climb to orbit. Both this incident and the near-disaster which hit STS-27 in December 1988 would take center-stage in 2003 as the Columbia Accident Investigation Board (CAIB) dug into the causal factors behind the STS-107 tragedy.

THE AGONY OF COLUMBIA

"If there ever was a time to use the phrase 'All good things come to people who wait', this is that one time," launch director Mike Leinbach radioed STS-107 commander Rick Husband on the morning of 16 January 2003. "From the many, many people who put this mission together, good luck and Godspeed."

"We appreciate it, Mike," replied Husband from his seat on Columbia's flight deck. "The Lord has blessed us with a beautiful day here and we're going to have a great mission. We're ready to go."

Without further ado, at 10:39 a.m. EST spectators lining the roads of the Space Coast were deafened by the roar of the three main engines and the harsh crackle of the SRBs as the 113th Shuttle mission got underway. Weather at KSC that morning was as close to perfect as it could possibly be. Columbia's scientific research mission had been delayed by more than a year, due to changes in the Shuttle manifest and fleet-wide technical troubles, but as Husband and his six crewmates – pilot Willie McCool, mission specialists Dave Brown, Kalpana Chawla, Mike Anderson and Laurel Clark, together with payload specialist Ilan Ramon, the first Israeli astronaut – headed to orbit, there was every reason to suppose that a spectacular flight was on the cards. After all, the lessons of Challenger had already taught everyone that once the SRBs were jettisoned, the most dangerous part of the mission was behind them. "Learning a lesson is one thing," said former chief flight director Milt Heflin, "but living by what you learned is sometimes difficult, I think."

Unsurprisingly, the launch made headlines in Tel Aviv and Israel's ambassador to the United States could not help but see parallels between the "lowest ebb" of the Jewish people, two generations earlier in the Second World War, and the "great achievements" on this turn-of-the-millennium space mission. For the next 16 days, STS-107 proceeded well, as the astronauts worked in two 12-hour shifts to support 80 experiments in life sciences, fluid and combustion physics, materials processing and Earth observations around the clock. Only the most minor issues troubled them: one of two heaters in a cryogenic fluid storage tank in the payload bay stubbornly refused to work, part of the intercom between Columbia's flight deck and Spacehab module hiccupped occasionally and an electrical spike in a dehumidifier left the astronauts working in slightly warmer conditions than normal. As the mission entered its final days, Ramon remarked on the fragility of Earth's atmosphere, which resembled a thin veil between the Home Planet and the ethereal blackness of space. "It saves our life," he breathed, "and it gives our life."

At the end of STS-107, the atmosphere (or at least Columbia's high-speed passage through it) would do neither.

Video footage taken during launch clearly showed a briefcase-sized chunk of SOFI falling from the ET about 81 seconds into the flight and impacting the left-hand wing with a resultant shower of debris. Such debris, of course, had been seen regularly on earlier missions and was not considered a safety-of-flight issue. Improvements to the TPS had been introduced over the years and in April 1994 Endeavour flew STS-59 with six new-specification tiles referred to as Toughened Uni-Piece Fibrous Insulation (TUFI). These afforded a varying density level from 'high' at the tiles' outer surfaces to 'low' at their inner surfaces, coupled with a greater capacity to withstand debris impacts. But the rationale for TUFI was to reduce post-flight maintenance, rather than to significantly beef-up the Shuttle's protective armor during re-entry. Tile damage, right up until STS-107, continued to be regarded principally as a maintenance issue, rather than a safety-of-flight concern. And nine years after STS-59, as Columbia circled the globe on what would be her last mission, the media quickly latched onto the news of the foam strike to such a degree that Husband was sent an email by Mission Control to advise him of the situation, lest he be caught unawares by a journalist's question. The email, from flight director Steve Stich, assured him the imagery had been painstakingly analyzed by experts and there was "absolutely no concern for entry". Husband thanked Stich and pressed on with his mission.

Early on 1 February, the crew donned their pressure suits and took their seats for the hour-long glide back to the Shuttle Landing Facility (SLF) at KSC. On Columbia's flight deck were Husband, McCool, Chawla and Clark, with Brown, Anderson and Ramon seated downstairs on the middeck. At 8:15 a.m. EST, as Columbia flew backwards and upside down some 270 kilometers above the Indian

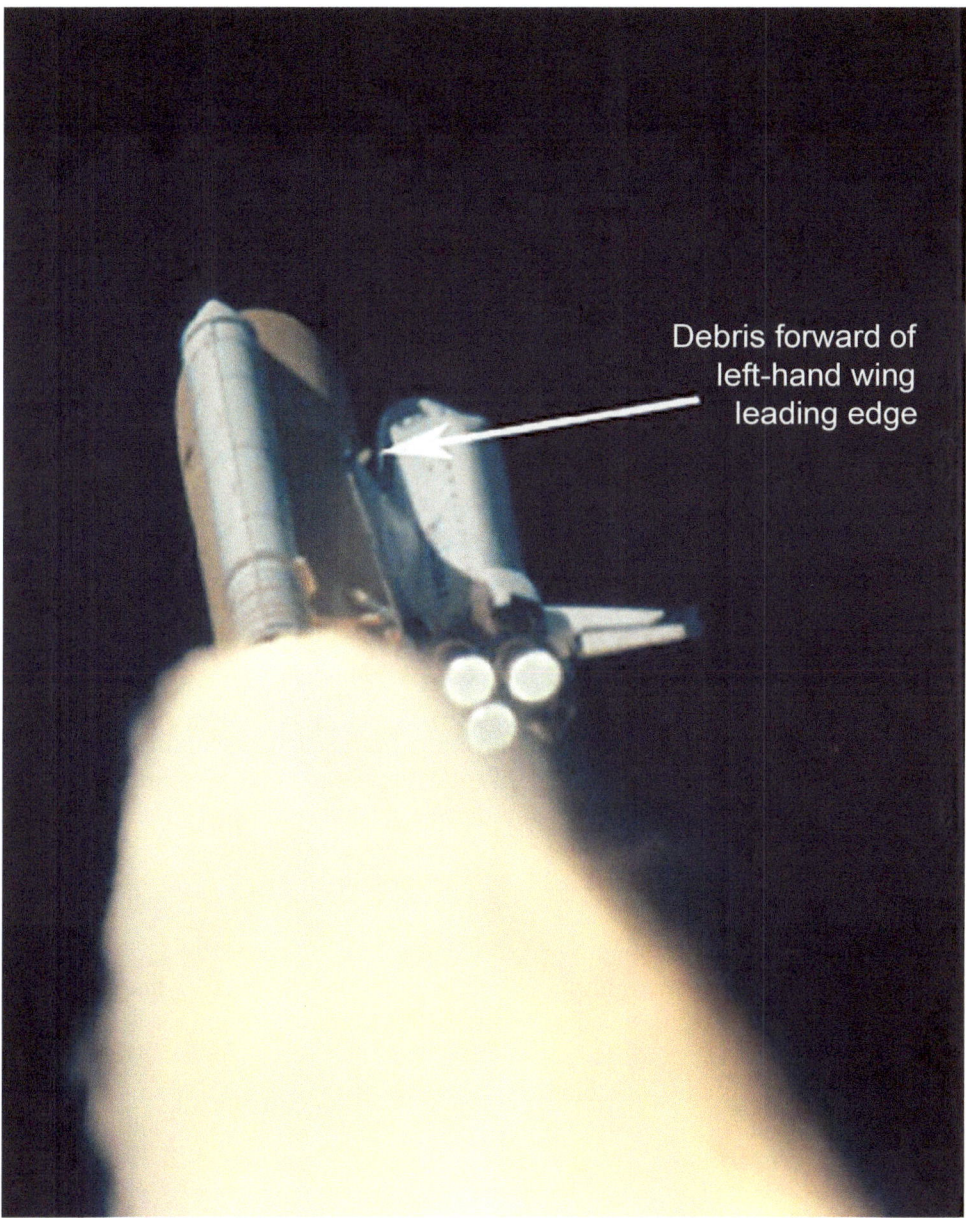

Debris forward of
left-hand wing
leading edge

Fig. 6.6 A chunk of debris falls from the External Tank (ET) Bipod Ramp and heads towards Columbia's left-hand wing at T+81 seconds into the STS-107 launch.

Occan, at an orbital velocity of 28,200 kilometers per hour, Husband was given the go-ahead to fire the OMS engines for the irreversible de-orbit 'burn' to commit to landing. For the next half-hour, the vehicle fell like a stone through orbital darkness, its General Purpose Computers (GPCs) guiding it smoothly towards a runway on the other side of the planet. The second half-hour, however, was always much more dynamic, as compression of the steadily thickening air at hypersonic speeds subjected the Shuttle's airframe to extreme temperatures and produced a spectacular light show through the flight deck windows. "Entry is fiery, just an amazing light show," STS-51L's Dick Scobee once remarked. "And the fires of hell are burning outside your window and you're sitting there, nice and comfortable, watching all this go on and it's just a neat feeling."

Seventeen years after Scobee's untimely death, the crew of Columbia were experiencing the onset of that same feeling. Little could they have possibly realized that it would soon turn into their doom. "That might be some plasma now," observed McCool, as he glimpsed the peculiar, salmon-pink glow which gradually replaced the pitch blackness of space to which he had become so acclimatized over the last two weeks.

"Think so, already?" queried Clark from her seat directly behind McCool.

"That's some plasma," confirmed Husband.

"Copy, and there's some good stuff outside," said Clark. "I'm filming overhead right now."

"It's kinda dull," added McCool.

"Oh, it'll be obvious when the time comes," Husband assured him. Minutes later, the glow steadily brightened and McCool told Brown, Anderson and Ramon that he could now see vivid orange and yellow flashes across Columbia's nose. In the chatter which followed, Husband likened it to being inside a blast furnace. The time was 8:44 a.m. EST and the Shuttle had by now dropped to an altitude of about 120 kilometers as it hurtled like a meteor across the eastern Pacific, nine minutes from the California coastline and a continent away from its scheduled 9:16 a.m. EST landing in Florida.

Flying with her nose angled upwards to subject the toughened RCC of her nose cap and WLEs to temperatures far above 1,500 degrees Celsius, Columbia descended to Earth at more than 24 times the speed of sound as aerodynamic stresses on her airframe doubled, then tripled, then quadrupled. Aware of the intensely dangerous situation that they were in (but unaware that their ship was, in fact, fatally stricken), the astronauts calmly donned their gloves and performed communications checks. Armed with a camera, Clark intended to videotape the entire re-entry. At 8:50 a.m. EST, with the GPCs still flying the vehicle, Columbia's right-hand RCS thrusters automatically pulsed to adjust the position of her nose. This was one of several pre-programmed maneuvers to bleed off her immense speed. Three minutes later, exactly on time, she crossed the California coastline and observers on the ground were able to run out into their back yards to see the Shuttle streaking overhead, "at incredible speed", according to freelance photographer Gene Blevins. It was around this time, however, that he and colleague Bill Hartenstein saw something else. It looked like a big red flare, underneath

Columbia, as if something had broken away from the spacecraft. Meanwhile, at KSC, the astronauts' families, Israeli dignitaries and high-ranking officials, including NASA administrator Sean O'Keefe and associate administrator for spaceflight Bill Readdy (a former Shuttle commander) had begun to gather in the viewing bleachers on a beautiful Saturday morning. At 9:05 a.m. EST, as Evelyn Husband and her two children smiled and posed for a photograph next to the famous countdown clock, they knew that in only 11 minutes they would welcome their husband and father home from his mission.

Fig. 6.7 View of Columbia's payload bay, acquired during the STS-107 mission. Note the light-grey-colored Reinforced Carbon Carbon (RCC) panels at the leading edges of the wings. The damage to Panel 8 on the Shuttle's left-hand wing (seen on the right of this photograph) is hidden from view by the payload bay doors and unseen by the crew.

Little could they have realized that Columbia had, by this point, already broken apart.

Despite managers' assurances that the SOFI strike on ascent had been inconsequential in terms of having caused critical damage, lingering doubts remained, to such an extent that on 31 January – the day before Columbia's homecoming – NASA engineer Kevin McCluney gave a hypothetical description of the kind of data 'signature' they might see if the foam had indeed punched a hole through the TPS and allowed super-heated plasma to enter the wing and the Main Landing Gear (MLG) wheel-well from the point of Entry Interface at an altitude of 120 kilometers down to an altitude of about 60 kilometers, covering a timeframe of not much more than 15 minutes.

"First would be a temperature rise for the tires, brakes, strut actuator and the uplock actuator return," McCluney wrote.

At 8:52:17 a.m. EST, nine minutes after Entry Interface, flight director Leroy Cain and his Mission Control team at JSC saw the first unusual data on their monitors. Cain had begun his shift that morning with a crisp "Let's go get 'em, guys", before giving Husband the green light to begin the journey home. Most of the re-entry had been conducted under GPC authority (as was the nominal procedure), but with only 23 minutes remaining before Columbia was due to land at KSC, Maintenance, Mechanical, Arm and Crew Systems (MMACS) engineer Jeff Kling saw something strange. In what is described in flight parlance as an "off-nominal event", Kling noticed that two downward-pointing arrows had suddenly materialized next to readings from a pair of sensors buried deep inside the Shuttle's left-hand wing. Their purpose was to measure hydraulic fluid temperatures in feedlines running back to the elevons. A few seconds later, two more sensors also failed. The attention of Kling and a pair of his colleagues was now riveted upon the sensors, which looked as if their wiring had been inexplicably cut. But despite their best efforts, none of the men could identify a common thread to explain the failures.

"FYI, I've just lost four separate temperature transducers on the left side of the vehicle," Kling reported to Cain. "Hydraulic return temperatures. Two of them on System 1 and one in each of Systems 2 and 3."

"Four hyd return temps," queried Cain.

"To the left outboard and left inboard elevon."

There surely had to be a common cause for all four failures to have occurred in such close temporal proximity to one another. But when Kling added that "no commonality" existed between them, Cain was perplexed. Immediately, his line of thought returned to the foam strike on the left wing. Were these four failures somehow linked to that event? He would later admit that his initial suspicion was that a hole had been punched in the wing and was indeed allowing plasma to wreak havoc on the sensors, cables and systems inside. But upon checking with Guidance, Navigation and Control (GNC) officer Mike Sarafin, his worries were assuaged, for Columbia's performance as she flew over the California-Nevada state line at 22.5 times the speed of sound remained right on the money. Cain then checked in a second time with Kling, who advised him that all other data was normal.

"...Tire pressures would rise, given enough time, and assuming the tires don't get holed," continued Kevin McCluney's chilling hypothesis from a day earlier, *"the data would start dropping out as the electrical wiring is severed..."*

Suddenly, at 8:58 a.m. EST, Husband made his first radio call to Mission Control since the onset of Entry Interface, a quarter of an hour earlier. But his transmission was abruptly cut off. A few seconds later came a loss of temperature and pressure readings for both the inboard and outboard tires of the Shuttle's left-hand landing gear. If the tires had been holed or were losing pressure, it would be a bad day, for Columbia was already carrying the heavyweight Spacehab research module and many doubted that a 'wheels-up' belly landing would be survivable. The criticality of the landing gear hardly requires explanation and in March 1994 debris had been observed falling from the Shuttle's underside during touchdown and rollout on the runway. And in November 1995, low pressures were recorded in Columbia's tires during STS-73. This required Mission Control to instruct the crew to put their ship into a variety of different orbital attitudes to raise temperatures on the belly, which served to slowly increase tire pressures to within their required minimum limits. But in the eventuality that the tires were losing pressure or 'holed', the astronauts would have to bail out and parachute to safety…which was feasible only during controlled, gliding flight in the lower atmosphere.

"…Data loss would include that for tire pressures and temperatures, brake pressures and temperatures," McCluney's chilling prediction concluded.

After hearing the report from Kling, Capcom Charlie Hobaugh, a veteran Shuttle pilot, called Husband to inform him of the tire pressure messages and ask him to repeat whatever it was that he had tried to say. But there was no reply. Meanwhile, Cain continued to press Kling on whether the messages – four hydraulic sensors gone, followed in short order by an apparent breach of Columbia's tires – were evidence of a real emergency or were faulty instrumentation. The reply was that all sensors were registering themselves as 'off-scale low', meaning that they had simply stopped functioning. Seconds later, at 8:59:32 a.m. EST, Husband again tried to communicate with Mission Control. They would be the final words ever received from Space Shuttle Columbia.

"Roger, uh, buh…"

It was clearly an acknowledgement of Hobaugh's call and an effort to say something else, but his words were broken in mid-sentence, together with all the other data flowing from the spacecraft. Communications were never restored. Thirty-two seconds later, a ground-based observer armed with a camcorder shot video footage of multiple trails of debris streaking across the skies over Texas.

With the flow of telemetry thus severed, the situation in Mission Control was becoming increasingly uncomfortable, as Kling told Cain that no common thread existed between the hydraulic sensors and the tire pressure messages, but also that instrumentation responsible for monitoring the positions of the Nose Landing Gear (NLG) and MLG was gone. As the period of radio quietness crept on and on, Instrumentation and Communications Officer (INCO) Laura Hoppe expected some "ratty comm" during re-entry but was surprised at how protracted and 'solid' the unnerving silence was.

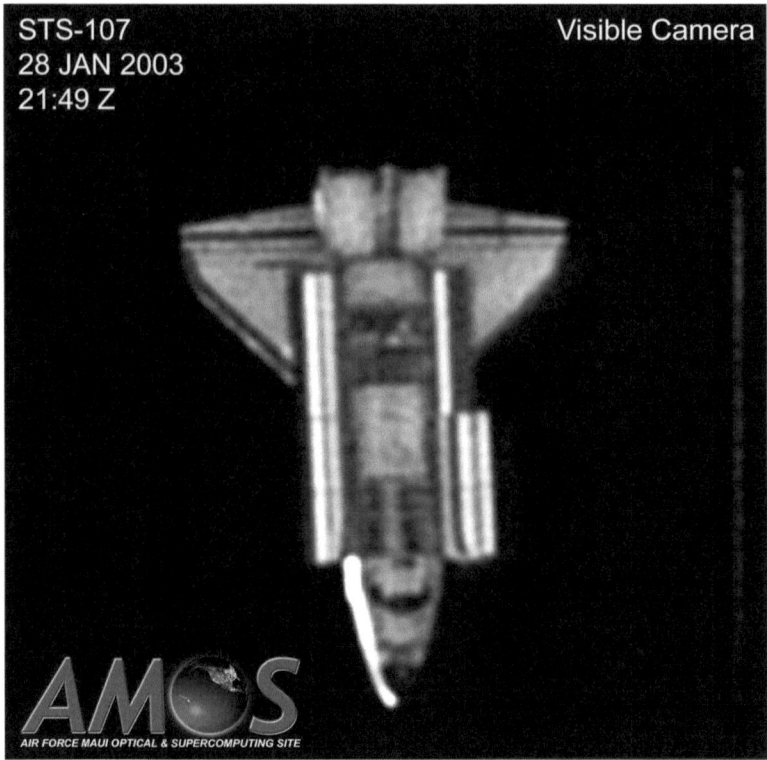

Fig. 6.8 Ground-based view of Columbia in orbit during STS-107, as seen by the Air Force Maui Optical and Supercomputing Site (AMOS) in Hawaii. The Shuttle's payload bay doors obscure the damage to Panel 8 on the leading edge of the left-hand wing.

From the Capcom's seat, at 9:03 a.m. EST Hobaugh attempted to call Husband. "Columbia, Houston, comm check." His words were greeted by static. A minute later, he repeated the call. But again, there was no reply.

Sixteen hundred kilometers away at KSC, astronauts Jerry Ross and Bob Cabana were chatting outside the recovery convoy commander's van at the SLF when they heard that communications with Columbia had been lost. At first, they were unconcerned, until they were advised at 9:04 a.m. EST that powerful radars – which were meant to lock onto the incoming Shuttle as it swept into Florida air-space and lined up for its final approach to the runway – could see nothing coming over the horizon. "That was the absolute black-and-white end," Cain later remarked, sadly. "If the radar is looking and there is nothing coming over the hori-zon, the vehicle is not there." Unlike aircraft, which can adjust their flight profiles to effect a second approach to a runway, the Shuttle fell from space with the aero-dynamic grace of a brick, with only a single shot at landing. Throughout re-entry, its blistering orbital velocity was gradually slowed through a series of sweeping,

S-shaped 'turns' to permit it to touch down at around 350 kilometers per hour. As such, the vehicle's trajectory could be plotted precisely from the point of entering the atmosphere to the instant of touchdown. The assembled crowds at KSC watched the famous countdown clock as it ticked towards a 9:16 a.m. EST landing. They braced themselves for the Shuttle's trademark double sonic boom, followed by their first glimpse of the tiny black-and-white spacecraft, rapidly descending from the west. But it never came.

Veteran Shuttle commander Steve Lindsey was responsible for taking care of the STS-107 families that morning and he recalled someone nearby innocently remark "they're always late". However, Lindsey knew better. Returning Shuttles were *never* late. When the double sonic boom did not reverberate across Florida, when Columbia did not appear on the horizon, and when her tires did not kiss the SLF runway at precisely 9:16 a.m. EST he instinctively knew that something was terribly wrong. So too did other astronauts. Ross bowed his head and said a brief prayer. At one point, Sean O'Keefe looked over at Bill Readdy – a tough, no-nonsense, ex-fighter pilot – and saw the color suddenly drain from his face. Readdy was visibly trembling.

Far to the west, in Texas, state police were inundated with 911 calls, reporting bright flashes in the morning sky, loud explosions and falling debris. In Mission Control, televisions were not tuned to outside broadcasts and it was left to off-duty NASA personnel, watching the unfolding CNN coverage from their homes, to call in with the devastating news. Engineer Michael Garske saw Columbia (what was left of her) hurtle overhead from the roadside south of Houston and he immediately ran to the nearest public telephone to call his colleague Don McCormack in Mission Control.

"Don, Don, I saw it," Garske yelled. "It broke up."

"Slow down," McCormack replied. "What are you telling me?"

"I saw the orbiter. It broke up!"

No one who witnessed the faces of those inside Mission Control that day can ever forget the wordless horror that came next. Seen more than three million times on YouTube, it is possible to see veteran flight director Phil Engelauf take a telephone call from off-duty flight director Bryan Austin, who – like Garske – had seen the end of Columbia with his own eyes. Although no one in the windowless control center had seen the destruction, they had resigned themselves to it. A minute or two earlier, Cain asked Flight Dynamics Officer (FIDO) Richard Jones when he expected to receive tracking data from the radars in Florida. He was told it ought to have happened at least a minute earlier. In one of the back rooms, a knot of former flight directors, including Lee Briscoe and Jay Greene, together with JSC director Jefferson Howell and deputy director Randy Stone, listened to the unfolding disaster. An ashen-faced Stone turned to Howell and spoke. "This," he said, "is going to be the worst day of your life."

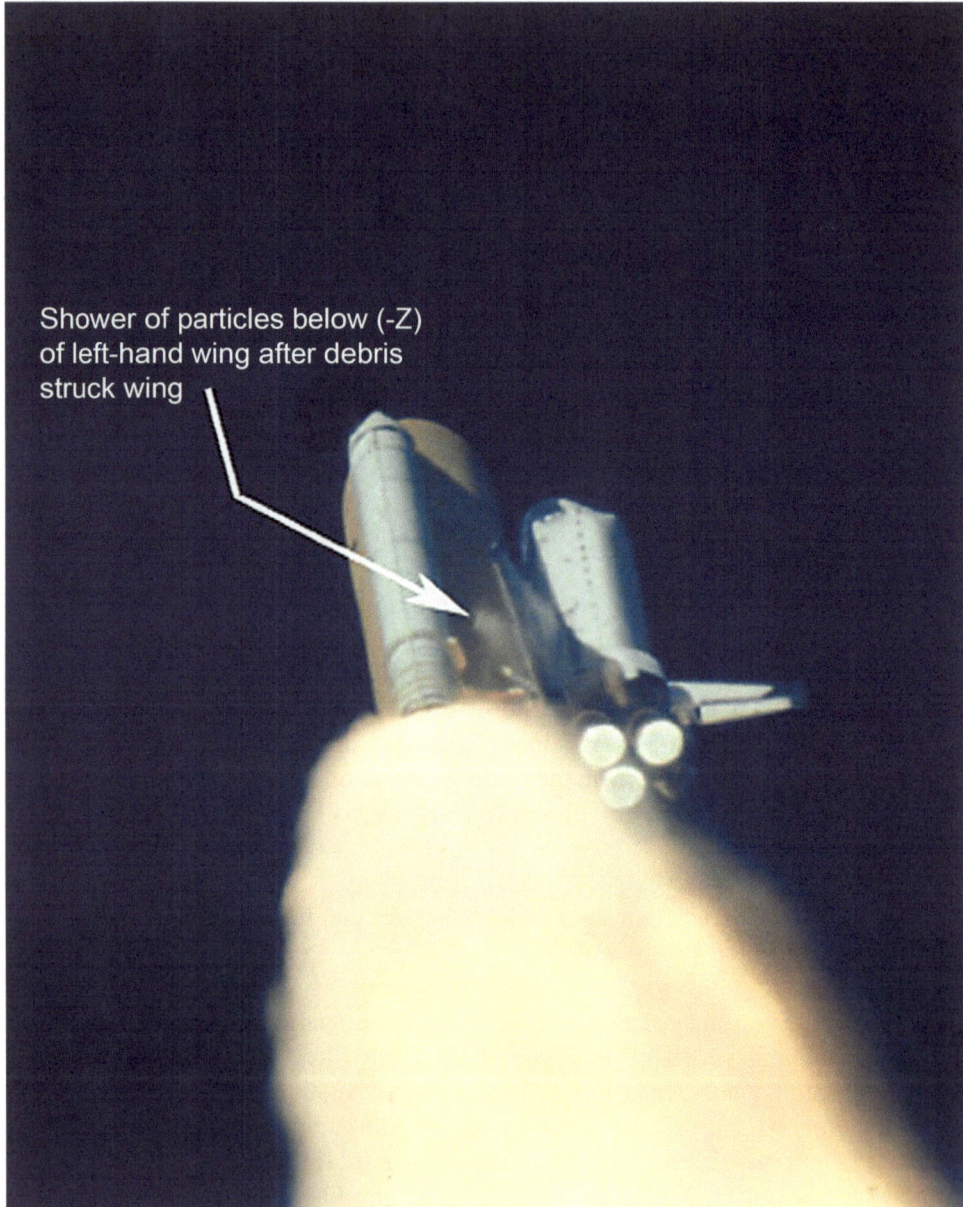

Shower of particles below (-Z)
of left-hand wing after debris
struck wing

Fig. 6.9 A shower of particles falls from Panel 8 on the leading edge of Columbia's left-hand wing following the foam impact.

Elsewhere at JSC, Milt Heflin was chatting to one of the division chiefs. All at once, John Shannon, a lead flight director, walked in and, without a word, pulled a white binder from a shelf and shoved it under his arm. Heflin knew the binder contained the contingency procedures to deal with a Shuttle disaster.

"John, what's happened?" asked Heflin.

Shannon did not stop walking. "We lost them," he said. Then he left the room.

In Mission Control, as Engelauf relayed Austin's emotional report to Cain, astronaut Ellen Ochoa (who had flown with Husband on an earlier mission) turned her head away, an agonized look of distress on her face. Cain composed himself for a moment or two and then declared a Space Shuttle Contingency. At 9:12 a.m. EST, he instructed Ground Control (GC) officer Bill Foster to "Lock the doors". It was an admission, if any were needed, that all hope was gone. No one was to leave Mission Control, no telephone calls were to be made and all personnel were to begin preserving their data files and writing up their logbook notes for the investigation that would inevitably follow. After checking with Jones that no further tracking data had been acquired, Cain referred his team to the contingency procedures in the Flight Control Operations Handbook.

Back at KSC, the STS-107 families were ushered away from the viewing site and were back in the crew quarters by 9:30 a.m. It was left to former chief astronaut Bob Cabana to break the dreadful news. Mission Control, he told them, had not received any radio beacon signals which would have been activated had the crew managed to bail out. And in any case, Columbia was traveling so high (about 60 kilometers) and so fast (23,400 kilometers per hour) at the instant of breakup that there was no possibility that anyone could have survived.

The crew's children suddenly started to scream.

TIMELINE TO TRAGEDY

In the days that followed, few could have imagined that a humble, lightweight chunk of foam could have crippled a Space Shuttle. "Was it something that happened after launch? Was it something that happened during the re-entry? Or was it something that happened during ascent and we didn't see it? Those are all possibilities," said Shuttle program manager Ron Dittemore in a 5 February press conference. "It just does not make sense to us that a piece of debris would be the root cause for the loss of Columbia and its crew. There's got to be another reason."

In mid-March, the Shuttle's flight data recorder was recovered in hilly terrain near Hemphill, Texas, and proved to be in remarkably good condition, considering that it had fallen to Earth from the very edge of space. It was shipped to a data-storage-tape specialist for cleaning and stabilization, before it was returned to KSC to be copied and from thence to JSC for a minute engineering analysis of its contents, which had taken and logged readings from more than 570 sensors scattered throughout Columbia's airframe. The tape had broken between its 'supply' and 'take-up' reels and a portion had become stretched, but after being hand-cleaned in filtered, de-ionized water, dried with lint-free cloth and nitrogen and wound back onto its original hub with new flanges, it was ready to tell the story of STS-107's final minutes. It provided a strong signal and valid data up to 9:00:18 a.m. EST, a full minute after Husband's last partial radio call. For the investigators of the CAIB, chaired by former U.S. Navy admiral Harold Gehman, it would represent both a treasure trove and a smoking gun. But more than that, it was a timeline to tragedy.

Another timeline came in the form of Laurel Clark's recovered videotape, filmed during the early stages of descent, which contained 13 minutes of footage of herself, Chawla, Husband and McCool in jubilant spirits, putting on their gloves and admiring the spectacular light show outside the flight deck windows. Little could they have known that even as they looked forward to their Florida homecoming, ionized atoms from the steadily thickening air were entering a gaping hole in their ship's left wing and soon would start to destroy it. The surviving tape

Fig. 6.10 The light-grey-coloured Reinforced Carbon Carbon (RCC) panels lining the edge of the Shuttle's left-hand wing are clearly visible in this post-landing view from a previous Columbia mission, as are the wheel-well doors to the Main Landing Gear (MLG).

ended at 8:47:30 a.m. EST. Clark clearly continued filming past this point, but whatever happened in the next few frightful minutes – stored on the outermost edge of the cassette – was burned away during its fall to Earth.

About a minute or so after the surviving tape ended, a 'strain gauge' on an aluminum spar just behind one of the RCC panels at the leading edge of the left-hand wing measured an unusual spike in structural duress. It seemed that the aluminum was getting hotter, expanding and beginning to soften. Each of the Shuttle's wings consisted of upper and lower surfaces, connected by means of an aluminum framework, with 22 RCC panels (11 per wing) wrapped in a 'U' shape around the forward edge. The panels were numbered, with Panel 1 closest to the fuselage and Panel 11 furthest away. Analysis of the STS-107 ascent footage showed that the foam liberated from the ET appeared to have hit Panel 8, a location which was not visible to the astronauts whilst they were in orbit since it was obscured by Columbia's open payload bay doors. The strain gauge which recorded the data was behind Panel 9 and it revealed that trouble began brewing only a few minutes after Entry Interface. When the CAIB published its report in August 2003, the authors were convinced that the strain-gauge data was proof-positive that Columbia had started her re-entry with a crippled RCC panel. That conclusion was later refined to one specific area (Panel 8), due to the strength of the gauge's readings. In fact, investigators would judge that the gauge must have been situated within 40 centimeters of the point where hot plasma was playing on the aluminum spar. Twenty seconds after the first indication from the strain gauge that something was not right, a sensor in a hollow cavity behind Panels 9 and 10 measured an unusual increase in temperature. Because this sensor was not only heavily insulated, but also some distance away from the breach, it suggested that the hole in Panel 8 was a large one, perhaps 15-25 centimeters wide. Anything smaller would almost certainly not have generated such strong observed readings.

At approximately 8:50 a.m. EST, Columbia's GPCs began to delicately guide the spacecraft towards Florida, swinging the nose to the right. A few seconds later, sensors attached to the left-hand OMS pod registered an unusual rate of temperature increase. Rather than steadily climbing, they rose slowly. Wind-tunnel tests later showed that hot plasma from the Panel 8 breach location was blowing metallic vapor from melted insulation through air vents on the uppermost face of the left wing, which interfered with the normal air flow around the vehicle and delayed the anticipated temperature increases. As re-entry temperatures worsened, Inconel – the heat-resistant alloy which sealed the RCC panels – started to spray Columbia's metallic skin. Then, just after 8:52 a.m. EST, the aluminum spar behind Panel 8 finally burned through. The hot plasma now had access to the interior of the wing and immediately cut sensor wiring and heated the aluminum trusses which supported its upper and lower surfaces. By this stage, temperatures on the RCC were twice as much as they were built to withstand. Under normal re-entry conditions, the Shuttle compressed the thin air to generate two shockwaves, forming a 'boundary layer' several centimeters thick, which would resist further compression and provide a natural insulator. But a smooth surface was critical for this boundary layer to take shape. With a ragged hole in Panel 8, the protective properties of the boundary layer were correspondingly disrupted.

When the aluminum spar finally burned through, the STS-107 crew were still flying high over the Pacific, oblivious to the evolving danger. However, their blissful sense of safety would not last for long. Fourteen seconds later, the hot plasma began to destroy three bundles of wiring that ran along the outboard wall of the MLG wheel-well, before entering the box-like enclosure through vents in the door hinges. It was at this point that the unusual data of elevated hydraulic fluid temperatures enigmatically appeared on Jeff Kling's display in Mission Control. The left wing was literally being eaten from the inside, and as the Shuttle penetrated deeper into the atmosphere the GPCs struggled to maintain effective control. Husband and McCool may have noted a slight tug as the nose pulled to the left under the effects of increased aerodynamic drag, but it was corrected by the computers commanding the elevons to balance out the discrepancy to the right.

By 8:53:28 a.m. EST, however, as Columbia crossed the California coastline, she was in severe distress. Not only were the aluminum trusses softening and melting, but so too was the aluminum skin of the wing itself and the adhesive material which held heat-resistant tiles and AFRSI blankets in place. It was almost certainly shards of this compromised TPS that ground-based observers saw falling from the spacecraft as they watched its glowing plasma trail cross the early

Fig. 6.11 Image of Columbia, as seen from the Starfire Optical Range at Kirtland Air Force Base in New Mexico. This photograph, which clearly shows material streaming away from the Shuttle, was acquired at 8:57 a.m. EST on 1 February 2003, just prior to loss of contact.

morning sky, traveling west to east. There were reports of a bright flash at around this moment, which probably represented the point at which the plasma finally burned through the upper surface of the left wing.

When the debris path was mapped, it became apparent that the most 'westerly' recovered tiles – those which fell away from Columbia early in its structural breakup – had originated from RCC Panels 8 and 9. At about 8:54:11 a.m. EST, the GPCs responded to further disruption of the normal flight regime, although on this occasion the forces trying to pull the nose to the left and roll the vehicle to the left suddenly reversed and the left-hand wing seemingly acquired additional 'lift'. The computers readjusted the elevons to counteract the undesired motion. On the flight deck, Husband and McCool may have noticed these changes on their instruments but did not contact Mission Control. Investigators would later blame this perceived lift on the now-weakened lower surface of the left wing, which was starting to bow inwards under the increased pressure. At about this time, ground-based observers videotaped a large piece of debris falling away from the Shuttle. It was probably a major chunk of the lower portion of the left wing. The CAIB investigators would speculate that by this point, the melted aluminum spars and trusses might have been lying "in pools on the bottom of the wing", with only the RCC panels holding them in position. By 8:56:16 a.m. EST, the plasma continued its furious rampage into the left-hand wheel-well and began to scorch the MLG struts as sensors registered increased pressure in the outboard tires. A concave depression on the lower surface of the wing gradually increased in size and the GPCs again responded by commanding the elevons to counteract this unwanted deviation from the flight profile. Columbia remained barely in control, but that changed shortly after 8:58 a.m. EST, when her flight characteristics suddenly and abruptly shifted.

It was not until the loss of tire pressure triggered an alarm on the flight deck that Husband finally radioed Mission Control. But his transmission was cut off in mid-sentence. Presumably, Clark continued videotaping as the first of many master alarms echoed through the cockpit in the next few minutes. We shall never know.

By this point, the commands to adjust the elevons were proving hopeless and at 8:59 a.m. EST a pair of RCS thruster firings, each lasting about 1.5 seconds, were commanded by the computers in an unsuccessful effort to regain control. A second master alarm sounded in the cockpit as the elevons' control circuitry failed and, seconds later, ground-based observers in western Texas saw a large piece of debris fall away from the Shuttle as it continued its descent. This was probably the remains of the left wing, finally detaching and vanishing into the plasma trail. By 9:00 a.m. EST, two other large fragments also separated, likely the vertical stabilizer and part of an OMS pod. As Columbia fell nose-first, with one wing missing and several other major structural elements gone, master alarms would have echoed in the cockpit. The cooling system failed, the electrical system experienced intermittent shorts and the Shuttle spun out of control at more than 20 degrees per second. For the astronauts, it must have seemed like a bad simulator run, but on this occasion even their years of training left them helpless. We can never know how they reacted in those horrifying final seconds, but the information gleaned from the recovered flight data recorder showed that it stopped working at 9:00:18 a.m. EST.

Just as her sister Challenger did 17 years before, Columbia fought vigorously and valiantly to the very end to save her humans.

AN ENDURING PROBLEM

It had long been recognized that flying the Space Shuttle through the furnace of re-entry and, in less than an hour, bleeding off an orbital velocity to accomplish a controlled runway landing was an extremely hazardous undertaking. But in addition to the near-misses of STS-27 and STS-87 and the tile damage experienced by so many other missions, there were also several 'close-calls'. In August 1989, during Columbia's STS-28 re-entry, an unusual aerodynamic 'shift' was noticed in her performance. Normally, the Shuttle's wings experienced a smooth (called 'laminar') air flow in the upper atmosphere. As the spacecraft descended deeper into the steadily thickening air, this flow became more turbulent and imparted greater heating and aerodynamic loads on the wings. As STS-28 returned to Earth, the transition from laminar to turbulent air flows occurred only 15 minutes after Entry Interface, a full five minutes sooner than it ought to have done, with the result that the left wing experienced a substantially higher heating environment than the rest of the vehicle.

The cause was traced to a textural 'roughness' of the wing, which had imparted a slight change in her flight characteristics. The issue was complicated further by protruding 'gap-fillers' between several of her tiles. Fortunately, STS-28 landed safely. Six years later, in November 1995, during her return from STS-73, a similar incident occurred: an earlier-than-expected increase in heating and atmospheric drag associated with the left wing. Post-mission analysis again revealed that surface roughness might have been a causal factor and, in the aftermath of STS-107, aerodynamicist John Anderson of the National Air and Space Museum (NASM) suggested that a toxic mix of surface roughness and critical RCC damage could have conspired to pull Columbia into a fatal 'sideways' orientation. Aerodynamic shifts like those experienced on STS-28 and STS-73 probably would have been insufficient in themselves to induce catastrophic structural damage but combined with the foam strike and the compromised RCC panels, it could have presented enough of a problem to tip STS-107's fortunes over the edge.

This underlined the reality that, far from being an 'operational' vehicle, the Shuttle – which had logged 113 missions by February 2003 – remained an experimental flying machine. "Any arrogance I may have had went out the window on 1 February," said NASA flight director and Shuttle program manager Wayne Hale. "I would have told you we understood what we were doing and we had mature processes and good hardware. I think all those assumptions have been shattered." Indeed, over-confident hubris was one key criticism leveled at NASA by the CAIB. "I think there's been a little bit of denial that NASA, at least in the Shuttle program, has modified its organizational structure over the years into one that no longer contains the attributes that they built their reputations on," said Harold Gehman. "The Board is absolutely convinced the management team they have right now is not capable of safely operating the Shuttle over the long term." Emphasizing costs and schedules and efficiencies was not the right way to run a space program. "The whole nation," added Gehman, "and Congress and the White House has an unrealistic view of how we do space exploration."

Fig. 6.12 Beautiful view of Earth, as seen through the overhead windows on Columbia's flight deck during STS-107.

The idea that falling SOFI was indeed responsible for the destruction of Columbia moved from hypothesis to rock-solid certainty one day in July 2003, when a similarly sized chunk of foam was fired at an RCC panel under test conditions at the Southwest Research Institute in San Antonio, Texas. It was blasted out of the institute's nitrogen-gas cannon at a comparable velocity of 850 kilometers per hour. A slight rotational spin was also imparted upon the foam to better mimic the aerodynamic conditions at 81 seconds into Columbia's ascent. The result of the test left onlookers open-mouthed in horror and amazement. For that tiny piece of debris delivered 1,000 kilograms of force and left a gaping hole, no less than 40 centimeters wide, in the RCC panel. A critical component of the Shuttle's TPS armor, which had for decades been regarded as virtually impregnable, had been crippled by a small chunk of foam. The 'smoking gun' had been definitively found.

In the aftermath of the CAIB report, NASA decided that the Bipod Ramp – the area from which the liberated foam originated – would be removed from all future ETs and replaced with rod-like heaters to prevent ice accumulation, as discussed in Chapter 4. "NASA has to cut down on the amount of debris that comes off," read one of the CAIB's key recommendations. "They have to toughen the orbiter, they have to be able to inspect and repair the orbiter and then they have also got to give the crew a better chance to survive. All four of these contribute to safer operations and not any one of them…is a fix."

Techniques and tools for repairing the Shuttle in space were considered long before STS-1, with repair kits designed for use by spacewalkers. It would have been tricky, said astronaut George 'Pinky' Nelson, who was involved in the development of the kit. "Imagine standing on perfectly slick ice and trying to work on a wall," he remarked. "Every time you touch it, you slide away."

But as mission after mission survived TPS maulings and returned to Earth safely again and again, the repair kit was put on a back burner and eventually canceled. But in the aftermath of STS-107, it was decreed that all future Shuttle crews would carry at least five different sets of tools, including a cure-in-place, caulk-like Non-Oxide Adhesive Experiment (NOAX) material and its hand-held applicator gun. On STS-114 in July 2005 – the first flight after the loss of Columbia – a Cure-In-Place Ablative Applicator (CIPAA) was used, but from STS-121 in July 2006 the smaller and less cumbersome Tile Repair Ablator Dispenser (TRAD) was introduced in its stead. It took the form of a static mixer, a 90-centimeter hose and an applicator gun, about the same size as a hand-held vacuum cleaner. Initially intended only for emergency use, it was evaluated in orbit on STS-123 in March 2008, when spacewalkers Mike Foreman and Bob Behnken practiced the application of a new ablator into intentionally damaged tile samples. The astronauts then smoothed the ablator into place using foam-tipped tools. The silicone-based material, designated Shuttle Tile Ablator-54 (STA-54), comprised a 'base' material and a catalyst, which, when mixed, resembled cake-frosting in consistency, but cured to the texture of a pencil eraser in 24 to 48 hours. During re-entry, it would act to dissipate heat by a process of charring to protect the damaged TPS tiles.

And for most of the post-Columbia missions, salvation also meant location, location, location. All but one of the last 22 Shuttle missions between July 2005 and July 2011 traveled to the International Space Station (ISS), which provided not only a 'safe haven' for astronauts in the event of significant tile damage, but also an improved capability to inspect the Shuttle's airframe in space with high-resolution 400-millimeter and 800-millimeter camera lenses. When Discovery flew STS-114, in addition to an army of photographic assets on the ground, her ET was equipped with cameras to monitor falling debris. So too were the SRBs fitted with cameras to observe the inter-tank region, with one sitting just below the nose of each booster and another close to the ET attachment ring. And instrumentation mounted behind the WLEs afforded mission controllers a steady stream of acceleration and temperature data. This consisted of 66 sensors to gather acceleration data and 22 others to monitor temperatures and calibrate debris impacts during ascent.

"Shortly before launch, the units enter 'trigger-mode' and at liftoff are activated via a G-force switch to begin storing and processing ascent accelerometer data stored at 20,000 samples-per-second, per channel, within the unit's internal memory," NASA pointed out. "Temperature and battery voltage data is stored every 15 seconds." Summary data was later downlinked to the ground for screening and comparison to threshold criteria to determine whether any potential impact events had occurred. However, these sensors caused hearts to leap into throats on several missions. "Excessive triggers" were reported during the STS-123 ascent, a pair of off-scale-high data 'spikes' were observed on STS-130 in February 2010 and intermittent communications dropouts from the sensors were recorded during the last Shuttle launch, STS-135 in July 2011. Hand-held wireless scanners were also tested on the tiles for the first time on STS-118 in August 2007.

Shuttle crews also used a 15-meter-long sensor-tipped extension to the RMS – known as the Orbiter Boom Sensor System (OBSS) – to conduct thorough inspections in orbit. The OBSS elements were originally built as RMS spare parts. When affixed to the RMS, it doubled the robotic arm's useful length to about 30 meters, enabling it to inspect previously unreachable areas of the Shuttle's underside. The Laser Dynamic Range Imager (LDRI) that was developed by Sandia National Laboratories and an Intensified Television Camera (ITVC) acquired three-dimensional imagery and could achieve resolutions good enough to observe RCC damage as small as a 0.5-millimeter-wide crack. Also attached to the OBSS was the Laser Camera System (LCS), provided by the Canadian vision-systems firm Neptec Design Group, which could scan the Shuttle's flight surfaces in three dimensions at a rate of 64 millimeters per second and record at resolutions down to a few millimeters. To ensure the quality of returned images, the OBSS sensors had to be positioned within 3.3 meters of the targeted area of inspection and this closeness required the astronauts to put the RMS into a combination of automated and manual modes to ensure there was no contact with the Shuttle surfaces.

Fig. 6.13 Spacewalker Steve Robinson works to remove gap-fillers from Discovery's undersurface during STS-114.

These tools also enabled experts on the ground to determine when it might be necessary to take action and effect prompt repairs. "We started with simple things like, what does damage *look* like on the outside of a Shuttle?" explained flight director Paul Hill, who led a vehicle inspection and repair team after the loss of Columbia. "In the flight control world, we didn't have a lot of experience with that. Our job was to bring the spacecraft back, unhurt. Whatever minor damage we tended to land with, a bunch of guys at KSC repair it and get the Shuttle ready to go again. We always were handed almost pristine Space Shuttles." Now the flight controllers were faced with questions of what damage to the exterior of the orbiter – its belly, its nose cap, its WLEs – actually looked like and how much damage quantified as 'too much'. To achieve those ends, the imaging capabilities of the OBSS and other tools came to the fore. 'Tolerable' damage before STS-107 was thought to be anything less than a hole a few millimeters wide, but by May 2003 it became clear that far smaller dings and dents could still harm the Shuttle. "And if there's any cracking around it," said Hill, "all bets are off."

During STS-114, the OBSS imagery revealed a pair of protruding 'gap-fillers' between tiles on Discovery's belly, which spacewalker Steve Robinson was able to easily remove with his gloved hands. Made from ceramic-impregnated cloth, gap-fillers fulfilled several functions. They helped to prevent the 'chattering' of tiles against each other during ascent and aided in the closure of openings between the tiles to ensure the thermal integrity of the TPS. Although the gap-fillers were not expected to present a major problem for the STS-114 re-entry, prudence on this first mission after the loss of Columbia required their removal. The first two flights after the disaster also featured extensive testing of tile-repair methods. Even the third mission, STS-115 in September 2006, required an additional 'late' inspection of the tiles after a video survey revealed the presence of a piece of debris near the Shuttle. The painstaking five-hour survey revealed the ship's TPS skin to be in fine shape, and STS-115 returned safely to Earth after a 24-hour delay.

Still, the occurrence of heat-shield damage remained with the Shuttle until the very end of its career. In May 2000, three years before STS-107, a damaged tile seam caused a TPS breach which permitted plasma to penetrate Atlantis' left-hand wing during re-entry. Fortunately, the mission landed safely. But in a haunting reminder of the pre-Challenger philosophy, Martin Cioffoletti, vice president for Shuttle integration at prime contractor Rockwell International, highlighted the difficulty in 'proving' that falling ET debris could induce catastrophic damage. "We couldn't stand up and say this is going to kill anyone," he said later, "because there was no evidence this was causing that kind of significant damage." Incremental improvements were made during the final few years of the Shuttle's operational lifetime. In July 2006, Discovery flew with hardened tiles on her NLG doors and new procedures were implemented to ensure that the gap-fillers remained secure and did not protrude from the airframe. (More than 5,000 gap-fillers were removed and replaced in the wake of STS-114.) Even on STS-114, the debris problem continued in the form of foam liberation from the Protuberance Air Load (PAL) Ramp on the ET, although no damage was caused to the Shuttle itself.

In June 2007, however, a 10-centimeter by 15-centimeter section of AFRSI on Atlantis' left-hand OMS pod managed to become detached during ascent and after extensive photographic inspection it was stapled and pinned back into place by spacewalker Danny Olivas. A similar situation arose on her next mission, STS-122 in February 2008, although in that case no repair was needed. Nor was any corrective action necessary when a small section of AFRSI became detached from Endeavour on STS-126 in November 2008. Other missions which incurred notable dings and dents included a small gouge below Endeavour's right wing on STS-118 in August 2007 and several scuffs on the orbiter's belly during STS-127 in July 2009. Three RCC panels that had sustained the loss of small quantities of their protective silicon carbide coating were also replaced before Discovery flew the STS-120 mission in October 2007.

Prior to the loss of Challenger, ship-to-ship transfers in an emergency would have been handled by Personal Rescue Enclosures (PREs), which took the form of 86-centimeter-wide spheres, made from three layers of Urethane, Kevlar and an outer thermal protection blanket, together with a small viewing port, fabricated from toughened Lexan. These could be zippered closed to provide airtight enclosures with an internal volume of 0.33 cubic meters to enable astronauts to be transferred from Shuttle to Shuttle by a spacewalking crewmate. The PREs contained a rudimentary oxygen supply and carbon dioxide 'scrubber' to keep its occupant alive for about an hour. "In order to get into it, you had to get into a fetal position, into a ball," remembered Shuttle commander Rick Hauck, "and the concept of the sphere was that it was just small enough that it could go through the crew hatch in the Space Shuttle in the event that you had to rescue people." The system was never used in actual flight, which seemed just as well, in the mind of Shuttle commander Loren Shriver. "It would not have worked as a crew-transfer vehicle at all, now that I look back on it. But I was pretty naïve back then on space-related things, so it was a nice test."

Shadowing each Shuttle flight from STS-114 onwards was the Launch On Need (LON) capability, in which the hardware – orbiter, ET and SRBs – for the next scheduled mission were assigned double-duty in executing a rescue in the event of irreparable TPS damage. Four-person subsets of the next scheduled crew (a commander, pilot and two mission specialists) would perform the LON and by using hardware which was already close to flight it would be possible to launch the rescue mission within 40 days of a Contingency Shuttle Crew Support (CSCS) decision. In the meantime, the astronauts of the damaged orbiter would remain aboard the space station as a 'safe-haven'. But it was recognized as a high-stakes and risky endeavor, not least for the astronauts embarking on the rescue, whose own vehicle could also incur TPS damage. "The CSCS is a non-certified capability," NASA explained, "that is utilized as part of an 'all-out

Fig. 6.14 High above Atlantis' left-hand wing, STS-117 spacewalker Danny Olivas prepares to staple-down a loosened section of insulation blanket on one of the Orbital Maneuvering System (OMS) pods.

effort' to provide the opportunity to safely return the STS crew home via an STS rescue vehicle in the event that the orbiter has been declared 'unflyable' and cannot perform a successful entry and landing."

Prior to STS-121, an unflyable Shuttle would be abandoned after its crew had been rescued and left to burn up during re-entry, with some scenarios having its payload bay doors open to promote structural disintegration in the upper atmosphere. But with STS-121, the Remote Control Orbiter (RCO) capability became possible. An 8.5-meter RCO In-Flight Maintenance (IFM) cable was delivered to the ISS to permit a remote-controlled re-entry and landing. This employed existing 'autoland' capabilities and afforded an electrical signal connection between the Ground Command Interface Logic (GCIL) and the flight deck instrumentation. With the cable installed in the crippled vehicle before it was set adrift, it became possible for Mission Control to autonomously command the activation and running of the Auxiliary Power Units, the deployment of the air data probes from the Shuttle's nose, the closure of fuel cell reactant valves and the arming and the deployment of the landing gear and drag chute.

Had such a dire eventuality arisen, Vandenberg Air Force Base in California was chosen as the landing site. Its location near the Pacific coastline meant that the orbiter could be 'ditched' in the ocean if problems threatened to make a remote-controlled landing attempt unsafe. As the LON vehicle was readied for launch, the astronauts of the crippled Shuttle would remain aboard the ISS with the station's resident crew, rationing their food supplies accordingly. "The number of days the station can support a stranded Shuttle crew would be determined, in part, by the consumables already on board, plus what is brought up with the Shuttle," NASA noted. "The level of consumables on-board the station, including food, water, oxygen and spare parts, will be reviewed…up until the day of the first Shuttle launch to define the CSCS capability for that particular mission." The damaged Shuttle would then be autonomously commanded to undock from the ISS "prior to violating the cryogenic redline requirements to perform an unmanned undocking and disposal burn".

Perhaps the two most significant missions after the loss of Columbia in terms of the LON capability were STS-125 in May 2009, the final servicing of the Hubble Space Telescope, and the very last Shuttle flight, STS-135 in July 2011. In the latter case (with no subsequent LON rescue vehicle available), a smaller-than-normal crew of four flew the mission, which would allow them to remain comfortably on the ISS and return to Earth, one by one, aboard Russian Soyuz vehicles over the span of several months. But any damage incurred by Atlantis on STS-125 posed a tougher nut to crack. As Hubble operated at an orbital inclination of 28.5 degrees, the ISS at 51.6 degrees was demonstrably out of reach. Therefore, a four-person crew trained for the 'STS-400' rescue mission.

Fig. 6.15 A pair of T-38 Talon jet trainers fly over Launch Complex 39 in April 2009, just before the launch of the final Hubble Space Telescope servicing mission. Atlantis is on Pad 39A in the foreground, being readied for STS-125, whilst her sister Endeavour can be seen in the distance on Pad 39B, filling backup duty for the STS-400 Launch On Need (LON) mission.

Under this plan, Endeavour would rendezvous with Atlantis about 23 hours after launch. Her pilot would use the RMS to snare a grapple fixture in Atlantis' payload bay. "The grapple occurs with the payload bay of each vehicle pointed toward each other," explained NASA, "with the orbiters perpendicular to one another for clearance." The RMS would then execute a 90-degree yaw maneuver. There would be no more than 7.3 meters of separation between the nose-to-tail vehicles at their closest point. This configuration would provide optimal conditions for Atlantis' seven-member crew to transfer to Endeavour. It would involve three spacewalks, totaling about nine hours, which promised to be among the most dramatic ever attempted. The first, lasting 4.5 hours, would involve two of Atlantis' spacewalkers stringing a cable along the length of the RMS to serve as a 'translational path' to Endeavour and then assisting one of their colleagues across. The next day, a 2.5-hour spacewalk would see two more of Atlantis' crew members undertake the ship-to-ship transfer. The final two crew members would do likewise after Atlantis had been reconfigured to ensure that the ground could command a disposal 'burn' that would safely destroy it during atmospheric re-entry. One publication labeled STS-400 as 'The Shuttle Mission No One Wants'. And it is indeed fortuitous that it was a mission that never had to be flown.

7

A Lifetime of Challenges

MIRACLES

Early in November 1984, STS-51A commander Rick Hauck received an unusual meeting request from a senior NASA public affairs official. Hauck was about to lead an eight-day mission on Space Shuttle Discovery, one of whose tasks was to retrieve and bring home a pair of communications satellites – Indonesia's Palapa-B2 and Western Union's Westar-VI – whose attached boosters had failed to properly lift them to geostationary orbit the previous February. Instead of positioning the satellites at 35,600 kilometers above Earth, the Payload Assist Module (PAM) boosters had ignited, spluttered, then died, leaving their expensive cargoes stranded in lopsided orbits at unacceptable altitudes which carried them no higher than 1,100 kilometers and as low as 240 kilometers. Although the fault lay with the boosters, and not with the Shuttle itself, it was an embarrassing setback as NASA sought to attract commercial clients to its reusable spacecraft. Moreover, Palapa-B2 was part of a sizeable $79 million deal with the Indonesian government and Westar-VI's operators had switched over to the 'safer' Shuttle when a European Ariane rocket exploded in September 1982. Options to retrieve the two satellites and bring them back to Earth aboard the Shuttle for refurbishment soon entered the realm of possibility and it fell upon the shoulders of Hauck and his crew to do it.

As outlined in Chapter 2, an increasingly over-confident attitude prevailed within NASA at that time; an attitude certainly not aided by the media-fueled public perception that the Shuttle was operational and its missions were 'routine'. "NASA was still in its halcyon days," one astronaut remarked, "still riding on the coat-tails of the successful Apollo missions, successful Skylab, successful Apollo-Soyuz, successful first tests of the orbiter. NASA continued to be bullish on itself."

© Springer Nature Switzerland AG 2021 201
B. Evans, *The Space Shuttle: An Experimental Flying Machine*,
Springer Praxis Books, https://doi.org/10.1007/978-3-030-70777-4_7

Hauck and his crewmates, of course, knew differently, as did everyone else in the astronaut corps. Finding Palapa and Westar in space, maneuvering close to these fully fueled monster satellites in weightlessness and at orbital velocities of 28,200 kilometers per hour and having spacewalking astronauts physically 'grab' them, secure them into the Shuttle's payload bay and bring them back to Earth was more than just science fiction; it seemed to be dangerous lunacy. And as if the dual retrievals were not enough of a challenge on STS-51A, Hauck's crew – pilot Dave Walker, robotic arm specialist Anna Fisher and spacewalkers Joe Allen and Dale Gardner – also had two communications satellites of their own to deploy. (Before launch, the pranksters in the astronaut office reminded them to make sure they launched the right satellites and brought the right ones back home.)

Hauck and his crew roared into orbit aboard Discovery on 8 November 1984 and successfully deployed their own haul of satellites. Over the next few days, in a remarkable feat of orbital mechanics and precision flying, they triumphantly rendezvoused with Palapa and Westar, captured them and stowed them in the Shuttle's payload bay for the journey to Earth. Allen and Gardner completed almost 12 hours of spacewalking to grab and secure the satellites, aided by their crewmates aboard the orbiter. At the end of their work, the spacewalkers displayed a hand-made placard, emblazoned with the legend: 'For Sale'. Beaming with pride, Hauck asked them to declare their success to Mission Control.

"Rick, that's the commander's job," one of the spacewalkers replied. "You should report that we have two satellites aboard." Then, with scarcely a pause, he added: "And you can also use the words 'Fucking Miracle'."

Hauck chuckled at this 'insider' joke. His thoughts cannot have failed to go back to the meeting a couple of weeks earlier. The astronauts had been in crew quarters at the Kennedy Space Center (KSC) in Florida and were already irritated by the media perception that STS-51A would be a piece of cake, with two 'easy' satellite deployments and two 'easy' satellite recoveries. It was with some trepidation that the crew entered the meeting room to see the high-ranking NASA public affairs official waiting for them. Hauck briskly asked for the agenda. The official replied that he had no specific agenda but wanted to wish them good luck before their mission. "We were all surprised," remembered Allen, "because this really was occupying a good chunk of our morning and time was very important to us right then." Hauck saw red. "There is something you can do," he told the official and proceeded to cite media reports alleging that members of STS-51A's crew considered the mission to be easy. "I can assure you that none of us said that, nor do we believe it. And I will personally tell you that my assessment is: if we successfully capture *one* satellite, it will be a miracle, and if we get *both* satellites, it will be a fucking miracle. You can quote me on that!"

Then, Hauck and his crew turned on their heels and left the room.

Fig. 7.1 Joe Allen (right) and Dale Gardner display a mock 'For Sale' sign over the Palapa-B2 and Westar-VI communications satellites, following their successful retrieval on STS-51A in November 1984. Such high-profile salvage missions, whilst spectacular, served to promulgate the myth that the Shuttle was an operational vehicle.

Two weeks later, in space, his language was considerably more refined. "Houston," he radioed, "we've got two satellites, locked in the bay!"

Yet Hauck's sense that flying a Shuttle mission 'successfully' was nothing less than minor miracle was not universally shared in the seemingly bulletproof era prior to the tragic loss of Challenger. Several highly dangerous flights sat on the Shuttle's manifest for 1986 and beyond, one of which Hauck would have commanded. The Ulysses and Galileo space probes – the former destined to explore the poles of the Sun, the other to study the giant planet Jupiter – were scheduled to launch five days apart in May 1986. As if flying aboard the Shuttle did not pose sufficient risk, both probes were nuclear-powered and both would be propelled on their journeys by a thin-skinned, liquid-fueled booster called the Centaur-G Prime. NASA's long-standing rule of thumb dictated that no single failure should be capable of endangering the Shuttle or its crew. But the Centaur threatened far more than that. Nicknamed a 'balloon tank', it required full pressurization to achieve structural rigidity in its walls and included many non-redundant critical systems. Loaded with over 16,500 kilograms of liquid oxygen and hydrogen, the risk of propellants sloshing in its tanks posed serious controllability problems if the crew had to perform a Return to Launch Site (RTLS) abort.

In such a dicey abort scenario, the astronauts would have been forced to dump the Centaur's propellants during an already highly dynamic phase of flight. "To make sure that it can dump safely, you need to have redundant parallel dump valves, helium systems that control the dump valves and software that makes sure contingencies can be taken care of," said Hauck, who would have commanded STS-61F, the Ulysses mission. "Then, when you land, you're sitting with the Centaur in the bay, that you haven't been able to dump all of it, so you're venting gaseous hydrogen out this side and gaseous oxygen out that side. Not a good idea." At the time of the STS-51L tragedy, both Challenger and Atlantis were set to be modified with extra plumbing to load and empty the Centaur, with emergency dump vents – capable of draining the whole booster within 250 seconds of an abort – on opposite sides of their aft fuselage. But the risk of propellant leaks or explosions remained dangerously high. NASA had also accepted recommendations to run the Space Shuttle Main Engines (SSMEs) at a hitherto-untried 109 percent of their rated performance to get the heavy Centaurs into orbit.

And as if that was not enough, in the months before January 1986 there were worrisome signs about sloppy working practices. One astronaut saw a technician climbing onto a Centaur with an untethered wrench in his pocket, whilst another tried to smooth out a weld only to inadvertently damage the tank in the process. "Here's a case, the only time I know of, where Headquarters was basically directly going ahead with the program, even though they weren't meeting the real safety requirements," remembered Gary Johnson, former deputy director of the space agency's Safety, Reliability and Quality Assurance Office. "We were still basically bulletproof," added astronaut Mike Lounge, one of Hauck's STS-61F crewmates. "Until Challenger, we just thought we were bulletproof and the things would always work." In the months following the disaster, the KSC safety office refused to process the Centaur, citing "insufficient verification of hazard controls" both by NASA and the booster's manufacturer, General Dynamics. Additional safety concerns and cost overruns ultimately precipitated the Centaur's cancellation in June 1986. For Hauck and the others who might have flown Centaur missions, it came as a great relief. He had already seen first-hand that flight safety was being compromised, but the launches of Ulysses and Galileo – the 'Death Star Flights', as chief astronaut John Young darkly called them – were not the only unacceptably hazardous missions under consideration at that time.

More than a year earlier, in October 1984, STS-41G spacewalkers Dave Leestma and Kathy Sullivan were assigned an engineering task to transfer hydrazine fuel between two spherical tanks in the Shuttle's payload bay. NASA hoped to visit, service and refuel satellites on future missions. Two such visits were Landsat-4 and the Compton Gamma Ray Observatory (CGRO), planned for later in the decade. And whilst safety mechanisms were in place – with triple-redundancy connectors in CGRO's case – to prevent the highly toxic hydrazine leaking from propellant tanks or getting onto spacewalkers' suits, there remained the acute danger that it might explode in direct orbital sunlight, taking off the back end of the Shuttle in the process. Furthermore, if the astronauts were to accidentally bring it back into the cabin, the tiniest ingestion of the lethal stuff or any contact with their skin could kill the entire crew. Even before the loss of Challenger, STS-41G commander Bob Crippen was adamant that his astronauts should not work with real hydrazine in space. "Crip and the safety folks were very concerned that we shouldn't do this with hydrazine; we should just do it with water," said Leestma, indicating their similar heat-transfer properties. "I think Crip thought I was a little too cavalier, because I insisted that we should do it with hydrazine." At length, Crippen was only persuaded to agree to the STS-41G tests when he watched Leestma safely demonstrate the entire process with triple-redundancy containment and satisfactorily evaluate clean-up methods in the event of a spillage. Procedures were set out for the spacewalkers to 'bake-out' hydrazine droplets on their suits whilst in direct orbital sunlight and then, once back inside, to seal their equipment in airtight bags, vent the airlock's atmosphere and pipe in fresh air prior to removing their helmets. Just like the Centaur, it was perhaps fortuitous that plans to refuel satellites with the Shuttle went away in the months after January 1986.

OF MANY HANDS AND ROBOT ARMS

Even some payloads which did get off the ground did not go well at first. Deployments of satellites, remembered astronaut Steve Hawley, were typically timelined to occur as soon after launch as possible, to hedge against the risk of a mission-critical problem which might require the Shuttle to return to Earth sooner than planned. In addition to Leestma and Sullivan's spacewalk, STS-41G included the deployment of a major Earth observation satellite, whose accordion-like solar arrays stubbornly refused to unfurl. Efforts to 'jostle' them open with Challenger's Remote Manipulator System (RMS) mechanical arm proved fruitless, as did attempts to thaw out potentially frozen latch hinges. Clearly, a more robust approach was necessary, but not whilst still within Mission Control's earshot. The crew waited until Loss of Signal (LOS), then sharply jolted the RMS to the left and right. After several tries, this seemed to do the trick and the arrays sprang open.

Fig. 7.2 STS-41G commander Bob Crippen monitors his displays during Challenger's fiery descent from space in October 1984. Note the pinkish orange color of ionized plasma outside the Shuttle's windows.

"What did you guys do?" asked the surprised Capcom when radio communications with Mission Control were re-established.

"We aren't going to tell you," came the response, "but just check it out and make sure that it's ready to deploy!"

Six years later, in April 1990, a far bigger beast – the Hubble Space Telescope – was also perched at the end of the RMS, ready for its own long-awaited deployment. Its high-gain antennas had been unfurled without incident, as had one of its two huge solar arrays, which opened like enormous kitchen roller-blinds. The second array, however, unrolled a few centimeters, then suddenly stopped in its tracks. Time was of the essence to get Hubble out of the payload bay. Umbilical power from the Shuttle had already been severed, as planned, and the $1.5 billion telescope had only two hours of lifetime in its internal batteries before its arrays had to be fully open and generating their own electricity. If the problem could not be fixed within that time limit, the most important scientific payload ever launched by the Shuttle would be dead. Astronauts Bruce McCandless and Kathy Sullivan donned their space suits, ready to go out and manually unwind the stubborn array, but the problem turned out to be an erroneous software indication and the heart-stopping glitch was promptly resolved.

Other payloads similarly stalled at the end of the mechanical arm. In the summer of 1992, the European Retrievable Carrier (EURECA) suffered intermittent data problems and its deployment was postponed by 24 hours while the issues were investigated. In February 1994, deployment of a free-flying payload called the Wake Shield Facility was called off due to a communications malfunction and an issue with its attitude-control system. Two years later, a crew recovered a Japanese satellite whose solar arrays had already been jettisoned, having refused to latch correctly into place. However, even after retrieval by the Shuttle, colder-than-expected fuel temperatures raised the specter of a possible hydrazine leak. And after the deployment of the Advanced Communications Technology Satellite in September 1993, it was realized that the primary and backup separation 'cords' of the payload separation mechanism had liberated a sizable quantity of sharp-edged debris which resulted in 36 'hits' on the Shuttle's fuselage and aft bulkhead and several tears, gouges and scratches; one area of penetration passed straight through the aft bulkhead itself. "None of the debris hits had any effect during the flight," NASA noted, "and all damage sites will be repaired during turnaround operations." But those words proved darkly prescient of the agency's attitude towards the Thermal Protection System (TPS), uncovered in the wake of the Columbia disaster, where such damage was regarded as a turnaround nuisance and not a safety-of-flight issue.

Other missions required more direct intervention from spacewalkers. During the recovery of EURECA in June 1993, two antennas refused to close properly and astronauts David Low and Jeff Wisoff were sent outside to manually wind them into place. Two years earlier, during efforts to deploy the long-delayed CGRO, fears that (like Hubble) one of its 20-meter-long solar arrays might get stuck proved unfounded when it opened as intended. The astronauts breathed a sigh of relief. "Well," said Jay Apt to crewmate Jerry Ross, "I guess everything's downhill from here!" Commander Steve Nagel, looking on with pride, was inclined to agree. But their words heralded a false dawn. Shortly afterwards, as the time came for the critical high-gain communications antenna to deploy, it refused to budge. The crew exchanged anxious glances. No fewer than six tries to unfurl it were made, with the pilots even trying to shake it open with bursts of the Shuttle's thrusters, but to no avail. Sitting on the end of the RMS, the observatory reacted to neither thruster bursts nor efforts to jolt its antenna open. A spacewalk by Ross and Apt to physically unfurl it was the only option.

Ross took off his wedding ring and handed it to Nagel.

"Steve," he said, "I'm going downstairs to get ready."

Nagel nodded. A minute or two later, the call came up from Mission Control, asking them to do just that.

Having suited up in Atlantis' tiny airlock, Ross and Apt ventured outside to begin their mission to save CGRO. "I didn't know if we could fix it or not," Ross later admitted, "and here we are, on the spot, to try to go out and fix this thing… and if we can't, then we've got this great big lead weight." Indeed, CGRO was one of the heaviest payloads ever lifted into orbit by the Shuttle, tipping the scales at more than 17,000 kilograms. Physically, it looked like a stout steam locomotive. "We may not be able to bring it home, because the solar arrays have already been deployed and the antenna's partly released," continued Ross. "Oh, man!" Quickly, the spacewalkers split up, as Apt set up tools and Ross looked at the task which faced him. The troublesome antenna was on the back side of the observatory, unseen from Atlantis' cockpit. Ross positioned himself to shake it open, but he knew that the antenna resided close to CGRO's hydrazine tanks and he did not want to damage them and risk a spillage on his space suit. He gave the antenna a couple of shakes, but it felt jammed fast. With a few more tries, gradually, the antenna began to move, then suddenly sprang open. It turned out that a thermal blanket had gotten hung up on a bolt. Ross let out a war whoop of joy. His intervention to save this multi-million-dollar scientific observatory had taken a mere 17 minutes.

Interestingly, Ross was also involved in another situation which may (or may not) have shared some similarities with the salvation of CGRO. In December 1988, aboard STS-27, he was part of a crew which deployed a classified reconnaissance satellite for the Department of Defense. Years later, one of the astronauts from this highly secretive mission made an offhand remark that they had experienced some 'problems' with the payload, which called for them to rendezvous with it and execute some sort of repair. Of course, a repair does not necessarily imply a spacewalk – a glitch with a balky antenna, for example, could be rectified robotically with the RMS – but the question of whether Ross and crewmate Bill Shepherd did a spacewalk on STS-27 has remained a tantalizingly unanswered rumor ever since. To this very day, none of the astronauts from that mission has been authorized to discuss what they did in December 1988. "We separated from it and it had a problem," was all STS-27 commander Robert 'Hoot' Gibson has ever admitted. "We re-rendezvoused with it and assisted with fixing it, separated again and left it." We will never know until this mission, still highly secretive more than three decades later, is officially declassified.

A year after Ross' triumphant salvation of CGRO, spacewalkers successfully retrieved the Intelsat-603 communications satellite, which had suffered a booster malfunction and been left in a far lower orbit than intended. The original plan called for astronauts Pierre Thuot and Rick Hieb to grasp the satellite with a specially designed 'capture bar', before lowering it into the Shuttle's payload bay, installing it onto a new booster and redeploying it back into space. This carefully choreographed plan seemingly fell apart when Thuot was unable to attach the capture bar. Every time he drew close to the slowly rotating satellite, his efforts induced unwanted oscillations and the final solution was to send a third spacewalker,

Fig. 7.3 STS-49 spacewalker Pierre Thuot stands on the end of Endeavour's Remote Manipulator System (RMS) mechanical arm during the Intelsat-603 recovery effort in May 1992.

Tom Akers, outside with them to physically seize it. This was the first (and only) occasion that as many as three humans made a spacewalk together. It worked and Intelsat-603 was captured, mounted onto its booster and sent on its way. Although there was a view within NASA that this satellite salvage, against many odds, spectacularly showcased the Shuttle's capabilities, it was also an unpleasant reminder and throwback to the pre-Challenger era of flying 'commercial' missions.

Recovering problematic payloads with astronauts' gloved hands came again to the fore in November 1997, when the crew of STS-87 deployed a Spartan solar physics satellite for what should have been two days of scientific observations. Astronaut Kalpana Chawla – who would later lose her life in the Columbia disaster – grappled the satellite with the RMS and set it free. The crew watched attentively, expecting Spartan to complete a pre-programmed pirouette a few minutes later to confirm that its systems were in good working order. But the pirouette did not occur. Chawla moved in to grapple the payload but did not receive a definitive 'Capture' indication on her instrument panel. As she retracted the RMS she inadvertently bumped Spartan and put it into a slight spin. "After I moved the arm back, I thought maybe Spartan was doing its maneuver," she said later. "That was my immediate reaction."

But the small satellite had entered a rotational spin of 1.9 degrees per second, forcing STS-87 commander Kevin Kregel to pulse Columbia's thrusters to match its rate. However, these efforts exhausted the fuel budget and flight director Bill Reeves called off the attempt to grapple Spartan. A spacewalk by mission specialists Winston Scott and Takao Doi was already planned for later in the flight and they were sent outside to manually grab it. Positioned on opposite sides of the payload bay, Scott and Doi waited as Kregel maneuvered to within a safe distance. He then gave them a 'Go' to capture the telescope-laden satellite.

"Winston," he radioed, "if we're patient, it looks like the telescope is rotating. The top end of the longitudinal axis looks like it will come down to you and the bottom one right up to Takao. If we just wait, it'll come right to us."

"Okay, copy that, Kevin," replied Scott. "We'll just be patient and see what happens."

A few moments later, the elation evident in his voice, Scott declared success as a very stable Spartan came perfectly into his gloved hands. Notwithstanding some difficulty latching the satellite back onto its berth in the payload bay, the recovery had worked. Hopes of deploying it a second time, later in the mission, came to naught, on account of the Shuttle's remaining propellant budget, and Spartan was flown again in October 1998 with great success. But STS-87 was not the first time that Spartan – a reusable spacecraft – had proven difficult to handle. On its very first flight in June 1985, it was successfully deployed, but when the astronauts returned to retrieve it at mission's end, they found that it was not oriented as it should have been. "It was supposed to be in an attitude which would be easy for us to just fly up to and grab," said STS-51G crewman John Fabian. "It turned out that the grapple fixture, instead of being out-of-plane to the vehicles, was on top." This required the astronauts to fly an out-of-plane maneuver, for which they had not trained, to pick it up. A decade later, in September 1995, another crew watched as a Spartan entered an unexpected attitude during retrieval operations, although they were able to grapple it without incident.

The criticality of the 15.2-meter-long RMS arm in effecting many of these space salvages cannot be underestimated and, from December 1998, it also played a fundamental role in building the International Space Station (ISS). However, the Canadian-built arm experienced more than its fair share of trials and tribulations over the decades. Just like a human arm, it comprised shoulder, elbow and wrist joints, linked by a pair of graphite epoxy booms, and had the capability to move payloads by means of its 'end effector': a mechanism which employed a snare with three tie wires to capture a prong-like capture fixture on deployable or retrievable payloads. Tests on the arm's first flight in November 1981 delivered encouraging performance results, but on STS-3 its wrist-mounted television camera failed and impacted the deployment of an interim environmental contamination monitor.

Camera failures, indeed, cropped up periodically during the arm's early career, but of significant concern were malfunctions in its joints. "It's electric motors that drive the individual joints and you can determine the rate at which the joints drive by how much current flows to the electric motors and how much current can flow can be determined by how you configure the software," said astronaut Steve Hawley. "It's all run by software in the Shuttle computers. And for a very big payload, in particular one that's going to be in proximity to the Shuttle, what you're worried about is a very remote failure case that the arm could fail in such a way that it drives by itself. The operator is always there, watching it, but obviously if it's close to structure you may not have time to react." Consequently, the RMS systems were limited in how fast the joints could 'drive' to give its human operator sufficient time to intervene. A wrist-joint failure in February 1984 precluded the deployment of an important free-flying satellite, whilst a camera malfunction in November 1994 meant that an icicle on the payload bay door could not be safely removed. And when Steve Hawley came to deploy Hubble in April 1990, he found that 'signal noise' in the RMS joints was imparting unanticipated oscillations in the form of a wobble-like motion. This prompted Hawley to advise his colleagues to take such oscillations into account when planning their movements of the robotic arm. Such issues proved to be far from insignificant nuisances, for the RMS formed a critical piece of operational hardware, both for payload deployment and, from 1998, in the assembly and maintenance of the International Space Station.

During STS-51I in August 1985, a protective sunshield that housed Australia's Aussat-1 communications satellite was commanded to open to permit routine health checks in advance of deployment. When one of the sunshield's clamshell-like doors failed to fully open, astronaut Mike Lounge used the RMS to push the balky door open and the satellite was deployed ahead of schedule to avoid thermally over-stressing both Aussat-1 and its PAM booster. Meanwhile, the RMS had sustained a failure in its elbow joint. Lounge had no option but to command it in 'single-joint' mode. "Instead of some co-ordinated motion," he said, it proved "a little awkward and took a while". Watching the proceedings, STS-51I commander Joe Engle could see the difficulty of Lounge's task in manually operating electrical switches, selecting individual RMS joints in turn, then moving them one at a time to properly position the huge arm.

Fig. 7.4 STS-61 spacewalker Story Musgrave rides the Remote Manipulator System (RMS) mechanical arm during work to service the Hubble Space Telescope.

Matters were worsened by the fact that STS-51I was also tasked with retrieving and repairing a giant Syncom communications satellite which had failed to enter its correct orbit a few months earlier. "That became a concern," Engle said later, "as to how much that was going to slow us down in the grapple and the capture and then the redeployment of the failed Syncom that we were going to repair." The repair underlined a growing 'seat-of-the-pants' attitude that NASA seemingly took in the pre-Challenger era. The Syncom had been launched by the crew of STS-51D in April 1985 and suffered a problem with its deployment switches. Attempts to resolve the issue with spacewalking astronauts and a jury-rigged 'flyswatter' ultimately proved unsuccessful and Syncom was left useless in a much lower orbit than intended. Watching this unfolding drama on the ground, STS-51I astronauts Lounge and James 'Ox' van Hoften began swapping ideas about a rescue. "Back then," said van Hoften, "there was a much more 'can-do' spirit at NASA and everyone felt like, hey, you can do anything." Clustered around the dining table in flight director Jay Greene's home, with Shuttle program manager Glynn Lunney also in attendance, the astronauts – aided by a bottle of Old Overholt whiskey – sketched out a plan to hand-grab Syncom, haul it into the Shuttle's payload bay, fix its faulty electronics and redeploy it. "We drank that bottle," remembered STS-51I pilot Dick Covey, "celebrating just the fact that we had hoodwinked the whole system into letting us think that we could go do this. It would not happen today!" By his own admission, van Hoften's previous mission to fix NASA's Solar Max observatory had been planned to the tiniest detail; on STS-51I, conversely, they would be winging it. The idea won the support of the president of Hughes Aircraft, who called NASA administrator James Beggs to invest several million dollars in the rescue effort. Fortunately, the STS-51I repair went flawlessly, but it would be one of the last times that the space agency would risk the Shuttle and its spacewalking astronauts to salvage a commercial customer's payload.

HAZARDS OF SPACEWALKING

From the first Shuttle-based Extravehicular Activity (EVA) on STS-6 in April 1983 to the last on STS-134 in May 2011, more than 160 spacewalks were performed by over a hundred astronauts from the United States, Russia, Japan, Switzerland, France, Canada, Sweden and Germany. None of those excursions was ever routine and even the space suits themselves – the legs and the Hard Upper Torso (HUT) – had endured a complex, tumultuous and almost tragic development. One astronaut intimately involved in the suit's evolution was Story Musgrave, who performed the first Shuttle spacewalk. "We had ten major replans," said Joe McCann, NASA's former head of EVA life support systems. "The problems started pretty early with the upper torso and putting pivots in that, because Story couldn't get through it. He couldn't don the thing, because his elbows couldn't get close enough together, so we ended up putting gimbals and bellows on the suit to allow him to get in and be able to operate. That was a significant challenge to get that done, but it remained a key worry point because we had these pivots buried in the fiberglass, and cases of them loosening up after a while. If they blew out, you would blow the bellows and be dead." A multitude of other nagging malfunctions also characterized the suit's early days, ranging from problems with carbon dioxide-scrubbing lithium hydroxide cartridges to clogged sublimators and cycling water pressure regulators to battery failures.

But perhaps the most serious incident occurred on 18 April 1980. Astronaut George 'Pinky' Nelson was at home, working in his garden, when he received a call that an accident had occurred with one of the suits. "They were doing some tests in one of the vacuum chambers and going through the procedures of donning the suit, flipping all the switches in the right order and going through the checklist," Nelson recalled. "There's a point at which you move a slider valve on the front and that pushes a lever inside a regulator and opens up a line that brings the high-pressure emergency oxygen tanks online. You do that just before you go outside. You don't need them when you're in the cabin, because you can always repressurize the airlock, but when you're going outside, you need these high-pressure tanks. It turned out that a technician threw that switch and the suit blew up! Not just pneumatically, but burst into flames, and he was severely burned. It was pure oxygen in there. The backpack is flammable in pure oxygen, it went up in smoke." Chester Vaughan, who sat on NASA's investigation panel following the accident, remembers a debate over whether to publicly release the pictures of the badly charred space suit, "because most people might believe that we had someone inside". Fortunately, the technician did not need to be inside the ensemble on that occasion, simply to reach across and activate switches on its chest panel. The resultant flash fire meant he spent that night in the Houston's Galveston Burns Center.

Fig. 7.5 Astronauts Story Musgrave (left) and Don Peterson participate in the first-ever Shuttle spacewalk in April 1983.

Almost three years later, on 15 November 1982, STS-5 astronauts Bill Lenoir and Joe Allen were preparing for the first Shuttle spacewalk when the gremlins of misfortune hit. As the two men proceeded through standard leak checks on their suits, a problem was noted with Allen's ventilation fan; it sounded, said the crew, "like a motorboat". In essence, it was starting up, running unexpectedly slowly, surging, struggling and finally shutting down. Nor was Lenoir's suit behaving any better. His primary oxygen regulator failed to generate sufficient pressure, regulating to 26.2 kilopascals instead of the necessary 29.6 kilopascals. Some of the helmet-mounted floodlights also failed to work correctly. After fruitless attempts at troubleshooting, the spacewalk was deferred to STS-6. "I guess I was the bad guy," reflected STS-5 commander Vance Brand. "I recommended to the ground that we [cancel] the EVA, because we had a unit in each space suit fail in the same way. It looked like we had a generic failure there. It was the first time out of the ship. We didn't want to get two guys – or even one guy – outside and then have [another failure]. We could've taken a chance and done it, but we didn't. I'm not sure Bill Lenoir was ever happy about that, because he and Joe, of course, wanted to go out and have that first EVA."

As the lead spacewalker, Lenoir was indeed upset at losing the chance to put months of spacewalk training to the test. He wondered if it was possible for just one of them to go outside. "We tried to talk them into it," he remembers, "but it was made more difficult by Joe being inop[erable], because now I'm trying to talk them into not only are we going to go EVA with a suit that you don't really understand, but there's only one person going to be out there, so you don't have a buddy system." In Allen's mind, it was more black-and-white: the suits could have catastrophically failed whilst they were outside in the vacuum of space. And in any case, STS-5 had accomplished its primary mission objective of deploying two communications satellites. Only two other important tasks remained: doing the spacewalk and achieving a safe landing. Allen observed that, out of the two, he would prefer the safe landing.

Following Columbia's safe return home, a task force led by Richard Colonna, manager of NASA's Program Operations Office, was established to investigate the incident. The fault in Lenoir's suit was traced to two missing 'lock-up' devices (each the size of a grain of rice) in his primary oxygen regulator. Their absence permitted a locking ring in the suit to open, which in turn triggered a pressure leak. According to paperwork provided by the suits' manufacturer, Hamilton Standard, the locking devices had been fitted and signed-off by a supervisor in August 1982 but had not been fitted at all and were left unchecked. The responsible employee and the supervisor were barred from further work for NASA. The absence of the locking device allowed the pressure in Lenoir's suit to creep back from 29.6 kilopascals to 26.2 kilopascals. And in Allen's case, the issue lay with a faulty magnetic sensor in the fan electronics. Colonna's final report found that "even with no improvements, if the regulator were fabricated properly, the PLSS [the backpack Portable Life Support System] would function properly." It listed ways to test and inspect future regulators and motors and recommended testing inside the Shuttle's airlock on the day before launch. Plans were also set in motion to provide sensors with better moisture resistance for future motors and new tests to enable defects to be identified at an earlier stage.

A spacewalk was conducted for the first time by STS-6 astronauts Story Musgrave and Don Peterson in April 1983 and, although successful, it turned up a few anomalies of its own. Most significant was a point at which Peterson received a 'High O2 Usage' warning during what NASA described as a "high metabolic period". The message cleared quickly and did not recur and was later attributed to flexure in his suit, coupled with a high work rate. "I was working with a ratchet wrench," Peterson said later. Earlier in the flight, the crew had launched a large communications satellites out of a big collar-like structure at the rear of the payload bay and Peterson's task was to crank it back into place. "We had foot restraints, but it took so long to set them up and move them around that we didn't do that, so I just held on with one hand and cranked the wrench with my other. My legs floated out behind me. As I cranked, the suit's waist-ring rotated back and forth, the seal popped out and the suit leaked badly enough to set off the alarms." Peterson stopped his work and called Musgrave. As circumstances transpired, the astronauts were out of direct contact with Mission Control at the time and by the time communications were restored, they were back in the Shuttle.

Evidently Peterson's alarm had been caused by overwork and excessively rapid breathing, which depleted his oxygen supply, forced an increase in feed-level and triggered the warning. Biomedical data revealed his heart was pounding at 192 beats per minute whilst cranking the wrench. After their return to the airlock, and with data from the suits' computers downloaded to the ground, flight controllers pored over the incident with a measure of dismay. "They were upset about it," said Peterson, adding that he and Musgrave would certainly have been directed to terminate their EVA earlier than planned had Mission Control known about the situation.

Over 28 years, spacewalking proved to be one of the Shuttle's most challenging missions. Astronauts selected to perform EVAs recognized the need to develop the physical strength and necessary stamina to handle the demands of working for eight hours or more in a bulky space suit in the most hostile environment known to humanity. Training on the ground was conducted in a variety of simulators, ranging from virtual-reality computer tools to air-bearing tables to the 'neutral buoyancy' afforded by working in a huge water tank. Before the first Hubble Space Telescope servicing mission in December 1993, spacewalker Kathy Thornton worked out in the gym, but crewmate Jeff Hoffman noted that most tasks did not demand immense physical strength and instead emphasized "technical co-ordination" and the need to be "very careful in how you moved things around and not messing anything up". In an interview with this author in November 2019, Story Musgrave added that the whole point of training was to remove as many of the physical demands of spacewalking as possible.

Fig. 7.6 George 'Pinky' Nelson (right) and James 'Ox' van Hoften work to repair Solar Max during STS-41C in April 1984.

That said, problems permeated almost every EVA. On the first excursion to repair Hubble, Musgrave and Hoffman hit an obstacle when the doors to the telescope's gyroscope compartment refused to close and seal properly. "The doors in the gyro compartment are the biggest doors in the whole telescope," Hoffman explained. "In fact, they're asymmetrical; one of the doors is bigger than the other. We had opened and closed those doors a hundred times in the water. We knew how they worked. There were several latches, but there was one big handle. You turned that handle and that basically closed the latches; then you just had to throw a couple of bolts and tighten up the bolts and the door is secured." Upon closer inspection, it appeared that two door bolts did not reset correctly and Hoffman suspected that the doors had somehow become 'warped', perhaps by uneven heating. If the doors did not close, Hubble would be lost, for its thermal control capability would be gone and light leaking into the telescope's innards would ruin its delicate instruments. Whenever he tried to close the top of the door, the bottom would refuse to close, and its height precluded Hoffman from holding both ends at once. He asked Musgrave to assist him. Unfortunately, Musgrave was tethered and floating freely and could only push the door with one hand, since he needed to steady himself with the other hand. "It was basically a five-handed job," said Hoffman, "and we only had *four* hands. We tried a few times." On one occasion, Musgrave even tried pushing with his helmeted head, to no avail. At length, the spacewalkers recommended to Mission Control the use of the payload restraint device – "kind of a webbing tool, with a ratchet" – to help bring the doors into a position of closure. Lead flight director Milt Heflin agreed that the crew were in the best position to make the decision and gave them the go-ahead. It worked and the doors were successfully closed and latched, but the exhausted astronauts returned inside after almost eight hours, far longer than anticipated.

Other missions, too, encountered their own problems. On STS-41C in April 1984, Pinky Nelson and Ox van Hoften were tasked with retrieving and repairing Solar Max, aided by the jet-propelled Manned Maneuvering Unit (MMU) backpack and crewmate Terry Hart operating the robotic arm. Donning their suits and venturing outside, Nelson later remembered, felt just like the simulator, but what could not have been envisaged was the difficulty in connecting his Trunnion Pin Attachment Device (TPAD) onto the solar observatory in order to recover it. The device simply would not clamp onto Solar Max as their training convinced them it should. "We didn't know what was wrong," Hart reflected, "but, being mechanical engineers, we said 'If a small hammer doesn't work, use a bigger hammer!'" Nelson drove the TPAD faster and harder, still to no avail. As a last resort he grabbed one of the solar panels, but that caused the satellite to tumble. The astronauts returned to the airlock convinced their mission was a failure.

It became apparent that a small grommet, only 20 millimeters high and 6.4 millimeters wide, had caused an obstruction. The grommet helped to hold part of Solar Max's thermal insulation blanket in place. "What no one noticed is that one of the blankets had been put on with a little fiberglass stand-off that the grommets would fit over," said Hart. "The engineering drawings did not specify where those stand-offs ought to be, so when they assembled the satellite the technicians just put one wherever the grommet was. They glued it onto the metal frame, then stuck the blanket on." Problems arose when it was decided, a year after Solar Max's launch, to send the

Shuttle up on a repair mission. But when engineers started to design the TPAD, no one noticed the presence of the grommet. When Nelson had tried to capture it in space, the grommet foiled his every effort. The TPAD clearly would not work. The only alternative was to use the RMS to grapple the satellite. Solar Max was commanded to stabilize itself and the next day the Shuttle returned, the RMS grappled the satellite and it was stowed in the payload bay ready for several days of repair.

This underlined a growing uncertainty over the criticality of understanding the relationship between spacewalkers, their tools and knowing every inch of the hardware they would work on. With work already underway to develop a permanent space station, a huge number of spacewalks (nicknamed 'The Wall of EVA') would be needed to construct such a complex edifice. In the early 1990s, several successful EVA demonstrations were performed to prepare for station construction. One such demonstration was scheduled for STS-80, with astronauts Tammy Jernigan and Tom Jones set to test a variety of tools and methodologies for what lay ahead. "Of all the space station assembly missions coming up, probably more than 80 percent of them" required EVAs, explained Jones. "They're going to depend on these concepts that we think we've gotten right, but we've got to prove." Late on 27 November 1996, a fully suited Jernigan and Jones were sealed inside Columbia's airlock for their first EVA. They had already spent several hours pre-breathing pure oxygen to rid their bodies of nitrogen and their suits showed no evidence of problems.

Earlier that morning, the crew had been awakened to the tune of Robert Palmer's 'Some Guys Have All the Luck'. It would prove to be a bad omen.

Fig. 7.7 Tammy Jernigan and Tom Jones prepare their space suits for STS-80's ill-fated pair of spacewalks.

As the airlock pressure crept close to zero, Jernigan reached out her gloved hand to crank the handle and open the hatch. But it did not move. She tried again. Again, nothing. "Initially, I thought we just had a sticky hatch and the fact that Tammy's initial rotation wasn't able to free it up was just an indication that we'd have to put a little more elbow grease into it," Jones said later. But no matter how hard they tried, the hatch handle came to a stubborn halt after no more than 30 degrees of rotation, rendering it impossible to release a series of latches around its circumference. An engineering team on the ground was quickly convened to investigate the mishap, but it later became apparent that a misalignment of the hatch against the airlock seal had occurred. After the Shuttle returned to Earth several days later, inspections revealed that a tiny screw had somehow worked its way loose from an internal assembly and become lodged inside an actuator. But the incident posed a far greater danger than just a canceled EVA. If a life-or-death emergency were to arise – a stuck payload bay door latch, for example, that might require spacewalkers to intervene to manually close it to ensure a safe re-entry – there simply had to be a way for them to get outside.

Even late in the Shuttle program, the problems with space suits and equipment continued. In August 2007, spacewalker Rick Mastracchio noted a tiny hole in his glove, although there was no evidence of a leak and the astronaut was not placed in any danger. Six months later, a crew medical issue obliged one spacewalker to be replaced with another before a critical EVA to install Europe's Columbus lab onto the International Space Station (ISS). And in July 2009, an issue with a carbon dioxide scrubber inside spacewalker Chris Cassidy's suit forced an EVA to be terminated sooner than intended.

SICKNESS, SMALL SPACES AND TOILET TROUBLES

In fact, it was spacewalking – or rather the lack of it – which brought the issue of space sickness, to the forefront of many minds, early in the Shuttle program. When Columbia launched STS-5, the first mission with commercial primary payloads in November 1982, her crew was also expected to perform the first EVA in the new-specification Shuttle space suits. It would be the first time that an American citizen had opened a spacecraft hatch and floated into the void since the Skylab era, almost a decade before, and STS-5 mission specialists Bill Lenoir and Joe Allen were tasked with evaluating their comfort, dexterity, ease of movement and communications and cooling functionality. They had also trained for many months to practice techniques for the planned repair of NASA's Solar Max observatory, testing fixed and adjustable bolts using a specialized wrench. But the spacewalk was postponed by 24 hours when Lenoir and STS-5 pilot Bob Overmyer were afflicted by space sickness. Akin to motion sickness, this nauseous malaise with its stuffiness of the sinuses and low energy levels is known to affect around half of all astronauts and cosmonauts and typically lasts for no more than a couple of days, but for the five-day STS-5 flight it carried the potential not only to affect half of Columbia's crew, but also a sizeable chunk of their work schedule.

Modern thinking postulates that the influence of the strange microgravity environment on the human vestibular system (the workings of the inner ear) may be a fundamental root cause for the ailment. A sense of disorientation arises when sensations from the eyes and from other sensory organs conflict with those from the vestibular system and with information from a brain which has spent a lifetime in the 'normality' of Earth's gravity. A 'repatterning' of the central memory network occurs over a few days, such that unfamiliar sensations from the eyes and ears begin to be correctly interpreted. Nor was it possible to accurately predict which crew members might be affected, for on STS-5 the hardened U.S. Marine Corps test pilot Overmyer and T-38 jock Lenoir fell victim, but commander Vance Brand and civilian physicist Allen did not. Some years later, Lenoir remarked that Overmyer's suffering began in the form of severe sickness on the second day of the flight. "He filled a couple of bags, but I never stopped giving Bob credit," he said. "Sitting up in the pilot's seat, taking data, doing this, that and the other, he never missed a step. He didn't feel worth a damn. He'd puke his guts out and he'd get back to work…and he felt crappy for two days." Shortly afterwards, it was Lenoir's turn. It felt, he said later, like a low-grade hangover; the malaise enabled him to function and work effectively, but it left a lingering sense of lethargy and a desire to curl into a ball and sleep it off. Delaying the EVA allowed him to 'sack out' for several hours to recover – which allowed his crewmates to raid Lenoir's stash of jalapeño peppers in Columbia's pantry – although as described above it was a hardware malfunction which ultimately put an end to the second and final opportunity to attempt this spacewalk.

Canceling EVAs due to crewmember sickness, though unavoidable, was highly undesirable in the early years when the Shuttle was expected to morph into an operational vehicle. And whilst astronauts completed in-flight medical questionnaires and tried medications in the form of Dexedrine and Scopolamine tablets and, in later years, shots from the intramuscular anti-nausea drug Phenergan, the condition remained unpredictable in terms of who would develop it. In December 1982, NASA assigned physician-astronauts Norm Thagard and Bill Thornton to two Shuttle missions to comprehensively analyze the ailment. "The principal experiments," Thagard explained in a correspondence with this author, "involved eye-tracking changes. Since space adaptation syndrome is thought to be caused by visual-vestibular conflict, looking at changes in eye motion during tracking studies was thought to be important. Also, the auditory portion of the eighth – vestibular – nerve was studied using audio-evoked potentials." However, he added, "no reliable predictors of susceptibility were uncovered by either Bill Thornton's or my work."

Fig. 7.8 Norm Thagard participates in space sickness research during STS-7.

Some astronauts, including Thornton's STS-8 crewmate Dan Brandenstein, experienced sickness during ground-based tests, but in space were able to backflip out of their seats without a single problem. "It certainly makes your mission more enjoyable if you don't have to deal with that, but NASA was trying to decide what made people sick and how to prevent it, and it turned out, after a while, they quit trying and there was no correlation," he said. "Some guys could ride the spinning chair until the motor burned up and didn't get sick and then got into orbit and, within ten minutes, they were as sick as could be." Fellow STS-8 crewman Guy Bluford also suffered no ill-effects – happily wolfing down a sandwich immediately after reaching space – and Kathy Sullivan once joked that weightlessness was simply too much fun to waste time getting sick over. George 'Pinky' Nelson had experienced difficulties in the microgravity research aircraft but felt fine when he reached space.

However, Dale Gardner (tasked with a critical satellite deployment only hours after launch on STS-8) did succumb to the ailment. In Gardner's case, it had little impact on his ability to execute his tasks. Nor had it impacted the efforts of STS-3 crewmen Jack Lousma and Gordon Fullerton, but the very real possibility existed that space sickness would so incapacitate other astronauts that mission success and safety might be impaired. Indeed, some members of the medical community, aware that the Shuttle would fly for only a week or so, were aghast at the possibility that some astronauts might be totally out of action for most of their missions. One such instance came on 6 April 1984, when Challenger reached orbit for a week-long mission to deploy the Long Duration Exposure Facility (LDEF) and repair the Solar Max observatory. In charge of the LDEF deployment was Terry Hart, a former fighter pilot who had never been bothered by stomach-churning aerial maneuvers of any kind. But within minutes of getting to orbit, Hart knew he was in trouble. "For the whole first day, I was pretty much out of it," he remembered. "I just felt awful and I was throwing up…every 30 minutes or so, for a day." He appeared, reluctantly, on camera several times, but the impact was that Hart "barely made it through" his task of activating and checking out the RMS. Medication provided no comfort, although by the second day of the flight – the day that LDEF was to be deployed – he felt well enough to complete the task successfully. "There's a totally different mechanism going on from the Earthbound kinds of motion sickness," Hart said. "Whatever that mechanism is, I had it in spades. It really triggered. But like everyone else, after a day or so, everyone recovers and your body acclimates and everybody does fine."

In April 1985, when Senator Jake Garn reached orbit, he was so affected by the malady that he unwillingly gave his name to a unit of severity. The 'Garn Unit' was created as a measure of nausea in astronauts and some reports indicated that Garn (an experienced jet pilot) was incapacitated for the first several days of the seven-day mission. "He has made a mark in the astronaut corps," remembered fellow payload specialist candidate Bob Stevenson, "because he represents the maximum level of space sickness that anyone can attain…and so the mark of being totally sick and totally incompetent is 'One Garn'. Most guys will get maybe a tenth of a Garn!" It was a humorous take on the condition, to be fair, but on STS-51F in July 1985 a carefully planned week of solar physics observations was thrown into disarray when payload specialist Loren Acton suffered sickness. "I got sick as a dog," he said later. "Thirty seconds after the main engines shut off, I felt like my stomach and my innards were all moving up against my lungs. I was sick for four days and learned very quickly that you cannot unfold your barf-bag as fast as you barf." When one of his crewmates poked his head into the Shuttle's middeck to tell him that a critical solar physics instrument had started to return valuable data, Acton was too ill to even move. In many cases, it was incumbent upon Shuttle commanders to recognize the impact on their crews. "You try to figure out how much extra time you need to add to the schedule, based on how many rookies you've got," said Shuttle commander Steve Oswald. "Some guys are semi-Velcroed to the wall, throwing up, while the folks you least expected to be heroes are just chugging along, executing the plan." And for missions with enormous work schedules and carefully laid timelines, the negative effects of space sickness could be profound. Nor did it hit astronauts alone. In April 1985, a pair of squirrel monkeys flying as part of a life science experiment suffered lethargy, lost appetite and their feces accidentally escaped the confines of their holding pens and ended up floating underneath the Shuttle commander's nose.

Fig. 7.9 Astronaut Jim Dutton illustrates the smallness of the Shuttle's aft flight deck on STS-131. Note the commander's and pilot's seats at the right of this image and the aft flight deck windows at left.

Living and working in the tight quarters of the Shuttle's flight deck and mid-deck offered its own challenges, particularly with large crews. "We had a crew of seven, living inside a volume about the size of a minivan," quipped STS-51G astronaut John Fabian, "and so we were good neighbors for a week!" Admittedly, the 'three-dimensionality' of operating in weightlessness made it somewhat easier, said Shuttle commander Loren Shriver, noting that the entire volume of the cabin – the walls, floor and ceiling – could be occupied with ease. But other difficulties manifested themselves in the form of toilet failures and blocked wastewater lines. On STS-35 in December 1990, the urinal proved totally unusable, leaving crewman Mike Lounge with the unenviable task of stowing waste into sealed plastic bags. "It was stinky," he said later, with more than a hint of understatement. "It was not glamorous spaceflight." By the end of the nine-day mission, as the crew mopped up urine with old socks, the aromas in Columbia's middeck must have been unbearable. Luckily, fluid shifts in weightlessness had impaired the astronauts' olfactory sense, but STS-35 mission specialist Jeff Hoffman pitied the technicians who had to clamber inside the Shuttle after landing to clean everything up.

Another mission in August 1984 suffered from the formation of an icicle on an external wastewater dump line. "Heaters on the exit nozzle are supposed to ensure the fluid separates cleanly," wrote STS-41D astronaut Mike Mullane, "and does not freeze to it." Inspections with the RMS revealed a 30-centimeter icicle of frozen urine protruding into space, which aroused very real concern that it might break away during re-entry and damage the TPS. A similar incident had occurred on an earlier flight. A contingency EVA was considered to remove the icicle, but Mullane's crewmate Steve Hawley was cautious. "There's no translation path down there," he noted later. Using makeshift tools to have one astronaut hang over the payload bay wall to knock the icicle away, in Hawley's mind, seemed like a bad plan. In the end, STS-41D commander Hank Hartsfield was asked to dislodge it using the tip of the robotic arm. It was an exercise fraught with difficulty and danger. "One of the rules in those days was you're not allowed to operate the arm in a place where you can't see what it's doing," said Hawley. "I was supposed to go down to the side hatch window and watch as best I could. [Hartsfield] would move it through a pre-determined trajectory and, if he did it properly, they knew from the ground simulations that it would hit the ice." The attempt worked and the crew, whose mission occurred only a few weeks after the worldwide release of the movie 'Ghostbusters', earned for themselves the moniker 'Icebusters' from their fellow astronauts. The inevitable consequence, though, was that the six-person crew could no longer use the Shuttle's toilet, lest another icicle form. It was fine for solid waste, but the astronauts had to use plastic bags for liquids. Hartsfield saw Judy Resnik's difficulty and insisted that she use the toilet as usual, with the male crew members using the plastic bags.

A week or more with up to seven people living, working, sleeping, eating, urinating and defecating in a volume no larger than a campervan can hardly have been a pleasant experience. "The rinseless soap and rinseless shampoo we use on the Shuttle actually does a super job of getting you clean," remembered astronaut Pierre Thuot. "Each of us took a bath every day in orbit…so I didn't notice any odors or anything when we were in space. But when we landed at Edwards Air Force Base in California, the NASA guys came to open the hatch. They equalize the pressure first, and when they do that, instead of the air just equalizing, it rushes out. It's sucking all this smelly air out of the wet trash area underneath the deck and it got really ripe. It smelled like you were standing in a garbage pile. I'm sure they thought it was us, but it was really the garbage!"

If living conditions could prove tough and times, a chronic lack of sleep also plagued many Shuttle crews. "Joe kinda ran himself out of gas the last day," recalled STS-5's Bill Lenoir of crewmate Joe Allen, "because he wasn't sleeping real well and some combination of that. He needed to get some sleep." Nor was sleeplessness and exhaustion a factor on STS-5. During several long-duration Spacelab flights, astronauts found themselves putting in long hours to complete complex scientific research programs, often with two crew shifts working around the clock. On STS-55 in April 1993, astronaut Jerry Ross was the payload commander, responsible for all of the experiments and he remembered insisting that the rest of the crew got adequate rest. "Even though I wasn't," he quipped later. "I probably didn't get more than five hours sleep a night for the whole time that we were up there and when I got back on the ground, I was just flat wiped out." Dual-shift missions included phonebox-sized sleep stations in the middeck, to afford the off-duty shift some privacy, but to Ross they felt like uncomfortable coffins and any chance to sleep fitfully was lost due to the incessant noise of the duty shift tapping and banging, eating and working. Even Spacelab missions which had single-shift operations fared no better. On the STS-90 Neurolab flight in April 1998, several rodents died and the astronauts' free time was given up to care for them. Payload commander Rick Linnehan averaged three or four hours of sleep per night and started calling their mission the 'Blurrolab' because he could never quite recall what day it was. In many ways, as Bill Lenoir remarked, sleeplessness and space sickness were simply part and parcel of adapting oneself to the unusual weightless environment, which invariably took time. Crews had barely enough time on a comparatively short Shuttle mission to adjust to the new conditions, as brains rewired themselves after several decades of acquiring data from the eyes and ears.

Other problems plagued Shuttle missions in the form of failures of the deployable Ku-band communications antenna and the need to perform Debris Avoidance Maneuvers (DAMs) when the Shuttle flew close to other satellites. In September 1991, Discovery's STS-48 mission hit the news when her crew became the first to

Fig. 7.10 STS-75 astronauts (from left) Franklin Chang-Diaz, Jeff Hoffman and Claude Nicollier eat breakfast in their sleep stations aboard Columbia.

dodge a piece of 'space junk', passing within 1.2 kilometers of Russia's defunct Cosmos 955 spacecraft. A short 'burn' of the Shuttle's thrusters established a wide berth round the old satellite. "We were all busy conducting experiments," said pilot Ken Reightler, "so no one gave this operation much time or attention. I didn't even take a look out the window to see if I could see the intruder go by. I just wanted to make sure we didn't make the news back home!"

8

Hard Road To Wheels Stop

THE PREDATIONS OF MOTHER NATURE

If schedule pressure weighed heavily upon the shoulders of the Shuttle fleet, then the predations of Mother Nature – who routinely manifested her wrath through hurricanes and rain-sodden runways, lightning and high crosswinds, thunderstorms and low cloud ceilings – hamstrung many of NASA's attempts to launch and land the orbiters with equal vigor. Across the 135 flights conducted between April 1981 and July 2011, around 40 percent were directly impacted by unacceptable weather either in Florida or at the Transoceanic Abort Landing (TAL) sites before launch and 30 percent found their launches delayed at least 24 hours as a result. Others were affected more benignly by just a handful of minutes, thanks to an initially threatening weather outlook which eventually gave way to fairer skies. Weather in Florida and California also caused several missions to be brought home earlier than intended or retained in orbit for substantially longer. In November 1994, Atlantis' planned return to the Kennedy Space Center (KSC) was switched to Edwards Air Force Base in response to predictions of gale-force winds, lashing rains and low cloud ceilings associated with Tropical Storm Gordon. And in July 1997, as the STS-94 crew prepared to return to Earth, forecasted rain in Florida dimmed their chances of an on-time landing. Against all the odds, however, conditions abruptly improved just two minutes before the deorbit burn and the Shuttle landed safely. Only weeks prior to the loss of Columbia, in December 2002, Endeavour's return from STS-113 was delayed an astonishing three consecutive days, due to clouds, rain and high winds in Florida. Forty missions between March 1982 and September 2009 were kept in orbit at least one additional day, whilst five others between November 1981 and April 1997 were shortened by at least 24 hours owing to weather or payload-related issues.

© Springer Nature Switzerland AG 2021
B. Evans, *The Space Shuttle: An Experimental Flying Machine*,
Springer Praxis Books, https://doi.org/10.1007/978-3-030-70777-4_8

And this posed an additional headache for the astronauts' families. A red carpet was laid out on the Shuttle Landing Facility (SLF) at KSC in June 1983, with STS-7 expected to be the first mission to return to Florida; unfortunately, concerns about poor visibility forced Challenger and her crew to be waved-off to Edwards instead. In April 1984, after a spectacular satellite repair mission on STS-41C, thunderstorms in Florida forced mission managers again to divert Challenger from KSC to Edwards. For astronaut George 'Pinky' Nelson, who flew STS-41C and later the many-times-delayed STS-61C in January 1986, both missions were originally targeted for KSC landings and ended up returning instead to Edwards. In the pre-Challenger era, the astronauts were personally responsible for getting their families out to the landing site and finding and reserving motel rooms for them. Before STS-61C, Nelson's wife spent three weeks in a Cape Canaveral condo and their young children missed a significant amount of school time awaiting both the launch and landing, although NASA took a greater role in the care of the families in later years.

Fig. 8.1 Lightning strike on the launch pad before STS-134.

Weather worries aside, weight and center-of-gravity constraints of the returning orbiters – some of which carried particularly heavy payloads or brought large satellites such as the Long Duration Exposure Facility (LDEF) back to Earth – also had to be taken into consideration when landing site selections were made. In January 1990, the STS-32 crew deployed one satellite and retrieved the giant

LDEF, returning it home after almost six years in orbit. The presence of LDEF in Columbia's payload bay pushed her landing weight up to 103,400 kilograms. Although she was baselined to land on the 13-kilometer-long dry lakebed of Edwards' Runway 17, NASA managers opted to switch this to 4.5-kilometer-long concrete Runway 22, hopeful that a sturdier surface might afford commander Dan Brandenstein and pilot Jim Wetherbee greater margins of error and improved controllability should problems arise during touchdown and rollout. But even landing on the 'safer' concrete surface presented its own difficulties. With LDEF aboard, the large payload relocated the Shuttle's center-of-gravity slightly 'forward' and necessitated exceptionally deft handling on the part of the pilots to keep Columbia on the runway centerline. To ensure success, Brandenstein and Wetherbee maintained their ship's speed after Main Landing Gear (MLG) touchdown to gently rotate the Nose Landing Gear (NLG) onto the runway.

COUNTING DOWN TO RE-ENTRY

Computer issues conspired against other missions too. Late on 7 December 1983, the six-man crew of STS-9 was readying Columbia for her sixth return home and packing away hardware after a highly productive international science mission. All at once, just five hours before the scheduled landing, as commander John Young and pilot Brewster Shaw configured the five GPCs for re-entry, the gremlins that forever lay hidden in the Shuttle system suddenly showed up. During a firing of the Shuttle's Reaction Control System (RCS) thrusters, one of the five General Purpose Computers (GPCs) failed. The prognosis was not good. The computers – four of which carried the primary 'load' of re-entry software and the fifth, a backup, a hedge against the harrowing possibility that the four primary computers should fail, had an alternate set of software – were essential for monitoring thousands of discrete functions during the hour-long hypersonic descent back to Earth.

But worse was yet to come. Six minutes after the first failure, a second GPC failed. Young and Shaw successfully brought it back online, but their efforts to restart the first one proved fruitless and it had to be powered down. "A ground review of the [second GPC] memory dump indicated that memory alterations had occurred," NASA noted. "However, it was reinitiated and placed back in the redundant set." The mood on Columbia's flight deck was tense. "My knees started shaking," Young said of the first GPC failure. "When the next computer failed, I turned to jelly. Our eyes opened a lot wider than they were before." At length, the computer previously restored by Young and Shaw (plus two others in the primary set) were used to guide the Shuttle through a safe re-entry. However, even more trouble was afoot. Later that same morning, an Inertial Measurement Unit (IMU) failed. Re-entry was postponed to allow ground specialists to analyze the issue.

No crew activities were scheduled, except that the six men were kept in a state of preparedness for re-entry.

After nearly eight hours, the de-orbit burn was performed to begin the irreversible descent to Earth. Overall, Columbia's performance was nominal...until about four minutes prior to touchdown on the concrete Runway 22 at Edwards, when the temperature of one of the three Auxiliary Power Units (APUs) began to rise sharply. "Then we had another lesson," Shaw said later. "Never let them change the software in the flight control system without having adequate opportunity to train with it. There were 'gains' in the flight control system and [these] changed depending upon what phase of flight you're in. When you're flying a final [approach], there are certain gains that make the vehicle respond a certain way to the inputs the pilot makes on the stick. When the main gear touch down, the gains change and the gains are set up so you can de-rotate the vehicle and get the nose on the ground in an appropriate way. We had done all our training in a simulator with a certain set of gains...and then they changed the flight software and these gains, so that when it came time for John to land the vehicle in real flight, the gains were different than he'd done all his training on. Certainly, when John started to de-rotate the vehicle, it responded differently than he had trained on." With Shaw calling airspeeds and altitudes, Young brought Columbia's MLG gently onto the runway and began the process of de-rotating the NLG down. Everyone on the middeck started cheering. "We were down to about 150 knots," continued Shaw, "when the nose hits the ground and it goes smash!" The changed gains in the flight software and the 'different' response of the vehicle presented them with a hard, but safe, landing.

It subsequently became clear that the first GPC failure was caused by a sliver of solder, which had somehow dislodged itself during an RCS thruster firing and then shorted out the computer. The second failed GPC, which Young and Shaw had earlier managed to revive, went on to fail again, only seconds after landing. And six and a half minutes after that, one of the APUs shut itself down, as did a second just four minutes later. Little did the crew know at the time, but one of the APUs had caught fire (due to a hydrazine leak) as the Shuttle raced down the runway. "The reason the first one shut down," said Shaw, "was [because] it was on fire and the fuel wasn't getting to the catalyst bed, so it 'undersped' and automatically shut itself down." In response to that situation, Shaw – who, as Columbia's pilot, was responsible for the APUs – reconfigured systems to prevent the third APU from shutting down. "The next one [APU-3] didn't shut down until we shut it down," he said. "We had a generic failure of a little tube of metal where the fuel went through and was injected into the catalyst bed and it cracked. When we shut [the APUs] down and shut the ammonia off to them, the fires went out. We had some damage back there, but the fires stopped. We didn't know anything about that until the next day!" Two APU failures during one landing, said former Shuttle director Arnold Aldrich, was "about as close to a very serious incident as anything I can imagine".

Fig. 8.2 STS-83 commander Jim Halsell works the Auxiliary Power Unit (APU) issues which plagued his mission in April 1997. Columbia was forced to return to Earth less than four days into a planned 16-day voyage.

Issues with the three APUs plagued other missions, keeping crews on their toes not just during re-entry and landing, but also during the countdown and launch. These hydrazine-fed, turbine-driven devices provided the necessary muscle to position actuators, move aerosurfaces, retract liquid oxygen and hydrogen disconnect umbilicals, deploy the landing gear and provide steering on the runway; to put it mildly, therefore, they were an essential element of the Shuttle system. They were activated by the pilot at T-5 minutes by no fewer than 15 switch-throws on the center instrument panel. Columbia's second mission, STS-2 in November 1981, had been scrubbed because lube oil outlet pressures were too high on two of the APUs. Their pressures sat at around 690 kilopascals, far above the maximum permissible 414 kilopascals. Subsequent inspections found that their oil filters had become clogged by pentaerythritol, a crystal formed when hydrazine penetrated their gearboxes. Both gearboxes were flushed and the filters were replaced. In March 1982, STS-3 had one of its APUs shut down during ascent. According to Terry Hart, the Capcom in Mission Control, this was "the first time we had a major failure on launch". In the years preceding STS-51L the units revealed worrisome cracks in their turbine housings and other maladies.

After the loss of Challenger and the extended down time which followed, an Improved APU (IAPU) was developed which increased their operating lifetime to

75 hours – three times that of its predecessor, roughly translatable to 50 Shuttle missions – but their nature as a mechanical device and their consumption of hazardous hydrazine left them increasingly vulnerable to fires, explosions and leaks. An APU problem was experienced during the STS-34 ascent in October 1989 and the launch of the Hubble Space Telescope the following year was scrubbed owing to abnormal pressures and turbine speeds in an APU. The first IAPU flew on STS-45 in March 1992 and an improved controller for the unit was introduced from STS-57 in June 1993, but the problems continued to materialize from time to time. Severe leakage from a fuel pump seal was noticed on STS-78 in June 1996, and three months later on STS-79 one of Atlantis' units shut down and her crew was required to fly a highly undesirable two-APU re-entry and landing profile. In STS-79's case, the failure prompted mission managers to draconically tighten the weather criteria for landing: crosswinds could be no higher than 18.5 kilometers per hour, cloud ceilings were revised and clear visibility of at least 11 kilometers was required. Even following the loss of Columbia, APU problems cropped up from time to time. In July 2006, a tiny leak was detected during flight control system checks prior to Discovery's STS-121 re-entry, but it proved inconsequential to overall mission success.

THE GOOD...

Although thousands of people saw the Space Shuttle launch and land over its three-decade operational career, it is an everlasting regret that the experience evaded this author. But in June 1983, during a school sports day in Birmingham, England, came a most unusual sight which continues to resonate. As fellow six-year-olds stumbled across the school playing field, heads were abruptly jerked upwards by the roar of a low-flying aircraft. Yet this was no ordinary aircraft. Dozens of pairs of eyes, including my own, were momentarily captivated by the sudden appearance of the world's largest biplane. It was Enterprise – the 'hangar queen' of NASA's orbiter fleet and alone among the sisters in never actually making it to space – secured atop a Boeing 747 Shuttle Carrier Aircraft (SCA). Glinting in the late afternoon sunshine of a glorious midsummer's day, this astonishing machine swept over a multitude of young heads, midway through her tour of Germany, Italy, the United Kingdom and Canada, before heading home to the United States. Enterprise had already stolen 1983's Paris Air Show.

Four decades later, she sits in pride of place in the Steven F. Udvar-Hazy Center at the National Air and Space Museum (NASM), near Dulles International Airport in Washington, D.C. And whilst her career was distinctly overshadowed by those of her spacefaring sisters – Columbia, Challenger, Discovery, Atlantis and Endeavour – she was, and will forever be, a trailblazing icon. For during the

Shuttle's genesis in the 1970s, there existed great uncertainty about how human pilots could control this delta-winged spacecraft as it knifed its way back through the sensible atmosphere at hypersonic velocities of 28,200 kilometers per hour, fell to Earth with all the aerodynamic grace of a brick and touched down like an enormous glider on a runway at just 350 kilometers per hour. As if that level of complexity did not weigh heavily enough, Shuttle pilots also lacked the luxury of airliner pilots in that they could not circle the runway to make another approach. They had to land this infinitely complex machine perfectly and on the very first attempt.

And even highly trained and experienced Shuttle commanders and pilots were surprised from time to time by the extreme dynamism of this phase of atmospheric flight. During the early stages of the STS-57 re-entry in July 1993, whilst traveling at Mach 21 – more than 25,000 kilometers per hour – a sudden *wham* echoed through the airframe. Seated at the back of Endeavour's tiny cockpit, flight engineer Nancy Sherlock looked up in surprise, but noticed commander Ron Grabe and pilot Brian Duffy seemed unperturbed and said nothing. Shortly afterwards, it happened a second time and Grabe turned to his pilot.

"Have you ever felt that before?"

"No," replied Duffy.

Fig. 8.3 Enterprise separates from the Boeing 747 Shuttle Carrier Aircraft (SCA) over Edwards Air Force Base in California.

Both astronauts were veterans, but the noise remained anomalous to them. It later became apparent that Endeavour had been shaken by a 'density shear': a difference in air masses at extremely high altitude, which the spacecraft had transitioned at hypersonic speeds. "Those things extend up at a high altitude and when you're going very fast and you change from one to the other, you get an instantaneous notification that you're somewhere else," Duffy said later. "It's like somebody really picks up and shakes the whole orbiter. The whole vehicle rings because it's not very sturdy. It *looks* sturdy, but it's about as stiff as a Twinky and, truth be known, during ascent, you can actually feel it flexing."

In view of the problems suffered by the Shuttle during those troubled early years, it is unsurprising that NASA's wish to launch the first orbital mission, STS-1, by 1977 faded to naught. When eight astronauts were announced in March 1978 to fly the four Orbital Flight Tests (OFTs), the long-awaited maiden voyage was anticipated no sooner than the following spring. By September of that year, it had slipped to the right and was still at least 12 months away. And there remained risks of added delays if further Congressional funding was not forthcoming; the Shuttle by this stage having consumed 8 percent above its original $5.15 billion budgetary projection. Despite vigorous support from the U.S. Air Force, the program still had powerful political enemies who fiercely questioned the need for a reusable manned launcher which, on the face of it, looked to be doing the same missions that expendable rockets were already doing.

It may seem odd, therefore, but in 1977 the Shuttle did, in fact, fly, albeit not into space, but over the California desert at Edwards Air Force Base for a series of Approach and Landing Tests (ALTs) to evaluate its handling and approach characteristics in the real aerodynamic conditions of the low atmosphere. Under the language of the original Shuttle contract, two vehicles were required: 'Constitution' (Orbiter Vehicle-101) and 'Columbia' (Orbiter Vehicle-102). Constitution – later renamed 'Enterprise', following a massive letter-writing campaign, principally by Star Trek enthusiasts – would perform the ALT flights, after which it would be refurbished for space missions. In February 1976, astronauts Fred Haise, Gordon Fullerton, Joe Engle and Dick Truly were named to fly the ALT missions. In addition to a pair of ground-based simulators, one other crucial training device was Grumman's twin-engine Gulfstream II aircraft, two of which had been specially modified as Shuttle Training Aircraft (STAs) and delivered to NASA's Johnson Space Center (JSC) in Houston, Texas, in June and September 1976. The STAs mimicked the subsonic flight characteristics of the Shuttle from an altitude below 10,600 meters down to the runway, with identical cockpit instrumentation, visual cues and handling qualities.

"The [orbiter] has a flight control system that is one of the early fly-by-wire systems, which means there is no physical linkage between the stick and the controls," remembered Shuttle commander Steve Nagel. "It's all electronic commands through a computer that then tells the flight control surface or the reaction control jets to do whatever it needs to do to control the airplane. It's not that you can't learn to fly it; it just takes a while to get used to the fact it has no engines and is a poor glider, so it's coming down at a real steep angle. The handling qualities and responses you get out of the Shuttle Training Aircraft are very close to what the real orbiter is. If anything, the real Shuttle is a little nicer; a little bit more responsive."

Added Shuttle commander and former NASA deputy administrator Fred Gregory: "They are flown using the same profile, the same speed, the same sensation of very high sink-rates, with a flare about a mile from the end of the runway." These close parallels to the idiosyncrasies of the real Shuttle were achieved through independent control of six degrees of freedom, using deflected flaps to increase and decrease lift, a fully deployed MLG and retracted NLG to set up the proper wing-load requirements and the use of reverse thrust whilst airborne. The STA's computer was programmed to have the precise flying characteristics of the fully laden Shuttle (including its weight and center-of-gravity constraints) to enable pilots to carefully simulate a return from space. Typically, an instructor sat next to an astronaut and flew the STA to cruising altitude with standard controls. At this stage, the astronaut took over the flight with a Shuttle-specific set of controls on his or her side of the cockpit, lowering the nose to increase speed and achieve a 20-degree angle-of-attack on the Outer Glide Slope, before executing the 'flare' maneuver to reduce the descent angle and transition to the Inner Glide Slope. Due to the anticipated height of the flight deck of the 'real' Shuttle above the runway, a green light would illuminate on the instrument panel at an altitude of 10 meters to simulate the instant of touchdown. At this point, the instructor pilot would de-select the simulation mode, stow the thrust reversers and return the STA to cruising altitude for another attempt. Before each mission, Shuttle commanders and pilots would perform hundreds of these practice approaches. "We had participated, in my particular case, in 500-700 landing approaches," said Gregory of his first mission as a pilot. Commanders flew perhaps double that number. "It's an excellent simulator, much better than the moving-base simulators we have, because you're moving," remembered Challenger's first commander, Paul Weitz. "You're not trying to 'fake' yourself out by emulating or simulating accelerations by just moving the fixed-base [simulator] up and down and around."

Another useful tool (and a strict requirement for Shuttle astronauts to remain on 'active' flight status) was maintaining proficiency in NASA's fleet of T-38 Talon jets. The ability of this sleek, supersonic dart to mirror the handling qualities of the orbiter was problematic, however, since its lift-to-drag ratio is quite dissimilar. To best simulate the Shuttle's steep-angled approach to the runway, astronauts typically opened the T-38's speed brakes as wide as possible, then deployed the landing gear at the very onset of descent. "I built a gadget to work on the T-38 that would allow you with any given weight to set the power with the speed brakes 'down' to simulate what the data said the orbiter would fly it at," said Fullerton, "so that we

Fig. 8.4 Enterprise heads for the runway at Edwards on her fourth free flight. Note the absence of the Shuttle's aerodynamic tail cone.

could go fly the pattern we intended to fly in T-38s, making steep descents, flaring and touching down." But aside from simulating Shuttle landings, these jets offered astronauts a chance to keep their situational-awareness and cockpit-management skills sharp. "There was an area, just outside Houston, over the Gulf of Mexico, where we could go out and do what we called 'turn-and-burn', which is to do aerobatics and loops and rolls and chase around clouds," recalled Shuttle commander and former chief astronaut Dan Brandenstein. "All the time, that's a way of maintaining your piloting skills. Obviously, it's a kick for people that had flown thousands of hours, but for somebody who had never flown before or had very little experience, it was a real kick, because you could go supersonic, pulling 7 G."

In readiness for its role in preparing for the first Shuttle landings, Enterprise was rolled out of prime contractor Rockwell International's Palmdale facility in California in September 1976 and early the following year was towed overland to Edwards for ALT. She was then hoisted atop the Boeing 747 SCA for three 'taxi-tests' on the runway. The 747 came from American Airlines and had been purchased by NASA in June 1974, partly because it was the largest available aircraft to carry the Shuttle. It was principally used to fly the orbiters across the United

States from Edwards to the Kennedy Space Center (KSC) in Florida, or during trips back and forth to Palmdale for maintenance and refurbishment. But before it could take on this new role, the old airliner had to go on a substantial diet – its passenger seats were removed, as were its galleys – and it received upgraded air-conditioning ducts, electrical wiring and plumbing and modifications to its four engines. Its upper fuselage was beefed up with more internal buttresses and, following wind-tunnel testing, vertical stabilizers were added to its horizontal tail. It flew a 90-minute test flight with a joint NASA-Boeing crew near Seattle in Washington State in December 1976, which cleared a final hurdle before ALT. All three taxi-tests occurred in the early morning hours, to reduce the heat on the SCA's brakes and tires, which – in addition to carrying the 180,000-kilogram load of the aircraft itself – also had to accommodate the 68,000-kilogram Enterprise on top. The combo's speed along the runway was steadily increased from 143 kilometers per hour to 250 kilometers per hour.

The success of these tests laid the groundwork for the first 'captive-inert' flights carrying Enterprise. Over a two-week period starting on 18 February 1977, the SCA crew of pilots Fitz Fulton and Tom McMurtry and flight engineers Vic Horton and Louis 'Skip' Guidry made five airborne runs with the unmanned Shuttle on top to assess their structural integrity and handling behavior in real flight conditions. Following the final captive-inert test, plans called for three 'captive-active' flights with astronauts aboard (in which Enterprise remained attached to the SCA) and up to eight 'free' flights (in which the Shuttle physically separated from the 747 to glide to an unpowered, 'deadstick' landing at Edwards). The goal, according to Engle, "was to place the vehicle in aerodynamic flight by itself and exercise all of the systems that we could: hydraulic systems, electronic systems, flight control systems and landing gear, in a real flight environment, and to gather as much flight test information as far as stability and control parameters and performance parameters and do it partly in an ideal environment".

After a week or more in the weightlessness of space, Shuttle commanders and pilots would have their hands full with flying the vehicle through a succession of highly dynamic flight regimes – hypersonic, supersonic transonic and subsonic – as it transitioned from an orbiting spacecraft into an oversized, heavyweight glider, without needing the additional worry of how to deal with an unexpected cross-wind during the final approach or a low cloud ceiling hanging over the runway. By Engle's admission, the approach, flare and landing was "a very small part of its mission, but a very critical part of its mission". Hence it was essential to gain confidence in knowing how to land the Shuttle before STS-1. Haise and Fullerton flew the first captive-active flight on 18 June 1977 and briefly assessed Enterprise's air surfaces, rudder and speed brake, and ten days later Engle and Truly made low-speed tests of her flight control system. On 26 July, Haise and Fullerton completed a thorough check of her avionics ahead of the free flights. They also tested the deployment of Enterprise's landing gear, whilst atop the SCA.

Fig. 8.5 STS-132 commander Ken Ham (left) and an instructor fly the Shuttle Training Aircraft (STA) over White Sands in New Mexico. Note Ham's cockpit displays closely mirror the Shuttle's flight deck instrumentation, whilst the instructor possesses standard aircraft controls.

On 12 August, more than 65,000 people gathered at Edwards to see Haise and Fullerton perform the first free flight. The SCA and Enterprise went airborne from concrete Runway 22 at 8:00 a.m. PDT and swept into the steadily brightening desert sky, accompanied by five T-38 jets. Forty-eight minutes later, Fitz Fulton nosed the SCA into a dive and Haise pushed the separation button on the Shuttle's instrument panel. Seven explosive bolts fired and 'popped' Enterprise away. But her first moments in flight almost immediately triggered a caution-and-warning alarm. Right after separation, Haise placed his ship into a right-hand turn and pitched upwards to open the separation from the SCA, as Fullerton worked to remove circuit breakers and reset switches because the shock of the separation had dislodged a tiny ball of solder and a transistor in one of the Shuttle's General Purpose Computers (GPCs) and triggered the alarm and red lights. "All your control of the airplane is through fly-by-wire and these computers," Fullerton said later. "I had a cue-card with a procedure if that happened, that we'd practiced in the simulator, and I had to turn around and pull some circuit breakers and throw a couple of switches to reduce susceptibility to the next failure. By the time I looked around, I realized, hey, this is flying pretty good because I was really distracted from the fundamental evaluation of the airplane at first."

As the Shuttle flew smoothly on its first independent flight, Haise banked 20 degrees to the right and headed for Edwards' dry lakebed Runway 17. Maintaining a nose-down attitude, he performed a pair of 90-degree turns and opened the speed brake. Mission Control erroneously advised Haise that he had a lower lift-to-drag ratio than predicted in wind-tunnel tests, to which the astronaut responded by flying his final approach at higher speed, thus conserving energy to extend the glide. In fact, Enterprise's lift-to-drag ratio was right on the money. Realizing that the Shuttle was actually 'high and hot', and that he would therefore land 'long', Haise opened the speed brake from 30 to 50 percent to lose energy and initiated the flare at a height of 275 meters. As the vehicle leveled-out, he deployed the landing gear and touched down 900 meters beyond the intended point at 340 kilometers per hour.

The Shuttle's first experimental test flight under her own steam had lasted 5.5 minutes. One unusual visual aspect was the shortness of the NLG, which caused the wings to dip noticeably downwards in a negative 4.5-degree angle-of-attack when all six wheels were on the runway. This placed extremely high pressures on the gear, to such an extent that on operational missions MLG tires could be used only once and NLG tires no more than twice. To Haise, it was "almost funny" as he 'de-rotated' the nose. "For a little bit, you almost think you don't have a landing gear, because it goes down so far!"

During subsequent free flights, the astronauts evaluated the autoland capability. In this test, Haise and Fullerton allowed Enterprise's computers to guide them down to the 270-meter flare point, then took over and hard-braked their ship to a smooth stop.

Original plans called for four flights with an aerodynamic tail-cone fitted over three dummy Space Shuttle Main Engines (SSMEs) and bulbous Orbital Maneuvering System (OMS) pods in Enterprise's aft fuselage. This would be followed by two final flights with the engines and OMS pods exposed, to assess the aerodynamic conditions a Shuttle might experience during approach and landing. However, NASA felt that the first three free flights were so successful that the fourth landing was done without the tail cone. In Engle's mind, this was the right thing to do. "The orbiter flew pretty benignly with the tail cone on a relatively shallow glide-slope," he told the NASA oral historian. "But that was not the configuration that we needed to really have confidence in, to commit for an orbital launch, because that re-entry and landing would be made with the engines exposed and required a much steeper glideslope, much more demanding profile, much more condensed time period from flare to touchdown, because the airspeed would bleed off much faster with the additional drag. With the tail cone on, from a performance standpoint and a piloting task standpoint, we really didn't have what we needed until we flew it tail-cone-off."

Enterprise's last ALT free flight, on 26 October 1977, saw the Shuttle perform better than predicted, although several issues did arise during final approach. As he emerged from the pre-flare maneuver, Haise noticed that he was descending much faster than planned. He opened the speed brake early, but Enterprise's speed increased. He deployed the landing gear and pitched the nose downwards to attain the desired touchdown point. Haise struggled as the wings dipped and the MLG hit the concrete 'hard', left the ground and finally touched down. The remainder of the rollout proceeded without incident.

The problem was traced to a 270-millisecond 'time-lag' in the flight control software. The delay between Haise's inputs on the stick and Enterprise's response prompted him to over-correct, resulting in 'pilot-induced oscillation', a porpoise-like nodding motion and a bouncy landing. It provided important data for engineers monitoring the performance of the landing gear. The overall ALT campaign yielded invaluable data on the autoland capability, Enterprise's airworthiness at subsonic speeds and thoroughly wrung out her systems. In Fullerton's mind, his training for a real mission would have been much harder if he had not flown Enterprise. As Haise explained, "It handled better, in a piloting sense, than we had seen in any simulation – either our mission simulators or the Shuttle Training Aircraft. It was tighter; crisper in terms of control inputs and selecting a new attitude in any axis and being able to hold that attitude. It was a better-handling vehicle than we'd seen in the simulation."

Fig. 8.6 Mission Control breathes a sigh of relief on 14 April 1981, as Columbia returns safely from the first Space Shuttle mission.

More than three years later, on 14 April 1981, STS-1 crewmen John Young and Bob Crippen were ready to bring Columbia back through the atmosphere after two days in space. Although the ALT series had successfully rehearsed the last few minutes prior to landing (at subsonic speeds), the 45-minute period from the de-orbit 'burn' of the OMS engines through the searing furnace of re-entry and a complicated sequence of S-shaped aerodynamic turns to 'bleed' away the immense velocity and prepare for a runway landing remained an unknown quantity.

To play the situation safe, NASA opted to use the wide expanse of dry lakebed at Edwards to land the four OFT missions. This would afford Young and Crippen a more forgiving runway environment and greater margins for error, although it was expected that when the Shuttle's aerodynamic characteristics were better understood and it became fully 'operational', precision landings on the narrow concrete runway of the Shuttle Landing Facility (SLF) at KSC would become the norm. Returning to Florida (notwithstanding its changeable weather) was expected to significantly reduce 'turnaround' times on the orbiter by perhaps a week or more and save NASA around a million dollars in SCA transportation and other costs.

Twenty minutes before the de-orbit burn, Young and Crippen oriented Columbia into a tail-first attitude and switched on two of the three Auxiliary Power Units (APUs) to control the flight surfaces and hydraulics throughout re-entry. High above the Indian Ocean, the OMS engines ignited in the vacuum, slowing Columbia sufficiently to begin her perilous glide to a landing strip on the other side of the planet. The 2.5-minute burn was reported with typical coolness by Young: "Burn went nominal."

"Nice and easy does it, John," replied Capcom Joe Allen from Mission Control. "We are all riding with you."

Minutes later, Columbia was turned around and her nose pitched 'upwards' at a 39-degree angle. Young and Crippen removed the safety pins from their ejection seats and the overhead escape hatches, then switched on the third APU. Traveling at close to 25,750 kilometers per hour (Mach 20.8), they hurtled on as the color of ionized atmospheric gases outside morphed from a pale pink into a deeper pinkish-red, then a reddish orange, creating a scene which would not have been out of place in Dante's 'Inferno'.

As a tense world waited, the NASA public affairs commentator reeled off a steady stream of updates. "We will be out of communication with Columbia for approximately 21 minutes." He was referring to the longer-than-normal period of radio blackout, caused by the accumulated plasma 'sheath' around the spacecraft. "No tracking stations before the West Coast…and there is a period of about 16 minutes of aerodynamic re-entry heating that communications are impossible…" Descending lower, the astronauts were, at length, able to receive Ultra-High Frequency (UHF) radio calls, crackling between Mission Control and one of the T-38 jets which would accompany the Shuttle down to the runway. Shortly after the orbiter crossed the California coastline, near Big Sur, Young took manual control of his ship. Long-range tracking cameras on Anderson Peak captured the first ground-based images of Columbia, flying at an altitude of more than 35 kilometers.

"What a way to come to California!" exulted Crippen.

Still traveling at well over four times the speed of sound, the Shuttle passed over Bakersfield, Lake Isabella and Mojave Airport, enabling the astronauts to verify by glancing through their windows that their ground track was "right on the money". Young then executed a sweeping, 225-degree turn to align his ship with the lakebed Runway 23 at Edwards. Dropping below 12 kilometers, he took Columbia's stick. Control was crisp and precise. Watching the arrival of America's first Space Shuttle from orbit were tens of thousands of people, including Larry Eichel of the *Philadelphia Inquirer*. His testimony perfectly encapsulated the anxiety and nervous excitement of everybody awaiting this historic event. "The Shuttle appeared far above the north-east horizon, a white dot against a cloudless blue sky," he explained. "That dot was dropping so fast that to an eye accustomed to watching the more gradual descent of commercial jets, it seemed inevitable that the Shuttle would crash to the desert floor." As Columbia drew closer, her speed brake was gradually retracted and was fully closed by the time the vehicle was 600 meters above the runway. Falling precipitously at a rate seven times steeper than a commercial airliner, and almost twice the speed, the reaction of Eichel that a crash was about to occur can, perhaps, be forgiven. It was at this point, however, that Young pulled back on the stick, lifted the nose and transformed his ship, in a split second, from a falling brick into a graceful flying machine.

Weather conditions in the California desert were near-perfect and surface winds were calm. As clocks ticked past 10:20 a.m. PDT, Crippen deployed the landing gear and all six wheels were down and locked into position within the mandatory ten-second time limit. Columbia touched down 22 seconds later, at a speed of 342 kilometers per hour, and rolled for almost 3,000 meters before coming to a smooth halt. The speed brake was opened and full-down elevons were applied, giving the astronauts an impression of considerable deceleration. "As it touched down," recalled Eichel, "the rear wheels nestled into the hard-packed sand, kicking a rooster-tail high into the air." The countdown to landing was echoed by both the public affairs spokesmen at Edwards and by the crew of one of the T-38s, who were first to welcome Young and Crippen back home with a resounding "Beautiful! Beautiful!"

Beautiful, indeed, but far from easy. In fact, many astronauts described the Shuttle as the most difficult 'aircraft' they ever had to fly. "It's not nearly as easy to fly as a big air transport, like a Boeing 707 or 757, and certainly a lot more difficult to fly than a little NASA T-38," said astronaut John Fabian. "You've got to stay on top of it all the time. You've got to be thinking well ahead of the vehicle, so this is not just a flying job for…the guy who really knows how to maneuver the airplane. This is a machine that is flown by people who are of great intellect as well as great skill." Post-landing analysis revealed that Columbia's right-hand inboard brakes suffered higher-than-anticipated pressure, which caused a slight tug to the right, just before the wheels stopped. Young compensated for this by

Fig. 8.7   Columbia touches down at Edwards on 14 April 1981.

balancing the total braking to either side of the Shuttle, maintaining a near-perfect course straight down the runway centerline, stopping at the intersection of Runways 23 and 15. One notable surprise was the sheer amount of lakebed debris – pebbles and grains of sand – kicked up by the wheels.

"Do I have to take it to the hangar, Joe?" asked Young.

"We're gonna dust it off first," retorted Allen with a chuckle.

Immediately after wheel stop, the astronauts unstrapped and began safing the Reaction Control System (RCS) and OMS switches before the arrival of the ground crew. When the latter arrived, they hooked up sensitive 'sniffer' devices to verify the absence of toxic or explosive gases and attached coolant and purging lines to Columbia's aft compartment to air-condition her systems and payload bay and dissipate residual fumes. Whilst this procedure was underway, the ground teams worked in protective suits, then moved an airport-type stairway over to the hatch. Young had remained totally cool throughout re-entry, but now allowed his excitement get the better of him. As soon as he got outside, about an hour after touchdown, he bounded down the steps, checked out the tires and landing gear and jabbed the air triumphantly with both fists. He even kicked the tires…which scared the life out of NASA's director of engineering, Henry Pohl. "I was really worried about that, because those tires have got 375 psi pressure in them…and I knew the brakes got hot," he explained. "I was afraid the tires were going to explode. It would have been a shame to do all that flying and a terrific landing and then have a tire blow up because you went over and kicked it!"

Young, of course, could be forgiven. He was over-excited and it showed. "I've often claimed that John calmed down" by the time he got outside, Crippen said later, but noted with a twinkle in his eye: "You should've seen him when he was inside the cockpit!"

THE NOT SO GOOD…

Seven months after Young and Crippen's perfect landing, old ALT buddies Joe Engle and Dick Truly flew Columbia into orbit for a second time as STS-2. In doing so, they became the first team of astronauts to ride a 'used' spacecraft. But whereas STS-1 had been a two-day test flight, STS-2 was scheduled to last five days and transport a fully-fledged scientific research platform – OSTA-1, provided by NASA's Office of Space and Terrestrial Applications – into orbit, as well as the first test of Canada's Remote Manipulator System (RMS), a 15.2-meter robotic arm that was under consideration for the deployment and retrieval of payloads and the construction of a future space station. OSTA-1 was located on a pallet in Columbia's payload bay and comprised seven experiments, visibly dominated by the giant Shuttle Imaging Radar (SIR), to assess the capability of the Shuttle to serve as a research laboratory. SIR was a side-looking 'synthetic-aperture radar' tasked with examining geological features for mineral and petroleum exploration. Other OSTA-1 experiments would make spectral observations of rocks and minerals, classify surface features such as water, vegetation, bare land and snow and clouds and ice and measure air pollution in the lower atmosphere, whilst two more in Columbia's crew cabin would examine lightning flashes and the growth of dwarf sunflowers. It promised a jam-packed five days for Engle and Truly, but since STS-2 was only the second OFT mission, many of the critical payload objectives were helpfully 'front-loaded' into the timeline just in case of problems.

And it did not take long for those problems to show up.

After many delays on the ground, Columbia launched on 12 November 1981, Truly's 44th birthday. This made him the first U.S. astronaut to celebrate a birthday with a launch. The Shuttle was inserted perfectly into orbit, and OSTA-1 was activated and began taking its first data. But just two hours into the flight, Mission Control noticed a high pH indication on one of Columbia's three fuel cells. Tasked with generating electrical power using cryogenically-stored liquid oxygen and hydrogen reactants, each cell was connected to independent electrical buses and located beneath the floor of the forward payload bay. During 'peak' and 'average' power loads, all three cells were operated in tandem, with only two used (and the third kept in standby mode) during periods of minimal power needs. Each cell comprised a 'power section' and a monitoring 'accessory section'. In the power section, where the oxygen and hydrogen were transformed into electrical power, water and heat, 96 cells were contained in three 'sub-stacks'. Manifolds which ran along the length of those sub-stacks were responsible for the distribution of oxygen, hydrogen and coolant.

It was the No. 1 fuel cell which began to exhibit trouble on the afternoon of 12 November 1981. Although its overall performance remained normal, the situation rapidly deteriorated and within a few hours it revealed a sharp voltage drop, indicative of the probable failure of one or more of its sub-stacks. If that was indeed the case, the cell's ability to generate electricity for Columbia and drinking water for Engle and Truly might be seriously compromised. With the corresponding likelihood of a contaminated water supply, the No. 1 cell was deactivated. But there existed another risk that its water was being 'electrolyzed', thereby forming a potentially explosive mixture, and Mission Control depressurized the cell as an added precaution. Under highly conservative flight rules, laid down long before STS-1, all three fuel cells had to be fully functional for a mission to continue. It soon became apparent that STS-2 would be returning to Earth much sooner than intended.

Early the following morning, Capcom Sally Ride told the astronauts the disappointing news that their landing had been brought forward to 14 November. A carefully choreographed five-day mission now required to be condensed into only two days. "That's not so good," was all a dejected Truly could say. But remarkably STS-2 would return home with almost 90 percent of its scheduled tasks successfully ticked off the check list. Years later, Engle remarked that their training on the ground enabled them to rapidly replan their mission. "We had trained enough to know precisely what had to be done and we prioritized things as much as we could," he said. "We only had the ground stations, so we didn't have continuous voice communication with Mission Control and [they] didn't have continuous data downlink from the vehicle either, only when we'd fly over the ground stations."

This allowed the astronauts to burn the midnight oil in terms of getting through their tasks. "When our sleep cycle was approaching, we did, in fact, power down some of the systems and we did tell Mission Control goodnight," Engle said. "As soon as we went Loss of Signal (LOS) from the ground station, *then* we got busy and scrambled and cranked up the [RMS] and ran through the sequence of tests for the arm, ran through as much of the other data that we could, got as much done as we could during the night." Neither man slept on 13 November, their last night in space. When Mission Control sent the customary wake-up call in the morning, they pretended they were still sleepy.

Fig. 8.8 Engle and Truly put the Canadian-built Remote Manipulator System (RMS) mechanical arm through its paces during the truncated two days of STS-2.

But not everyone was fooled. After STS-2 landed, flight director Don Puddy pulled Engle aside. Mission Control knew that both men were awake through the night through their data. "We could see," he pointed out with a wizened grin, "that you were drawing more power than you should've been if you were asleep!"

Columbia returned safely to Earth the next day, touching down at Edwards Air Force Base. But even landing on this second mission was by no means smooth sailing. The astronauts were tasked with no fewer than 29 maneuvers during re-entry which spanned the entire hypersonic, supersonic, transonic and subsonic flight regime. It made Engle the only Shuttle commander to fly his ship under manual control all the way from the de-orbit burn to touchdown. "The rationale behind the maneuvers was [that] we were very anxious to see how much margin the Shuttle had in the way of stability and control authority, how much muscle the surfaces had at different Mach-numbers and angles-of-attack," he said later. "Also, in the event that a de-orbit had to be made on an orbit that had excessive cross-range to the landing site, in order to get more cross-range rather than S-turn back and forth to deplete energy, the technique was to just leave the vehicle in the bank in one direction and keep flying toward the landing site, off your straight ground-track toward your landing site. You could increase that cross-range ability by actually decreasing the angle-of-attack. It allowed the leading edge of the wing to heat up a little more and would cut down on the total number of missions that a [particular] Shuttle could fly, but it would allow you to get that extra performance…to make it to the landing site.

"How much the leading edge would heat up and just how much more lift-to-drag that would give you – turning ability, cross-range ability – was theoretically known and had some wind-tunnel test data, but the wind tunnels are very susceptible to a lot of variables, so you really want to know for sure what you have in the way of capabilities if you ever have to use them, and that's what our purpose was. During the entry, I would pulse the vehicle in all three axes to see what the effectiveness of the surfaces were during entry and how quickly the vehicle would damp out after being disturbed. Getting that data to verify and confirm the capabilities of the vehicle was something that we wanted very much to do and, quite honestly, not everyone at NASA thought it was all that important. There was an element in the engineering community that felt that we could always fly it with the variables and the unknowns, just as they were from wind tunnel data, and always come down the chute. Then there was the other school that felt you just don't know when you may have a payload you weren't able to deploy, so you have maybe the [center of mass] not in the optimum place and you can't do anything about it, and just how much maneuvering will you be able to do with that vehicle in that condition? How much control authority is really out there on the elevons and how much cross-range do you really have if you need to come down on an orbit that is not the one that you really intended to come down on?"

The Shuttle's re-entry profile as she hurtled through the atmosphere, bound for Edwards, was also markedly different from even these ambitious plans, thanks to the shortened mission. The RCS thrusters in the aft fuselage were commanded to fire over 1,000 times – consuming more than 800 kilograms of propellant, far more than planned – because the predicted fuel-consumption rate after two days differed from estimates for a five-day mission. Shortly before Entry Interface, a large quantity of propellant was dumped out of Columbia's forward RCS to allow more precise control of the ship's center-of-mass during descent. A series of flight tests were also conducted, the most important of which was a 'push-over/pull-up' exercise by Truly. He pushed the nose 'down' from a 40-degree angle-of-attack to 35 degrees, then lifted it to 45 degrees, prior to returning it to the original 40 degrees. This provided additional data on the vehicle's aerodynamic performance throughout re-entry. It enabled the astronauts to evaluate, at 35 degrees, how much more cross-range capability it afforded the Shuttle and, at 45 degrees, how they could pull up to a higher angle-of-attack if the demand arose to lower the heat on the leading edges of the wings. But performing such a complex re-entry made their lack of sleep a problem. "The fact that we were up all night," said Engle, "may not have been a good plan in retrospect." An additional problem arose with Columbia's water supply, when hydrogen from a burst fuel cell membrane caused their drinks to bubble, which led the men to avoid drinking. They returned to Earth tired and severely dehydrated.

Fig. 8.9 STS-42 commander Ron Grabe (left) and pilot Steve Oswald monitor their displays during Discovery's descent from orbit on 30 January 1992.

But there was an element of humor to complete STS-2. Engle had a long history as a test pilot at Edwards and both he and Truly had spent many weekends there during training, flying simulated approaches and landings in the STA. On one occasion, half-jokingly, the control tower told Engle to give him a call on Columbia's final approach, "and I'll clear you to land". It was not a normal thing to do, of course, because all Shuttle communications during landing ran through Mission Control in Houston. However, for a bit of fun and an acknowledgement of the important role played by his friends at Edwards in the success of the mission, Engle made the call.

"Eddy Tower, it's Columbia, rolling out on High Final," he radioed. "I'll call the gear on the flare!"

The Edwards control tower team loved it and came back with split-second timing. "Roger, Columbia, you're cleared No. 1. Call your gear!"

Touchdown was picture-perfect, but the absence of a nose-wheel steering capability meant Engle had to apply differential braking to maintain a straight and true course along the runway centerline. He also explained in the post-mission debriefing that a fluctuating indicator on his instrument panel made it difficult to maintain a constant deceleration rate. Yet, as with STS-1 before it, STS-2 had landed safely and another box seemed to have been ticked as NASA strove to land these vehicles like airliners. But as mission after mission landed without incident, every so often problems and close calls would bring into stark relief just how hazardous it really was landing these oversized, super-heavy gliders.

Nor were the problems with fuel cells yet done with the orbiter. The STS-2 problem was ultimately traced to a deposit – perhaps only a speck – of aluminum hydroxide in an aspirator, which prevented the proper removal of water from the cell. In the aftermath of the Challenger disaster, the system was extensively overhauled to improve its reliability and maintainability. End-cell heaters on each fuel cell power plant were deleted to preclude the risk of electrical failures and were replaced with Freon-21 coolant-loop passages. The hydrogen pump and water separator in each cell was improved to reduce excessive hydrogen entrapment in the power plant and new sensors were installed to afford greater visibility into potential overloads and unacceptable thermal conditions.

Following the loss of Challenger, the concept of the Minimum Duration Flight (MDF) was developed to respond to major contingencies in orbit whose severity demanded a prompt return. "We analyzed the plan to determine what could cause you to want to do something different," reflected Shuttle program manager Tommy Holloway. "Situations that resulted in minimum-duration mission were defined in flight rules. Failures that resulted in the Shuttle being at higher risk than was acceptable for a full-duration flight would be shortened to four days. That was pretty straightforward. For other situations, you'd determine what might go wrong that would result in a different plan and warranted a significant effort to build a contingency plan."

The fuel cells would return to haunt several other missions in the post-Challenger era, too, throwing up hydrogen pump anomalies, degraded transducers, flowmeter failures and on some occasion 'erratic' functionality. But more than a decade after STS-51L – and for the second time in the Shuttle's operational lifetime – the fuel cells reared their heads and necessitated an MDF and the shortening of a mission to four days. This was far more problematic than STS-2 (which had lost only a few days of its flight time), for Columbia's STS-83 mission was planned to run for a marathon 16 days, with an intricate series of fluid physics and materials science investigations and a crew of seven working in two shifts, around the clock. Soon after Columbia reached orbit on the afternoon of 4 April 1997, commander Jim Halsell and pilot Susan Still reported erratic behavior in Fuel Cell No. 2. The voltage output differential between the two 'banks' of one of its sub-stacks revealed a sharp increase. It had been noticed on the ground, before launch, but Columbia had nevertheless been cleared to fly. Halsell and Still, assisted by flight engineer Mike Gernhardt, adjusted the electrical system to reduce the load placed on the ailing cell. This intervention appeared to have the desired effect, and the rate of change in the cell slowed from 5 millivolts per hour to around 2 millivolts. However, it still exhibited a slight upward trend.

"There's always a difference between the two halves of the stack, but we're noticing a changing difference," explained Mission Operations representative Jeff Bantle. "Actually, that changing difference has leveled-off a lot, so the degradation was greater the first 12 hours of the mission." Bantle's main worry was that if the differential between the two banks increased to 300 millivolts – and it could, having already reached 250 millivolts by the evening of 5 April – the crew might be forced to shut down Fuel Cell No. 2 in its entirety. In that eventuality, the flight rules required a landing at the earliest possible opportunity. "The concern is degradation in a single cell. If it degrades enough, rather than getting power out from the cell, you would have power output *into* the cell. You could actually have crossover and localized heating, exchange of hydrogen and oxygen within the cell and could even have a localized fire. That's the very worst case. That's why we have flight rules that are very conservative to try to avoid and try to shut down and 'safe' a fuel cell before you would ever get to that point." Early on 6 April, Halsell and Still performed a manual purge of the cell, but as the situation worsened an MDF was called later that afternoon.

In the meantime, the ailing Fuel Cell No. 2 was shut down, together with several other non-critical elements of hardware to wring as much power to run the science payload for as long as possible. For a time, even the lights in the pressurized Spacelab module were dimmed and the crew worked by flashlight to perform experiments. According to one fluid systems engineer, Fuel Cell No. 2 had displayed a 500-millivolt discrepancy between its two stacks before it was even switched on, some 12 hours before STS-83 launched. Similar abnormal cell behavior had been noted on two earlier missions by Atlantis, but on both occasions the discrepancy leveled-off to well within safety guidelines shortly after it had been switched on and started to support its full electrical load. With that prior experience in mind, engineers activated Columbia's fuel cells for STS-83 and, sure enough, No. 2 settled down to 'normal' levels. It performed normally throughout ascent, but the discrepancy reared its head again in orbit. As a precaution, one of the fuel cells assigned to Atlantis' STS-84 mission was removed for checks after displaying a similar signature.

Fig. 8.10 STS-83's Mike Gernhardt performs Earth observation photography through the overhead flight deck windows of Columbia in April 1997.

On Columbia, the crew reacted with "shock and disbelief", according to mission specialist Don Thomas. Scientists scrambled to reprioritize their schedules to make the most of the one or two more days available prior to coming back to Earth. Already, efforts were underway to lobby NASA to stage a reflight of the mission later in the year. In fact, even before Columbia landed, plans were afoot to fly STS-83 with the same crew. "There were rumors already flying that after fixing the problem, NASA would be re-flying our crew in a few months to complete our Spacelab science mission," recorded Thomas on his website, OhioAstronaut.com. "That definitely helped ease the sting of coming home early."

So it was that when Halsell departed Columbia after completing a picture-perfect landing on 8 April, he was approached by KSC director Roy Bridges with a handshake and a pledge that his crew and vehicle would be given "an oil change and send you back". Three days later, Shuttle program manager Tommy Holloway authorized plans for a reflight in July. Internally designated 'STS-83R' (for 'reflight'), it was later assigned the numerical designator of 'STS-94' and launched on 1 July. NASA normally spent around $500 million per mission, although a substantial proportion of that figure was devoted to hardware testing, processing, training, planning and simulations, much of which did not require repetition. Holloway quoted about $60 million for STS-94 and stressed that flying Columbia within three months offered "a very good test of a capability we should have in place for the station, to bring an element of the station back, for whatever reason, and turn it around in as reasonable time as practical". True to the predictions, the reflight was cheaper: $55 million to process Columbia herself, plus $8.6 million for expenses associated with the turnaround of the Spacelab.

"Our approach," said STS-94 flight director Rob Kelso, "has been to treat this flight as a launch delay. The crew is exactly the same, the flight directors are all the same and the flight control team is almost identical. It's a mirror-image flight in many respects." In fact, even the embroidered patch worn by Halsell's crew was the same except for a different-colored border: red for STS-83, blue for STS-94. The Spacelab pressurized module remained in Columbia's payload bay at the Orbiter Processing Facility (OPF), although the tunnel adaptor was removed to give technicians better access to its interior. Ordinarily, between flights, the modules were transferred to the Operations & Checkout Building, but during the short turnaround technicians were able to accomplish many critical tasks, including replenishing fluids for the experiments. Normally, the Shuttle processing team supervised an orbiter for 85 days, but just 56 days in the OPF were required for the reflight. To ensure that necessary work – including the replacement of two Auxiliary Power Units (APUs) and several Reaction Control System (RCS) thrusters in Columbia's nose – was completed, several structural inspections were deferred until her next mission. Fuel Cells No. 1 and 2 were removed and returned to their vendor, Connecticut-based International Fuel Cells, for analysis. Although the exact cause was not identified, it was believed to have been an isolated incident. Engineers took steps to develop monitors to provide better performance data. Meanwhile, Columbia was rolled into the Vehicle Assembly Building (VAB) for attachment to her External Tank (ET) and Solid Rocket Boosters (SRBs) on 4 June and from thence to Pad 39A on the 11th. Aiding the early July launch target, the orbiter was fitted with three SSMEs 'borrowed' from Atlantis and two SRBs 'borrowed' from Discovery's forthcoming STS-85 mission.

In fact, the landing-to-launch turnaround time between STS-83 and STS-94 totaled only 81 working days, the shortest ever achieved in the post-Challenger era. Before the loss of STS-51L in January 1986, a total of 16 missions had eclipsed this figure, with the turnaround between two Atlantis flights in late 1985 holding the empirical record. After landing from STS-51J on 7 October and launching again for STS-61B on 26 November, Atlantis was reprocessed in 46 working days. Across the entire Shuttle program, turnaround times gradually improved as systems and processes matured, with the 668 working days needed to prepare Columbia for STS-1 dropping to just 187 working days for STS-2, then 97 working days for STS-3 and only 77 working days for STS-4. 'Operational' missions in the pre-Challenger period came in at a mean average of 89 working days, whereas after the loss of Challenger this increased to a mean average of 147 working days. After the loss of Columbia on STS-107 it rose to a mean average of 326 working days.

An MDF also came within a hair's breadth of being declared during Columbia's last fully successful mission, STS-109. Eighty-five minutes after launch on 1 March 2002, the payload bay doors were opened and the radiators lining their interior faces commenced the important task of dumping excess heat from the vehicle's electronics into space. It soon became apparent that one of two Freon-21 coolant loops was behaving in a sluggish manner. Initial suspicion centered upon a piece of welding slag or solder, which might have broken loose during ascent and become stuck inside the loop. Proscriptive mission rules forbade continuing the flight with only one loop functioning. But gradually the ailing Loop 1 stabilized and, in any case, it became clear that Loop 2 was perfectly healthy. The prospect of MDF receded. "The shake, rattle and roll of ascent really is a very dynamic test," said Shuttle program manager Ron Dittemore. "Not only do you get a lot of vibration, but you get a lot of acoustic vibration. If there was anything in Loop 2, it would have broken loose by now. That's why we believe that we saw the debris in Loop 1. We believe it's going to remain stable and support the remainder of the flight."

Strictly speaking, keeping STS-109 in orbit with one working coolant loop infringed mission rules, but Dittemore was quick to stress that Loop 2 remained "rock-solid" and even the contaminated Loop 1 would hold its own during re-entry. "The flow-rate we see on cooling Loop 1," Mission Control told the crew, "is large enough that it would be able to support a full nominal entry if called upon to do it all on its own." Certainly, flight director John Shannon was unwilling to apply the flight rules rigidly. "The flight rules are in black and white," said fellow flight director Jeff Hanley. "They say loss of cooling to those particular boxes… you should come home right away; you should not do a normal mission. John was one of the finest ascent/entry people that we've had. He also knew his team and he knew that you can't just take the rule at face value. You have to interpolate and read between the lines." As circumstances transpired, Shannon and his team worked out an approach to manage the hardware and the mission successfully went on to service the Hubble Space Telescope.

In addition to STS-2 and STS-83, one other mission did fall victim to an MDF and had to return sooner than planned. In November 1991, Atlantis' STS-44 crew were a week into a ten-day flight for the Department of Defense and had already deployed an important infrared early-warning satellite. On the morning of the 30th, one of the Shuttle's three Inertial Measurement Units (IMUs) – a critical element of the navigation hardware – failed. The crew attempted to cycle power to the device, in hopes of reviving it, but to no avail. An MDF contingency was declared. However, with Atlantis originally scheduled to touch down on the SLF in Florida her landing site was changed to Edwards, whose larger expanse of runway offered greater margins of safety for an incoming orbiter with a degraded navigational capability. They landed without incident on the morning of 1

Fig. 8.11 To protect against the risk of technical problems which might require a Minimum Duration Flight (MDF), primary payloads were typically deployed as early as possible after launch. This was particularly fortuitous on STS-44, whose Defense Support Program (DSP) early-warning satellite was launched only hours into the mission.

December, three days early, although a significant disappointment was that their families were in Florida. "It would have been spectacular to watch, because we landed on Runway 5, which meant we came right over the top of the buildings," said STS-44 pilot Tom Henricks. "Someone in the control tower could have looked in the Shuttle windows as we went by and, if you had been on the ramp, where NASA keeps its planes, you practically could have jumped up and touched our wheels."

The IMUs had previously exhibited trouble on STS-32 in January 1990 and a 'transient' problem was also noticed during Endeavour's STS-108 mission in December 2001. In the first instance, the commander reset the unit and no further action was necessary, whilst in the second case Endeavour's other two IMUs remained healthy and the mission was unaffected. Several other flights saw IMUs fail during pre-flight self-tests and the launch of STS-40 in June 1991 was scrubbed when one of the units failed to calibrate properly with its siblings. Certainly, in the case of STS-44 the astronauts had completed most of their critical work by the time MDF was called, but a hastily changed landing site would go on to blight several missions and leave families on one U.S. coast as their loved ones touched down thousands of kilometers away on the other.

THE 'INTERESTING'…

On no fewer than 26 occasions between August 1983 and July 2011, astronauts guided the Space Shuttle to land in the hours of darkness. And despite hundreds of hours of training in the simulator and the T-38 and the STA, landing this highly dynamic machine – whether by day or night – remained one of the most challenging trials for any commander or pilot. But on STS-8, Challenger would launch and land at night for the first time, a peculiarity driven by the needs of her primary payload, an Indian communications satellite. Commander Dick Truly and pilot Dan Brandenstein were both naval aviators who had flown hundreds of missions from aircraft carriers under cover of darkness. The assignment (whilst novel) fazed neither of them. But they were under no illusions that landing this super-sized glider at night would be a walk in the park. To provide additional margins for safety, NASA arranged for Challenger to land on the wide runway at Edwards, rather than risking the narrower, swamp fringed SLF in Florida. "In other words," remarked Brandenstein, "if we had some problem and ran off the side of the runway, we wouldn't go into the moat!"

An added difficulty in the early Shuttle era was how to properly illuminate the runway for the returning astronauts at night. Concrete Runway 22 at Edwards was selected for STS-8, since it was feared that should Challenger land on the dry lakebed Runway 17, her tires would kick up an enormous rooster-tail of dust which risked attenuating the light. "We felt it was safer to take the approach and land on the concrete, rather than the lakebed," said Brandenstein. The lighting devised to aid the Shuttle pilots was known as the Precision Approach Path Indicators (PAPI) and kept the vehicle on the correct outer glide path with a band of half-white and half-red lights. The PAPI system was located about 2.3 kilometers from the end of the runway and 3 kilometers from the predicted touchdown point. The correct flight path was determined by the pilots by centering the white light onto the 'band' of red lights. Transition and area lighting, consisting of 800-million-candlepower xenon floodlights, illuminated the entire area and green marker lights indicated the near and far ends of the runway.

As a hypersonic re-entry vehicle, the Shuttle could not be equipped with external landing lights of its own and everything had to be situated behind the Thermal Protection System (TPS) and inside the mold-line. "Here we are, not wanting to *not* be able to land at night because – to be a fully operational program – we were going to eventually land at night somewhere," said Shuttle commander Loren Shriver, who worked on developing PAPI. "Without landing lights, we needed some kind of illumination on the runway, in addition to the normal runway lights. There were lots of cues for the pilots, but there was nothing illuminating the touchdown zone. We had to figure out a way to supply some of that lighting onto the touchdown zone and far enough ahead that the commander could get the visual cues that he would normally have to fly in and land. We experimented with a number of methods to fly the glide slope on and then, after the pre-flare, to fly the shallow glide slope. We used various combinations of other high-powered lighting systems and ended up zeroing in on xenon lights. We found that certain arrangements of these lights in groups of two or four, and angled across the touchdown zone, not only headed the pilots in the right direction, but supplied the light. Then it became apparent that, once the pilots came in, if the light sources were behind them and they were trying to land on a lakebed, the wing-tip vortices and the Shuttle's rollout would produce a huge amount of dust, which would start to cut out the light in the rest of the touchdown zone. So it's maybe not a good thing to try to land on a lakebed at night, because the dust is soon going to block out all the light. Very soon after that, we put all that stuff on the concrete runways and decided if we were going to land at night, we wanted to land on a hard surface. It was an evolving process."

One case in which the PAPI system demonstrated its usefulness came in November 1990, when poor weather conditions at Edwards obliged NASA to divert Atlantis to KSC at the end of her STS-38 mission. "This is the fall of the year," remembered STS-38 commander Dick Covey, "and one of the things that they do in Florida during the fall is burn the underbrush in their pine forests; a very controlled type of burn, just to get everything down. They were doing that over on the west side of the [Banana] River…and the winds were predominantly from the

Fig. 8.12 Ghostly view of Atlantis, drag chute deployed, touching down on the Shuttle Landing Facility (SLF) at the Kennedy Space Center (KSC) in Florida on 21 July 2011. This was the final voyage of the 30-year Shuttle program.

north-east, so they were blowing that smoke out over Central Florida, towards Orlando." Based upon this visibility prediction, Covey was advised to land on the south-eastern end of the SLF (designated Runway 33), rather than the north-western end (Runway 15).

But by the time Atlantis commenced her hour-long re-entry, the winds shifted and the smoke began to obscure the southern half of the SLF. Additionally, STS-38 would land at sunset and the refractive effect of the dying afternoon light made the smoke appear thicker. In the final minutes before touchdown, Covey and pilot Frank Culbertson could see little but murk through their windows. Fortunately, as they rolled out on final approach, the PAPI lights glimmered into view, barely visible through the pall-like shroud of smoke. Although this gave a measure of visual guidance, as Atlantis dove through the smoke Covey and Culbertson could still see nothing but the lights; the runway itself was invisible to them. At length, the smoke cleared, and with seconds to spare, the SLF appeared right in front of them. Years later, Covey joked that he logged one of the Shuttle's very few 'instrument-only approaches' on STS-38, so poor was his visibility.

The evolution of runway-mounted illumination continued with the introduction of a system of 52 halogen lights, positioned at 60-meter intervals along the runway centerline and first used by STS-82 in February 1997. The modification had been requested by the astronauts to better support an increased cadence in night-time landings, which became increasingly commonplace at the end of Hubble Space Telescope, Mir and International Space Station missions. "It's a little bit tougher, at night, to judge our line-up with the runway," remarked STS-82 commander Ken Bowersox.

Additional instrumentation in the cockpit to assist the astronauts included the Heads-Up Display (HUD), first used on Challenger's maiden voyage, STS-6 in April 1983. On that flight, commander Paul Weitz and pilot Karol 'Bo' Bobko deemed the HUD an exceptionally useful landing aid, projecting instantaneous data on their velocity, descent rate, altitude and other critical flight parameters onto transparent viewing-glass over the cockpit windows. The HUD enabled Shuttle pilots to assimilate data from the 'heads-down' world of instrument-based flying with the 'heads-up' domain of looking directly through the windows at the approaching runway. John Blaha, who piloted two Shuttle missions and commanded two others, worked on the development of the HUD at Kaiser Electronics in San Jose, California, and remembered the early designs to be overly cluttered with data and disliked by many astronauts. "The biggest challenge was the older, established astronauts had not flown military aircraft with a HUD. The younger guys had all flown aircraft with a HUD, so there was some resistance," Blaha recalled. This resistance, to be fair, was perfectly understandable, for the earliest iteration of the Shuttle HUD proved overly cluttered with data and part of Blaha's role was to reduce its content to four or five critical parameters the commander and pilot would need. Implementation of HUD technology also proved beneficial in achieving pinpoint landings on the SLF in Florida, which, unlike Edwards, had a far less forgiving runway.

Another technology which was canceled early during the Shuttle's evolution was a braking parachute for deployment on the runway. The concept re-entered consideration following a particularly severe landing in April 1985, when Discovery touched down in a crosswind and sustained seized brakes and a burst tire. In the aftermath of the Challenger tragedy, options to enhance landing safety were investigated in greater depth. These included the installation of a specialized 'skid' on the landing gear, which (in the event of a burst tire) would preclude the chance of another blow-out by effectively providing a 'roll-on-rim' capability for a predictable rollout pattern. An arresting barrier was also installed at the end of the concrete runways at Edwards and KSC and at the Transoceanic Abort Landing (TAL) sites. Design requirements for the 'drag chute' included an ability to bring a fully loaded 112,500-kilogram orbiter to a complete halt following a TAL abort in less than 2,500 meters of runway, even with a tail wind of 18.5 kilometers per hour and maximum braking applied at a relative ground speed of 260 kilometers per hour. Housed in a cylindrical structure, just beneath the vertical stabilizer, the chute was manually deployed by the pilot after MLG touchdown and ahead of NLG touchdown. It would then be jettisoned when the rollout speed had slowed to around 110 kilometers per hour. During re-entry, the SSMEs adopted a lower-than-normal configuration to minimize the risk of damaging the drag chute.

Fig. 8.13 Discovery alights on the Shuttle Landing Facility (SLF) on 9 March 2011 to wrap up her 39th and final mission. Note the colossal Vehicle Assembly Building (VAB) in the background.

Air trials took place at NASA's Dryden Flight Research Center (DFRC) at Edwards Air Force Base in the summer of 1990, in which the hardware was fitted to a modified NASA B-52 Stratofortress aircraft and successfully tested at landing speeds of between 260 kilometers per hour and 370 kilometers per hour. These trials enabled engineers to determine that the drag chute would help to reduce the Shuttle's landing rollout distance by 300-600 meters, as well as alleviating stress on the vehicle's brakes and tires. In January 1991, the agency modified its procurement contract for the newest orbiter, Endeavour, by adding $33.3 million of funding to design, fabricate and install the drag chute. On 16 May 1992, at the end of Endeavour's maiden mission, STS-49, the drag chute was deployed for the first time. To ensure the safety of this first experimental test, the chute was deployed when all six wheels (MLG and NLG) were on the ground. When commander Dan Brandenstein and pilot Kevin Chilton issued the command to deploy the drag chute, pyrotechnics blew the door away from the chute compartment and a mortar fired, driving out firstly the 3-meter pilot canopy, then the main canopy, which 'reefed' to 40 percent of its total diameter for a few seconds to lessen the structural loads on the Shuttle itself. The drag chute trailed Endeavour by 27.2 meters on a 12.6-meter riser. Following the successful operation of the reefing line cutter, it blossomed to its fully inflated diameter of 11.8 meters. Photographic analysis of STS-49's landing illustrated that the reefed chute rode at a somewhat 'higher' angle than anticipated and the trajectory of the jettisoned compartment door differed from the B-52 tests. Its behavior and 'closeness' to Endeavour's centerline was later attributed to the effect of the aerodynamic flow for the fully-open speed brake on the vertical stabilizer.

The performance envelope was pushed further on subsequent missions. Two months later, Columbia – the second orbiter to receive the modification – returned from STS-50 and her pilots deployed the chute between MLG and NLG touchdown for the first time. However, an early observation was that it tended to 'drag' the vehicle slightly to one side during the rollout. This became apparent during the STS-52 landing in November 1992. In that case, the pilots deployed the chute a few seconds prior to NLG touchdown and noticed a just-perceptible 'tug' into the wind as the canopy billowed into its fully reefed configuration. In fact, it pulled the Shuttle about 4.6 meters to the left of the SLF runway centerline. Fortunately, the 90-meter width of the SLF made it a relatively minor controllability issue for STS-52 commander Jim Wetherbee. "It didn't cause me any concern," he remarked after the flight. "If we were landing on a very narrow runway, like over in Africa, and it pulled even more, then it would be a cause for a little bit more concern." With drag chutes fitted to Discovery in 1992 and Atlantis in 1993, all four orbiters were eventually equipped with these very visible and highly effective landing aids. Over time, as the system matured, teething difficulties were ironed out and mission after mission landed without incident.

But on 29 October 1998, as Discovery launched on STS-95, an incident occurred that no one could have foreseen. It was already a high-profile mission because John Glenn – the first American in orbit, a serving U.S. senator and the oldest human ever to fly in space at the age of 77 – was aboard, and STS-95 correspondingly drew many spectators. Several seconds after liftoff, the small aluminum panel covering the drag chute somehow detached and fell away. In the launch video imagery, the 5-kilogram panel was observed bouncing off one of Discovery's SSMEs and vanishing from view. Shortly after the Shuttle reached orbit, the astronauts were advised of the incident, but because it remained unknown if the drag chute had been damaged, partially melted or even destroyed, few NASA managers were willing to risk deploying it on the runway at the end of the mission. Over the next few days, a range of contingency plans were devised. Commander Curt Brown and pilot Steve Lindsey were asked not to deploy the chute after touchdown and were given specific details for how to jettison it if it inadvertently came open. At speeds of less than Mach 2.8, it would probably not inflate at all, whilst at lower speeds it might inflate or tear away entirely. Brown was told that should it deploy at an altitude of less than 15 kilometers, he might notice an upward pitch of the nose. If that happened, he would have taken his hands off the control stick and Lindsey would have immediately hit the Arm, Deploy and Jettison switches on his side of the cockpit to jettison the drag chute. Should an accidental deployment have occurred at less than 50 meters above the runway, it would have required an even faster response from Lindsey. In that eventuality, if the crew had done nothing, it was possible that the drag chute would inflate, pull Discovery's nose up and increase the sink rate to such a degree that landing would be far 'harder' than intended.

Fig. 8.14   During liftoff on 29 October 1998, the drag chute compartment door fell away from Discovery. As a precaution, the drag chute was not deployed on touchdown, nine days later, with the commander and pilot given instructions to discard it if necessary.

As such, in the morning mail on their landing day, 7 November 1998, Lindsey was instructed to maintain his hands on the control panel, with the covers for the Arm, Deploy and Jettison switches 'up' to afford him sufficient time to flip all three before the drag chute could unreef and inflate. But Brown, who piloted three Shuttle missions and commanded three others in his career, was unruffled by the issue. "I have practiced that out in the simulator…where we do our landing and rollout training," he observed. "It was more of an engineering evaluation; it wasn't really training, but that's kind of semantics. We have done that kind of thing." In the aftermath of this incident, the cause was traced to a failure of aluminum pins holding the drag chute door in place and strengthening modifications were incorporated for future missions. The next flight, STS-88 in December 1998, saw its drag chute capability disarmed, as investigations into the problem continued. But when Discovery launched on STS-96 in May 1999, it was with drag chute door hinges made from Inconel, far tougher than their aluminum predecessors.

Many other unusual (and even comical) anecdotes about landing anomalies cropped up over the years. On 13 August 1989, commander Brewster Shaw and pilot Dick Richards returned Columbia to Earth after her STS-28 mission. While landing on the dry lakebed of Edwards' Runway 17, Shaw noticed that although there were 'stripes' to outline the runway, its perimeter was not nearly as well-defined as the concrete runway and correspondingly affected his depth-perception. "When we came down and I flared the orbiter, I didn't know how high we were," he said later. "Looking at the photographs, we weren't very high, but I basically leveled the vehicle off and then it floated." The result was that Shaw allowed Columbia to 'float' on her MLG for a substantial part of the deceleration, before de-rotating the nose. "We got a lot of great data about low-speed flying qualities on the orbiter, but it wasn't supposed to work out that way!" For Richards, he remembered a point during re-entry – about Mach 10 – when superheated air streamed across the Shuttle, creating and depositing blobs of white-hot plasma, and one globule stuck to his cockpit window. After Columbia landed, Richards asked a technician to look at the substance, which was still in liquid form, and it was duly scooped into a coffee cup and taken away for analysis. Several days later, Richards caught up with Don Puddy, the head of NASA's Flight Crew Operations Directorate.

"What did they say about what that material was?"

"You're not gonna believe this," replied Puddy, before recounting the story. The technician had taken the cup containing the sample and put it on a counter. Another technician had then grabbed the cup, poured coffee and slugged it back. "That," he said with thinly veiled disgust, "is the worst-tasting coffee I've ever had."

To this day, no one knows what deposited itself on Richards' window at the edge of space and ten times the speed of sound in August 1989. But in a new (and somewhat dubious) 'first' for the space program, someone had at least tasted it.

THE 'UNREAL'...

In March 1979, two years before the first launch of the Space Shuttle, NASA took the unusual step of selecting a great white blotch of compacted salt and gypsum in New Mexico's Tularosa Valley as a potential landing strip for the new reusable spacecraft. Although it had long been planned for the primary End of Mission (EOM) landing site to be Edwards – at least for the four Orbital Flight Tests (OFTs) and early 'operational' flights – and subsequently KSC, the area popularly known as 'White Sands' offered near-perfect weather conditions, all year round, together with an enormous runway which provided the requisite margins of safety to land these heavyweight orbiters. Its vast size and color against the crystalline whiteness of the valley floor made it readily visible to pilots, even from space. It lay in a mountain-ringed area nicknamed 'Alkali Flats' and was first employed by Northrop Aviation in the 1940s to test military target drones. It acquired the moniker of 'Northrop Strip', which, following a typo in a press release, became known as 'Northrup Strip' and the new (mistaken) name stuck. By 1952, it was part of White Sands Missile Range and gained a pair of 10,600-meter runways, crossing each other in an X-like shape.

During the first two Shuttle missions in April and November 1981, White Sands was held in reserve to be used if Columbia needed to make an emergency return to Earth after a single orbit, a contingency known as Abort Once Around (AOA). "Should the orbiter not be in a safe orbit, the spacecraft would be slowed down by a de-orbit burn, high over the South Pacific, east of Samoa," outlined NASA. "The flight path would cross Baja California and the Mexican state of Sonora, until the spacecraft was in the denser atmosphere and the crew would fly it, 'deadstick', into Northrup Strip." Until the very end of the Shuttle program in July 2011, White Sands remained on NASA's list of active contingency sites as an AOA option, and astronauts regularly honed their flying skills there in the Shuttle Training Aircraft (STA). In December 2006, following a spate of bad weather at both KSC and Edwards, the crew of STS-116 came closer than any other flight (bar one) to executing a touchdown at White Sands. Eventually, the weather in Florida cleared and Discovery landed safely on the SLF.

One other mission, however, was not so lucky. On 22 March 1982, Columbia launched on STS-3, originally planned to spend seven days in space. Commander Jack Lousma and pilot Gordon Fullerton had trained to land at Edwards, but unseasonal rain had left its runways under several centimeters of water. As a result, four days before Columbia lifted off, NASA formally requested White Sands be activated as an additional landing site. But all was not well in the New Mexico desert either. Despite having 90-percent good weather throughout the year, White Sands was battered by its worst wind and sandstorm in a quarter of a century...on the very day that Lousma and Fullerton were to land. "It was just unbelievable how bad two or three days were, because it was in the spring of the year, which is

Fig. 8.15   Columbia lands at White Sands on 30 March 1982.

the high-velocity wind time for that missile range," remembered former Johnson Space Center (JSC) operations director Kenneth Gilbreath. "And, of course, it's a white powder. It is just blinding when the wind whips it up."

One astronaut detailed to White Sands on 29 March 1982 was future Shuttle commander and NASA administrator Charlie Bolden. "This dust storm was unlike anything I'd ever seen," he said later. "It's gypsum and it's very fine, like talcum powder. Everything was covered with plastic; the windows were sealed, but it didn't make any difference. That was a hint that this was not a good place to land the Shuttle."

Blissfully unaware of the poor weather at White Sands, Lousma and Fullerton prepared for landing. Then, less than 30 minutes before the OMS de-orbit burn to bring them home, Mission Control advised that conditions were unacceptable and that they would make a second attempt the next day, the 30th. Lousma and Fullerton became the first of 37 Shuttle crews to have their mission extended by at least 24 hours owing to weather or payload-related issues. The reason given in the case of STS-3 was higher than allowable gusting surface winds, but in fact high-altitude winds were also unacceptable for Columbia to land safely. The astronauts welcomed the extra time in space. "It was terrific," said Fullerton. "We got out of our suits and then we got something to eat and watched the world and I wouldn't have had it any other way. In fact, we flew right over White Sands, with the nose pointing straight down, and I could see this monster storm going on there. It looked like it was headed for Texas. It looked really bad down there." Chief astronaut John Young had flown weather reconnaissance sorties over the site and reported that conditions were far from adequate, with 50-centimeter-deep drifts of sand even blown into the public affairs areas of the site. "The runway got eroded by the wind, so we had people driving a road grader that night to grade it, compact it and get it ready for landing the next morning," said Grady McCright, the facilities manager at White Sands. "The wind didn't quit blowing until dark that night."

By dawn on the 30th, the sandstorm subsided and Columbia re-entered the atmosphere, bound for the New Mexico desert. At an altitude of 3 kilometers, Lousma tested the Shuttle's autoland capability – which was under consideration as a future capability for operational missions – then assumed manual control for landing. The deployment of the landing gear would be cued to use airspeed, rather than altitude, and the wheels began to lower about 30 meters above the runway. However, they took longer than anticipated to fully deploy and were only locked into position a couple of seconds before NLG touchdown. To observers on the ground, the return of STS-3 was a nail-biting sight, as Columbia streaked towards landing at over 320 kilometers per hour, with her gear still in the process of unfolding. Although touchdown was successful, the incident led NASA to determine that future missions would use altitude, rather than airspeed, as a landing gear deployment cue. (In fact, subsequent Shuttle flights typically deployed their gear about 75 meters above the runway, whilst traveling at a ground speed no higher than 550 kilometers per hour.) The effect was that Lousma touched down 1.2 kilometers past the runway threshold and had to apply differential braking to keep the spacecraft close to the centerline. The vertical impact velocity of both the NLG and MLG was within flight-rules, but the touchdown was far harsher than predicted and caused a gash-like scrape in one tire, a cracked brake rotor and extensive contamination by billowing clouds of gypsum dust.

So fine was the dust that it saturated the spacecraft and caused extensive damage that was not fully resolved in time for her next flight, STS-4, or indeed for the rest of Columbia's career. "I flew it several flights later, on my first flight and when we got on orbit, there was still gypsum coming out of everything," remembered Bolden. "They thought they had cleaned it…but it was just unreal what it had done!" As the gypsum-coated Shuttle sped down the strip, her forward gear still in the process of coming down, the nose pitched suddenly and unexpectedly back up into the air, again giving observers a moment in which their hearts leapt into their throats.

Even the NASA commentator's calm voice was laced with surprise as he counted down the number of feet to nose gear down and full weight on wheels: "Touchdown…Nose Gears…ten [feet]…five…four…three…", at which point the nose rose. He paused for a moment, repeated himself – "…three…" – and then, when the nose jolted harshly down and slapped the runway, "…touchdown!"

Fig. 8.16 Her Main Landing Gear (MLG) firmly on the ground, Columbia's Nose Landing Gear (NLG) is in the process of being de-rotated to the runway during the STS-3 touchdown. Note the scorching of the Thermal Protection System (TPS) and the large quantities of gypsum dust kicked up by the Shuttle's tires.

The effect, as Fullerton would relate, was "a kind of wheelie". The astronauts were trying to prevent what they thought might be a premature touchdown of the NLG. "It pointed out another flaw…in the flight software," said Fullerton. "The gains between the stick and the elevons – that were good for flying up in the air – were not good when the wheels were on the ground. [Jack] kinda planted it down, but then came back on the stick and the nose came up. A lot of people thought this is a terrible thing, but we improved the software and so people don't do that anymore, but we discovered a susceptibility." STS-3 also became the unwitting record-holder for the longest rollout: almost 4,200 meters. But despite the problems, the achievement was that Lousma and Fullerton discovered the problem before the Shuttle became operational, and additional simulator runs by the STS-4 crew would result in a decision to use an altitude of 60 meters, rather than an airspeed of 500 kilometers per hour, as a cue to deploy the landing gear. The vital point to be made about the off-nominal STS-3 landing is that it was successful, safe and instructive.

From his vantage point, Bolden watched the landing attentively. "Everything seemed to be going well until just seconds before touchdown, when all of a sudden we saw the vehicle kinda pitch up and then kinda hard-nose touchdown. We found out that, just as Jack Lousma had trained to do, you need to move [the stick] an appreciable amount [to disengage the autoland]. We didn't realize that. The way he had trained was just to do a manual download with a stick. When he did that, he disengaged the roll axis on the Shuttle, but he didn't disengage the pitch axis, so the computer was still flying the pitch as he was flying the roll. Gordon Fullerton just happened to look at the eyebrow lights and he noticed that he was still in auto in pitch. He told Jack, and so Jack just kinda pulled back on the stick and it caused the vehicle to pitch up. Then he caught it and put it back down and he saved the vehicle." As the servicing vehicles encircled Columbia, the spacecraft sat motionless on the runway, in Fullerton's words, "surrounded by white gypsum". So severe was the damage that the flow rate from the purge units attached to the forward fuselage had to be increased and the aft compartment's vent doors were closed to prevent further contamination. However, despite sterling efforts to remove the gypsum, the powdery stuff remained in small quantities, hidden in nooks and crannies, for the rest of Columbia's life. "It was unreal what it had done," lamented Bolden.

...AND THE OUTRIGHT UGLY

Re-entering the atmosphere after his first Shuttle mission afforded a spectacular perspective for STS-51D astronaut Jeff Hoffman. Seated on Discovery's flight deck on the morning of 19 April 1985, he watched mesmerized as the fires of hell burned outside. "You're surrounded by this red, then orange, then yellow, then white-hot plasma around the front windows. Behind you, there's this flickering wave, just like the wake behind a motorboat, but it's fiery and it's just awe-inspiring." After ten minutes of feeling as if he were riding through the innards of a neon tube, the spectacle diminished and Hoffman gradually felt his Earthly weight starting to return. He could let go of a pencil in mid-air and, for the first time in seven days, watched as it gracefully tumbled like a snowflake to the cockpit floor. At length, the sprawling outline of Florida and the straight-and-true outline of the SLF runway came into view as commander Karol 'Bo' Bobko and pilot Don Williams prepared for a routine landing. What happened next, however, was anything but routine.

Despite being 4.5 kilometers long – one of the longest runways in the world – and 300 meters wide, Shuttle commander Jack Lousma once remarked that he would have preferred the SLF to be only half as wide, but twice as long, in view of the challenges involved in landing the spacecraft. Built from extremely high-friction concrete and with a paving thickness of up to 40.6 centimeters at its center, the SLF is fringed by alligator-infested moats; an incentive, one astronaut joked, for Shuttle pilots to stay on the runway. (The alligators, in fact, live territorially in these waters and provide an effective deterrent for other animals that may want to cross the moat and reach the runway. Sometimes the alligators even bask in the sun on the Shuttle Landing Facility itself, earning it the alternate nickname of 'Gator Tanning Facility'). The SLF was not built perfectly flat, due to a 60-centimeter 'slope' from the centerline to the edge that helps to facilitate drainage along more than 8,000 small grooves, each 6.3 millimeters wide and deep, cut into the concrete. This grooving was intended to avert the risk of vehicles hydroplaning on the runway under wet operating conditions. By the morning of 19 April 1985, four missions had touched down on the SLF without incident, although on the STS-41G landing Challenger had sustained damage to her MLG tires due to rough runway conditions.

Fig. 8.17 One of Discovery's shredded tires after the STS-51D landing in April 1985.

Original plans called for Bobko and Williams to perform the first landing of the Shuttle with the 'autoland' engaged and the crew even created a mocking Latin motto for themselves: *Vide, mater, sine minibus* ("Look, Ma, no hands!"). But it would have been a risky test to execute. For Bobko, it required him to define a 'box' in performance during the final approach at which he could recover from the autoland in the event of a failure, assume effective manual control, and accomplish a safe touchdown. "The problem," he said later, "was to try and define how to recognize when the auto system was diverging and not let it get so far that I couldn't take over and make a safe landing."

"The crews were very concerned that they had everything that they can at their control to make sure it goes well," recalled Arnold Aldrich. "What they worried about was not that the autoland system wouldn't fly the vehicle right, but if there was some glitch in the autoland at a critical point of approach and they had to take control back over. Getting off the autoland and back onto manual control might be something they couldn't deal with." Charlie Bolden was equally unhappy that autoland should even be tested so close to landing. "We developed the procedures that we would use for autoland: how they would manually take over at the very last second and go ahead and land the vehicle," he recalled. "We recommended this was not a good thing to do. You're asking a person who's been in space to take over in this dynamic mode of flight and land the vehicle safely. Their physical gains, their mental gains, their balance; everything's not there. Not a smart thing to do." When common sense won out, Bobko's mission was retargeted to land at Edwards, where there was greater flexibility.

But by April 1985, with the Shuttle manifest gathering pace and NASA keen to avoid the $750,000 cost and a week of wasted processing time of flying the vehicle atop the SCA from California to Florida after each landing, STS-51D was directed to touch down instead on the SLF. And no one in their right mind would have considered putting autoland through its very first 'real-world' test on the moat-fringed, alligator-infested SLF. As such, Bobko would land Discovery manually. Unfortunately, as it alighted on the concrete the Shuttle was hit by a crosswind of 15 kilometers per hour, gusting at up to 22 kilometers per hour. This required Bobko to apply the right-hand brake and rudder more vigorously than the left, to maintain the spacecraft on the runway centerline during a long rollout. This 'differential' braking caused the inboard right-hand brake to lock up, followed shortly afterwards by its outboard counterpart. From his seat on the right-hand side of the cockpit, Williams remembered the incident vividly. "We're down to maybe just about walking speed and there's this big *bang, thump, thump, thump*," he said. "I knew right away what it was; it was a blown tire. "We're almost stopped anyway, so it turned out not to be a big deal and not an issue. Of course, the only thing to worry about is – since this tire is blown – there could be some debris problems, which might cause a puncture or might cause some reason to have to evacuate."

"AS GOOD AS IT WAS GOING TO GET"

Evacuating the orbiter in a post-landing emergency (though far less dynamic than a Mode 1 Egress on the launch pad) remained a harrowing prospect for the crew, not least because of the risk of fire or explosion and the fact that the Shuttle's still-hot tires were fully pressurized and the vehicle still carried a substantial load of highly volatile propellants. Eight emergency escape 'modes' existed for the Shuttle. Modes 1-4 covered emergencies in the pre-launch phase on the pad, in which the astronauts would exit the vehicle under their own steam or with support from ground personnel. Mode 8 provided for a high-altitude bailout and parachute-assisted return of the astronauts over land or water. That left three other modes, which dealt with how to escape an orbiter after an unhappy landing. In a Mode 5 Egress on the runway, the astronauts would escape unaided, primarily by jettisoning the hatch on the port side of the middeck and deploying a neoprene-coated nylon fabric inflatable slide, which would have yielded a drop of 3 meters to the ground.

But if the side hatch somehow refused to jettison or was otherwise unusable, the crew would evacuate the vehicle by means of a Sky Genie descent control device, deployed through one of the overhead flight deck windows. However, the temperature of the TPS tiles on the orbiter's cockpit roof was estimated to reach 48 degrees Celsius for a few minutes after landing, whilst the window glass could reach 87 degrees Celsius. This would have demanded great care on the part of the astronauts. The Mode 6 and 7 profiles envisaged assistance from ground-based fire crews, helicopter-borne Search and Rescue (SAR) forces or pre-positioned recovery convoy on the runway. Their remit covered the safe recovery of the astronauts in the event of a 'mishap' which resulted in the Shuttle coming down up to 45 kilometers from the intended landing site.

Had STS-51D's brake seizure and tire blowout occurred at higher speeds, or had controllability of the vehicle been seriously impaired, a crash was a realistic possibility. "It was a minor miracle," wrote Mike Mullane in *Riding Rockets*, "that Discovery didn't experience directional control problems...and careen off the runway". On the Shuttle's middeck, Hoffman was convinced from the noise that one of the fuel tanks had exploded, whilst crewmate Charlie Walker wondered if they had run over an alligator. "We didn't think anything more about it," recalled Walker, "until we got off the vehicle." Mission Control quickly advised the astronauts that the tire blowout had left a trail of debris stretching for some distance along the runway. As a result, for safety reasons, they would not be permitted to perform the customary walk-around inspection of Discovery after landing, lest one of the other tires should explode. The incident precipitated an immediate cancellation of plans for Shuttles to land on the SLF. Five days after STS-51D's landing, NASA announced that the next mission – Challenger's STS-51B – would land instead at Edwards. "The decision will provide more safety margin for the Challenger's tires and brake system," it reported, "because of the availability of the unrestricted lakebed and the smoother surface. The decision to land at [Edwards] for the next flight will enable engineers to determine what corrective actions are appropriate before returning to KSC for normal end-of-mission landings." But even STS-51B suffered severe brake damage, with the inboard rotors of her left-hand MLG being destroyed during touchdown and rollout.

Fig. 8.18 Pictured here surrounded by servicing vehicles on the Edwards Air Force Base runway after STS-1, the Shuttle remained an experimental flying machine from the very start of its career to the very end.

During this timeframe, of course, the development of the drag chute was several years into the future. "We were almost at a stop, luckily, because if the wheel had blown at high speed it could have been a lot more serious," reflected Hoffman. "I think that again got management's attention. They put a lot more work into getting the new brakes and getting the nose-wheel steering fixed." In the next few months after STS-51D, brake improvements were implemented and when Challenger returned from STS-61A in November she successfully trialed a new nose-wheel steering capability. Prior to that mission, the left-hand and right-hand wheel brakes were applied to steer the orbiter on the runway, at the risk of increasing the likelihood of wear and structural damage. On STS-61A, commander Hank Hartsfield had the ability to depress either the left or right rudder pedal, which signaled Challenger's GPCs to direct a hydraulic actuator to turn the NLG and steer the Shuttle precisely onto the centerline. As the vehicle slowed to around 170 kilometers per hour, he deliberately steered a few meters away from the centerline, then returned to normal and braked smoothly to a halt. "It went very well," he noted afterward. "I didn't get very far off the centerline." With this success, SLF landings were expected to resume with STS-61C, but this mission was diverted to Edwards due to poor weather. When Challenger lifted off to begin STS-51L on 28 January 1986, her tragic mission was scheduled to return to Florida, but Hartsfield did not consider one steering test as proof of the operability of a system. It was a moot point, in any case, for after Challenger's destruction NASA reverted to Edwards as the primary End of Mission (EOM) landing site for several years thereafter.

The difficulties posed by the brakes were nothing new. On the first 'operational' Shuttle mission, STS-5 in November 1982, Vance Brand undertook a test of Columbia's maximum braking capability seconds after touchdown. The left-hand inboard wheel locked-up during the rollout, owing to a brake failure. "We completely ruined the brakes," Brand recalled. "I had to stomp on them as hard as I could, which points out that we had a lot of flight test on the mission. Although it was the first commercial flight, I think we had 50-50 test objectives. That braking test was just one of them."

The damage was well-evidenced by skid marks on the runway, cracked stators and severe damage to the brake rotors. "Vance really got on the brakes, smoked the brakes," added Brand's STS-5 crewmate Bill Lenoir. "But we landed nice and sharply and got off and it was over. We walked down the stairs and 'machoed' it outta there!" The next mission suffered cracks in three stators in the right-hand inboard brake and in June 1983, after the return of STS-7, a chattering noise was heard from one of Challenger's wheels. Detailed inspections revealed that her right-hand inboard brake had suffered major structural damage to two rotors, including the beryllium heat-sink and carbon liner. Also, the right-hand outboard brake had two loose carbon pads and missing retainer washers. Cracked retaining washers were found in all brake assemblies and it was discovered that a similar situation might have occurred on previous Shuttle missions with no adverse effects. None, however, had previously been positively identified. It became clear that the washers were probably cracked during their manufacture or pre-flight assembly, with structural and thermal analyses confirming that neither the flight nor landing could have caused the damage. And the problems continued. On STS-8, a nose-gear thruster piston detached from the vehicle and ended up on the runway. In April 1984, Challenger landed with cracked rotors, chipped carbon brakes, missing washers and contamination by debris from the runway surface. These issues were described as "normal" and were not classified as safety-of-flight concerns. Nevertheless, as astronaut Sherwood 'Woody' Spring once said, in the pre-Challenger era, the brakes were being torn to shreds "on almost every landing".

Following the STS-51D incident, serious concerns were also raised about the quality of the runway surface and the SLF's lateral cross-grooves were ground out on the first 1,000 meters at each end to reduce friction and abrasion levels on the Shuttle's tires. By the end of 1994, the entire surface had been abraded to a smooth texture to reduce tire wear even further. Birds, too, were kept at a distance by using pyrotechnics, blank rounds fired from shotguns and propane cannons arranged around the runway perimeter. However, several other incidents in the post-Challenger timeframe made for unhappy landings. In April 1991, due to an incorrect call from Mission Control on high-altitude winds, STS-37 commander Steve Nagel landed Atlantis a full 190 meters 'short' of the runway threshold. Fortunately, he was returning to Edwards' vast dry lakebed and the error was not apparent to most observers. But if the mission had been targeted for the SLF, Atlantis would have touched down on the paved underrun prior to the runway and the 'low-energy' landing would have been much more apparent.

"It wasn't a real great day to land at Edwards," Nagel said later. "There were high-altitude winds aloft and a big wind-shear." As he prepared to put Atlantis through a 270-degree overhead turn to align her with the runway, by his own admission Nagel allowed the Shuttle to depart from the turn and "was not really

Fig. 8.19 Discovery deploys her drag chute during the STS-131 rollout in April 2010.

aggressive about correcting back". He rolled onto final approach a little lower than normal, but the presence of the windshear caused him to lose a lot of airspeed. "If I had been real aggressive in how I flew," Nagel continued, "really slowed it down and stretched it a little bit, I could've been back on the glide path okay, but I wasn't that aggressive with that." Less than a month later, on STS-39, Discovery's right-hand landing gear hit the runway a full 60 meters before its left-side counterpart, resulting in severe shredding of the outboard 'shoulder' of one of her tires.

As Shuttle operations matured, improvements gradually mitigated the risks which plagued the brakes and tires. Following the STS-61A test, improved nose-wheel steering systems were installed in Columbia, Discovery and Atlantis – and were a production feature on Endeavour – which permitted their safe high-speed engagement and effective lateral directional control of the Shuttle during rollout, even in cases of high crosswinds and blown tires. Improved brakes with thicker carbon-lined beryllium stator disks were installed to enhance the available braking energy and a longer-term campaign to develop all-carbon brakes was set in motion. This was expected to achieve a higher braking capacity by increasing maximum energy absorption. The carbon brakes, first flown by Discovery in April 1990, could endure maximum operating temperatures of 1,150 degrees Celsius, far higher than the 950 degrees Celsius capacity of their predecessors. And whilst earlier brakes had to be replaced after a single mission, the carbon brakes were designed to support up to 20 Shuttle flights and could support landing speeds of up to 415 kilometers per hour, a significant increase over the 330 kilometers per hour attainable with the beryllium brakes. As late as July 2006, when Discovery flew STS-121 with larger, smoother tires to better withstand higher loads, improvements to the landing gear continued to hold a high level of priority.

Yet it was only five years later, in the pre-dawn darkness of 21 July 2011, when Atlantis swept into Florida like an enormous bird of prey and came home to a smooth touchdown for the final time, that a collective sigh of relief could be definitively breathed by all. Having lost fourteen human lives in two unspeakably dreadful and wholly avoidable tragedies, and with so many other issues ranging from debris shedding during ascent to issues with brakes and tires, malfunctions with space suits, damaged TPS tiles and torn blankets, payload difficulties and various technical and human troubles on the ground, the surviving members of the Shuttle fleet earned themselves a dignified retirement with the safe landing of STS-135.

Fundamental lessons have been learned about how to keep future crews safe. As the United States moves towards SpaceX Crew Dragon vehicles atop Falcon 9 rockets, Boeing Starliners on Atlas V rockets and Orion deep-space missions on NASA's mighty Space Launch System (SLS), it seems doubtful that asymmetrical vehicles like the Shuttle will be seen again. And if they are, it is almost certain their abort capabilities will contain none of the survivability 'black zones' that so hamstrung the Shuttle.

In the deeply soul-searching months after the loss of Columbia in February 2003, veteran astronauts Scott 'Doc' Horowitz, John Grunsfeld and Marsha Ivins participated in a detailed analysis of the Shuttle's failure modes and the projected likelihood of crews escaping from a disaster to return alive to their families and loved ones. "The basic realization is we had built a very, very complex vehicle," said Horowitz, who piloted three Shuttle missions and commanded a fourth. "No matter how hard we worked on it, chances of making it or not making it on a mission was on the order of about one in a hundred." Far from the early claims that the risk was about one in several hundred, this reality was as shocking as it was unsurprising.

"And that," Horowitz said sadly, "was as good as it was going to get."

Bibliography

'Task Groups to Handle Efforts on the Manned Space Station and Space Shuttle.' NASA Headquarters News Release, 7 May 1969

'MSC Establishes a Space Station Task Group.' NASA Manned Spacecraft Center News Release, 14 May 1969

'Two Aerospace Firms Selected to Undertake 11-Month Contracts to Design a Reusable Space Shuttle.' NASA Manned Spacecraft Center News Release, 13 May 1970

'Phase A Study Contracts with Grumman, Lockheed and Chrysler for Alternate Space Shuttle Concepts.' NASA Manned Spacecraft Center News Release, 15 June 1970

'Request for Proposals for the Development of Landing Gear for the Space Shuttle.' NASA Manned Spacecraft Center News Release, 4 May 1971

'Request for Proposals for Development of Space Shuttle Thermal Protection System.' NASA Manned Spacecraft Center News Release, 17 May 1971

'Advantages and Disadvantages of Using Phased Approach for Developing the Space Shuttle. NASA Headquarters News Release, 16 June 1971

'Shuttle Preliminary Design Contract Extension.' NASA Manned Spacecraft Center News Release, 1 July 1971

'McDonnell Douglas Selected for Contract for Study of Space Shuttle Auxiliary Propulsion System.' NASA Manned Spacecraft Center News Release, 6 July 1971

'Selection of Rocketdyne for Space Shuttle Main Engine Contract.' NASA Headquarters News Release, 13 July 1971

'Three Contracts Awarded for Development of New Surface Materials for Space Shuttle.' NASA Manned Spacecraft Center News Release, 14 July 1971

'Textron and Rocketdyne Awarded Contracts for Auxiliary Propulsion System on Space Shuttle.' NASA Manned Spacecraft Center News Release, 12 August 1971

'Space Shuttle Contract Extensions.' NASA Headquarters News Release, 7 October 1971

'Nixon/Fletcher Shuttle Statements.' NASA Manned Spacecraft Center News Release, 5 January 1972

'Space Shuttle Decisions.' NASA Manned Spacecraft Center News Release, 15 March 1972

'NASA Releases Space Shuttle RFP.' NASA Manned Spacecraft Center News Release, 17 March 1972

'Space Shuttle Operational Site Selected.' NASA Manned Spacecraft Center News Release, 14 April 1972

'Four Companies Submit Proposals for Space Shuttle Program.' NASA Manned Spacecraft Center News Release, 12 May 1972

'Shuttle Aircraft Contract.' NASA Manned Spacecraft Center News Release, 5 July 1972

© Springer Nature Switzerland AG 2021   277
B. Evans, *The Space Shuttle: An Experimental Flying Machine*,
Springer Praxis Books, https://doi.org/10.1007/978-3-030-70777-4

'MSC Awards $540,000 Shuttle RCS Contract to Ball Aerospace.' NASA Manned Spacecraft Center News Release, 10 July 1972

'Shuttle Contractor Selection.' NASA Manned Spacecraft Center News Release, 26 July 1972

'NASA/NR Sign Shuttle Letter Contract.' NASA Manned Spacecraft Center News Release, 9 August 1972

'Initial Space Shuttle Hardware Procurement Action Initiated by NR.' NASA Manned Spacecraft Center News Release, 7 November 1972

'NASA Signs Definitive Space Shuttle Contract with Rockwell International Corporation.' NASA Johnson Space Center News Release, 16 April 1973

'JSC to Remodel Orbiter Reproduction Facility in California.' NASA Johnson Space Center News Release, 22 June 1973

'Martin Marietta to Develop Space Shuttle Tank.' NASA Johnson Space Center News Release, 16 August 1973

'Space Shuttle Facility Construction to Begin in 1974.' NASA Johnson Space Center News Release, 23 September 1973

'Contract Funds Shuttle Orbiter Assembly Site.' NASA Johnson Space Center News Release, 26 September 1973

'NASA Awards Shuttle Solid Rocket Motor Contract to Thiokol.' NASA Johnson Space Center News Release, 20 November 1973

'Grumman Selected for Shuttle Training Aircraft.' NASA Johnson Space Center News Release, 13 December 1973

'747 Selected for Space Shuttle Orbiter Ferry Flights.' NASA Johnson Space Center News Release, 17 June 1974

'NASA to Award SRM Contract to Thiokol.' NASA Johnson Space Center News Release, 27 June 1974

'Initial Space Shuttle Flights to Land at Edwards.' NASA Johnson Space Center News Release, 18 October 1974

'Shuttle Orbiter Wing Delivery.' NASA Johnson Space Center News Release, 30 April 1975

'Canada to Build Shuttle Remote Manipulator.' NASA Johnson Space Center News Release, 9 May 1975

'Space Shuttle Approach and Landing Test Crews Named.' NASA Johnson Space Center News Release, 24 February 1976

'Shuttle Space Suit and Rescue System.' NASA Johnson Space Center News Release, 24 March 1976

'Space Shuttle Landing System Components Delivered to NASA.' NASA Johnson Space Center News Release, 18 May 1976

'Successful Orbiter Heat Shield Test.' NASA Johnson Space Center News Release, 21 May 1976

'Shuttle Training Aircraft Delivery to JSC.' NASA Johnson Space Center News Release, 8 June 1976

'Second of Two Shuttle Training Aircraft Delivered to JSC.' NASA Johnson Space Center News Release, 15 September 1976

'Orbiter Crews Escape System Tested.' NASA Johnson Space Center News Release, 11 January 1977

'NASA Names Astronaut Crews for Early Shuttle Flights.' NASA Johnson Space Center News Release, 16 March 1978

'Shuttle Maneuver Engine Tested at White Sands.' NASA Johnson Space Center News Release, 7 September 1978

'New Mexico Lakebed Airstrip Named as Shuttle Backup Landing Site.' NASA Johnson Space Center News Release, 1 March 1979

'First Shuttle Crew Trains for Parachute Water Landings.' NASA Johnson Space Center News Release, 11 April 1980

'NASA Signs Canadians to Build Shuttle Robot Arm.' NASA Johnson Space Center News Release, 14 April 1980

'Shuttle Columbia's Flight Engines to be Retested.' NASA Johnson Space Center News Release, 1 May 1980

'Martin Marietta to Build Space Shuttle Orbiter Tile Repair Kits.' NASA Johnson Space Center News Release, 2 July 1980

'Space Shuttle to Carry Space Toolbox.' NASA Johnson Space Center News Release, 14 October 1980

'Space Shuttle Columbia Requires Only Minor Work Before Final Orbital Test Flight.' NASA Johnson Space Center News Release, 19 April 1982

'STS-5 Space Suit Inquiry.' NASA Johnson Space Center News Release, 19 November 1982

'Team Reports on STS-5 Space Suit Failures.' NASA Johnson Space Center News Release, 2 December 1982

'Fifth Crew Member Named to STS-7 and STS-8.' NASA Johnson Space Center News Release, 21 December 1982

'Inquiry Team Reports on Space Suit Failures.' Johnson Space Center News Release, 1 February 1983

'IUS Investigation Board Members Named.' NASA Johnson Space Center News Release, 7 April 1983

'51K Crew Announcement.' NASA Johnson Space Center News Release, 14 February 1984

'NASA Announces Updated Flight Crew Assignments.' NASA Johnson Space Center News Release, 3 August 1984

'NASA Changes 51B Landing Site to Edwards Air Force Base.' NASA Johnson Space Center News Release, 24 April 1985

'Space Shuttle Challenger Tapes Being Analysed.' NASA Johnson Space Center News Release, 17 July 1986

'NASA Awards Contract for Orbiter Arresting System.' NASA Johnson Space Center News Release, 2 June 1987

'Astronaut S. David Griggs Killed in Air Crash.' NASA Johnson Space Center News Release, 17 June 1989

'Partial Shuttle Crew Assignments Announced.' NASA Johnson Space Center News Release, 29 June 1989

'NASA Awards Space Shuttle Orbiter Drag Chute Contract Mod.' Johnson Space Center News Release, 4 January 1991

'NASA Awards Space Shuttle Orbiter 14-Inch Disconnect.' NASA Johnson Space Center News Release, 6 February 1991

'Space Shuttle Discovery's Flight on STS-39 Delayed, Atlantis on STS-37 Next Up.' NASA Johnson Space Center News Release, 28 February 1991

'Astronaut Seddon Injured During Training.' NASA Johnson Space Center News Release, 4 May 1993

'Astronaut Story Musgrave Injured During Training.' NASA Johnson Space Center News Release, 1 June 1993

'Super Lightweight External Tank to be used by Shuttle.' NASA Headquarters News Release, 28 February 1994

'Contract Signed for Glass Cockpit Shuttle Upgrade.' NASA Johnson Space Center News Release, 10 May 1994

'NASA Receives First New Shuttle Engine.' *Flight International*, 4 January 1995

'New Space Shuttle Main Engine Ready for Flight.' NASA Headquarters News Release, 21 March 1995

'NASA Managers Set Launch Dates for Discovery and Endeavour.' NASA Johnson Space Center News Release, 7 June 1995

'NASA Delays Launch of Space Shuttle.' NASA Headquarters News Release, 12 July 1996

'Shuttle Super Lightweight Fuel Tank Completes Test Series.' NASA Headquarters News Release, 18 July 1996

'Shuttle Super Lightweight Fuel Tank Completes Tests.' NASA Headquarters News Release, 11 September 1996

'Astronaut Cady Coleman Begins Training as Backup Mission Specialist for STS-83.' NASA Johnson Space Center News Release, 18 February 1997

'Rominger to Replace Ashby as STS-85 Pilot.' NASA Johnson Space Center News Release, 18 March 1997

'Shuttle's New Lighter, Stronger External Tank Completes Major Pressure Tests.' NASA Headquarters News Release, 28 March 1997

'Shuttle Program Reviewing Reflight of STS-83 Mission.' NASA Johnson Space Center News Release, 11 April 1997

'Microgravity Science Laboratory Mission Set for July; Remaining 1997 Shuttle Manifest Adjusted Slightly.' NASA Headquarters News Release, 25 April 1997

'New Space Shuttle External Tank Ready to Launch Space Station Era.' NASA Headquarters News Release, 15 January 1998

'Sen. Glenn Gets a 'Go' for Space Shuttle Mission.' NASA Headquarters News Release, 16 January 1998

'Astronaut Class of 1998 Reports for Duty.' NASA Johnson Space Center News Release, 19 August 1998

'Astronaut Crew Assignments Build on Space Station Experience.' NASA Johnson Space Center News Release, 15 August 2002

'NASA Announces Backup Commander for STS-134 Mission.' NASA Johnson Space Center News Release, 13 January 2011

'Astronaut Steve Bowen Named to STS-133 Space Shuttle Crew.' NASA Johnson Space Center News Release, 19 January 2011

Abbott, Matthew R. (2009) *NASA Johnson Space Center Oral History Project*

Aldrich, Arnold D. (2002) *NASA Johnson Space Center Oral History Project*

Allen, Joseph P. (2004) *NASA Johnson Space Center Oral History Project*

Blaha, John E. (2004) *NASA Johnson Space Center Oral History Project*

Bolden, Charles F. (2004) *NASA Johnson Space Center Oral History Project*

Brand, Vance D. (2000) *NASA Johnson Space Center Oral History Project*

Brand, Vance D. (2002) *NASA Johnson Space Center Oral History Project*

Brandenstein, Daniel C. (1999) *NASA Johnson Space Center Oral History Project*

Cleave, Mary L. (2002) *NASA Johnson Space Center Oral History Project*

Covey, Richard O. (2006) *NASA Johnson Space Center Oral History Project*

Covey, Richard O. (2007) *NASA Johnson Space Center Oral History Project*

Creighton, John O. (2004) *NASA Johnson Space Center Oral History Project*

Duffy, Brian (2004) *NASA Johnson Space Center Oral History Project*

Engle, Joe H. (2004) *NASA Johnson Space Center Oral History Project*
Fabian, John M. (2006) *NASA Johnson Space Center Oral History Project*
Fullerton, Charles G. (2002) *NASA Johnson Space Center Oral History Project*
Garman, John R. (2001) *NASA Johnson Space Center Oral History Project*
Gavin, Joseph G. (2003) *NASA Johnson Space Center Oral History Project*
Gibson, Robert L. (2016) *NASA Johnson Space Center Oral History Project*
Gilbreath, Kenneth B. (2003) *NASA Johnson Space Center Oral History Project*
Greene, Jay H. (2004) *NASA Johnson Space Center Oral History Project*
Gregory, Frederick D. (2006) *NASA Johnson Space Center Oral History Project*
Griffin, Gerald D. (1999) *NASA Johnson Space Center Oral History Project*
Hanley, Jeffrey M. (2016) *NASA Johnson Space Center Oral History Project*
Hart, Terry J. (2003) *NASA Johnson Space Center Oral History Project*
Hauck, Frederick H. (2003) *NASA Johnson Space Center Oral History Project*
Hauck, Frederick H. (2004) *NASA Johnson Space Center Oral History Project*
Hawley, Steven A. (2002) *NASA Johnson Space Center Oral History Project*
Heflin, J. Milton (2017) *NASA Johnson Space Center Oral History Project*
Hill, Paul S. (2015) *NASA Johnson Space Center Oral History Project*
Hoffman, Jeffrey A. (2009) *NASA Johnson Space Center Oral History Project*
Holloway, Thomas W. (2015) *NASA Johnson Space Center Oral History Project*
Holt, John D. (2005) *NASA Johnson Space Center Oral History Project*
Howell, Jefferson D. (2015) *NASA Johnson Space Center Oral History Project*
Hutchinson, Neil B. (2000) *NASA Johnson Space Center Oral History Project*
Hutchinson, Neil B. (2004) *NASA Johnson Space Center Oral History Project*
Hyle, Charles T. (1999) *NASA Johnson Space Center Oral History Project*
Jackson, Bruce G. (2009) *NASA Johnson Space Center Oral History Project*
Johnson, Caldwell C. (1999) *NASA Johnson Space Center Oral History Project*
Johnson, Gary W. (2010) *NASA Johnson Space Center Oral History Project*
Kehlet, Alan B. (2005) *NASA Johnson Space Center Oral History Project*
Leestma, David C. (2002) *NASA Johnson Space Center Oral History Project*
Lenoir, William B. (2004) *NASA Johnson Space Center Oral History Project*
Lind, Don L. (2005) *NASA Johnson Space Center Oral History Project*
Lounge, John M. (2008) *NASA Johnson Space Center Oral History Project*
McBarron, James W. (2000) *NASA Johnson Space Center Oral History Project*
McCright, Grady E. (2000) *NASA Johnson Space Center Oral History Project*
Nagel, Steven R. (2002) *NASA Johnson Space Center Oral History Project*
Nelson, George D. (2004) *NASA Johnson Space Center Oral History Project*
O'Connor, Bryan D. (2006) *NASA Johnson Space Center Oral History Project*
Peterson, Donald H. (2002) *NASA Johnson Space Center Oral History Project*
Pohl, Henry O. (1999) *NASA Johnson Space Center Oral History Project*
Richards, Richard N. (2006) *NASA Johnson Space Center Oral History Project*
Ride, Sally K. (2002) *NASA Johnson Space Center Oral History Project*
Ross, Jerry L. (2003) *NASA Johnson Space Center Oral History Project*
Ross, Jerry L. (2004) *NASA Johnson Space Center Oral History Project*
Seddon, Margaret R. (2010) *NASA Johnson Space Center Oral History Project*
Seddon, Margaret R. (2011) *NASA Johnson Space Center Oral History Project*
Shaw, Brewster H. (2002) *NASA Johnson Space Center Oral History Project*
Shelley, Carl B. (2001) *NASA Johnson Space Center Oral History Project*
Shriver, Loren J. (2002) *NASA Johnson Space Center Oral History Project*

Silveira, Milton A. (2006) *NASA Johnson Space Center Oral History Project*

Smith, Emery E. (2006) *NASA Johnson Space Center Oral History Project*

Stevenson, Robert E. (1999) *NASA Johnson Space Center Oral History Project*

Stewart, Troy M. (1998) *NASA Johnson Space Center Oral History Project*

Stone, Brock R. (2006) *NASA Johnson Space Center Oral History Project*

Sullivan, Kathryn D. (2007) *NASA Johnson Space Center Oral History Project*

Sullivan, Kathryn D. (2008) *NASA Johnson Space Center Oral History Project*

Templin, Kevin C. (2012) *NASA Johnson Space Center Oral History Project*

Thompson, Robert F. (2000) *NASA Johnson Space Center Oral History Project*

Van Hoften, James D.A. (2007) *NASA Johnson Space Center Oral History Project*

Vaughan, Chester A. (1999) *NASA Johnson Space Center Oral History Project*

Walker, Charles D. (2004) *NASA Johnson Space Center Oral History Project*

Walker, Charles D. (2005) *NASA Johnson Space Center Oral History Project*

Walker, Charles D. (2006) *NASA Johnson Space Center Oral History Project*

Weitz, Paul J. (2000) *NASA Johnson Space Center Oral History Project*

Whittle, David W. (2006) *NASA Johnson Space Center Oral History Project*

Williams, Donald E. (2002) *NASA Johnson Space Center Oral History Project*

Wren, Robert J. (2007) *NASA Johnson Space Center Oral History Project*

Burrough, Bryan (1998) *Dragonfly: NASA and the Crisis Aboard Mir*. London: Fourth Estate

Cabbage, Michael and Harwood, William (2004) *Comm Check*. New York: Free Press

Cooper, Henry S.F., Jr. (1987) *Before Liftoff*. Baltimore, Maryland: The Johns Hopkins University Press

Evans, Ben (2005) *Space Shuttle Columbia*. Chichester: Praxis

Evans, Ben (2011) *At Home in Space*. Chichester: Praxis

Evans, Ben (2012) *Tragedy and Triumph in Orbit*. Chichester: Praxis

Evans, Ben (2013) *Partnership in Space*. Chichester: Praxis

Evans, Ben (2014) *The Twenty-First Century in Space*. Chichester: Praxis

Galison, P. and Roland, A. (2013) *Atmospheric Flight in the Twentieth Century*. Berlin: Springer Science and Business Media

Glenn, John, with Taylor, Nick (1999) *John Glenn: A Memoir*. New York: Bantam Books

Heppenheimer, T.A. (1999) *The Space Shuttle Decision: NASA's Search for a Reusable Space Vehicle*. Washington D.C.: NASA History Office, Office of Policy and Plans

Hilmers, David, with Houston, Rick (2013) *Man on a Mission*. Grand Rapids, Michigan: Zonderkidz

Husband, Evelyn, with Vanliere, Donna (2003) *High Calling*. Nashville: Thomas Nelson, Inc.

Jenkins, Dennis R. (2001) *Space Shuttle: The History of the National Space Transportation System*. Hinckley: Midland Publishing

Jones, Tom (2006) *Sky Walking*. New York: HarperCollins Publishers

Kelly, Scott J. (2017) *Endurance*. London: Penguin Random House

Lenehan, Anne E. (2004) *Story: The Way of Water*. New South Wales, Australia: The Communications Agency

Linenger, Jerry M. (2000) *Off the Planet*. New York: McGraw-Hill

Massimino, Mike (2016) *Spaceman*. London: Simon & Schuster

Melvin, Leland (2017) *Chasing Space*. New York: Amistad

Morgan, Clay (2001) *Shuttle-Mir: The United States and Russia Share History's Highest Stage*. Houston, Texas: NASA History Series

Mullane, Mike (2006) *Riding Rockets*. New York: Scribner

Nelson, Bill with Buckingham, Jamie (1988) *Mission.* San Diego: Harcourt Brace Jovanovich

Portree, David S.F. and Trevino, Robert C. (1997) *Walking to Olympus: An EVA Chronology.* NASA Johnson Space Center: NASA History Series

Reichhardt, Tony (2001) *Space Shuttle: The First 20 Years.* Washington, D.C.: Smithsonian Institution

Young, John W. with Hansen, James R. (2012) *Forever Young.* Gainesville, Florida: University Press of Florida

About The Author

Ben Evans is a British writer with an interest in human space exploration that stretches back further than he can remember. He wrote his first article for the British Interplanetary Society's magazine *Spaceflight* in March 1992 and has also contributed to *Astronomy*, *BBC Sky at Night*, *Astronomy Now*, *All About Space* and *Countdown*. He has written nine books for Springer-Praxis since 2003: a survey of the Voyager mission to the outer Solar System, the careers of Columbia and Challenger and a six-volume History of Human Space Exploration. The first volume of this series, *Escaping the Bonds of Earth*, was shortlisted for the Eugene M. Emme Award in Astronautical Literature. Since 2012, he has written space news and history stories for AmericaSpace.com.

© Springer Nature Switzerland AG 2021
B. Evans, *The Space Shuttle: An Experimental Flying Machine*,
Springer Praxis Books, https://doi.org/10.1007/978-3-030-70777-4

Index

© Springer Nature Switzerland AG 2021
B. Evans, *The Space Shuttle: An Experimental Flying Machine*,
Springer Praxis Books, https://doi.org/10.1007/978-3-030-70777-4

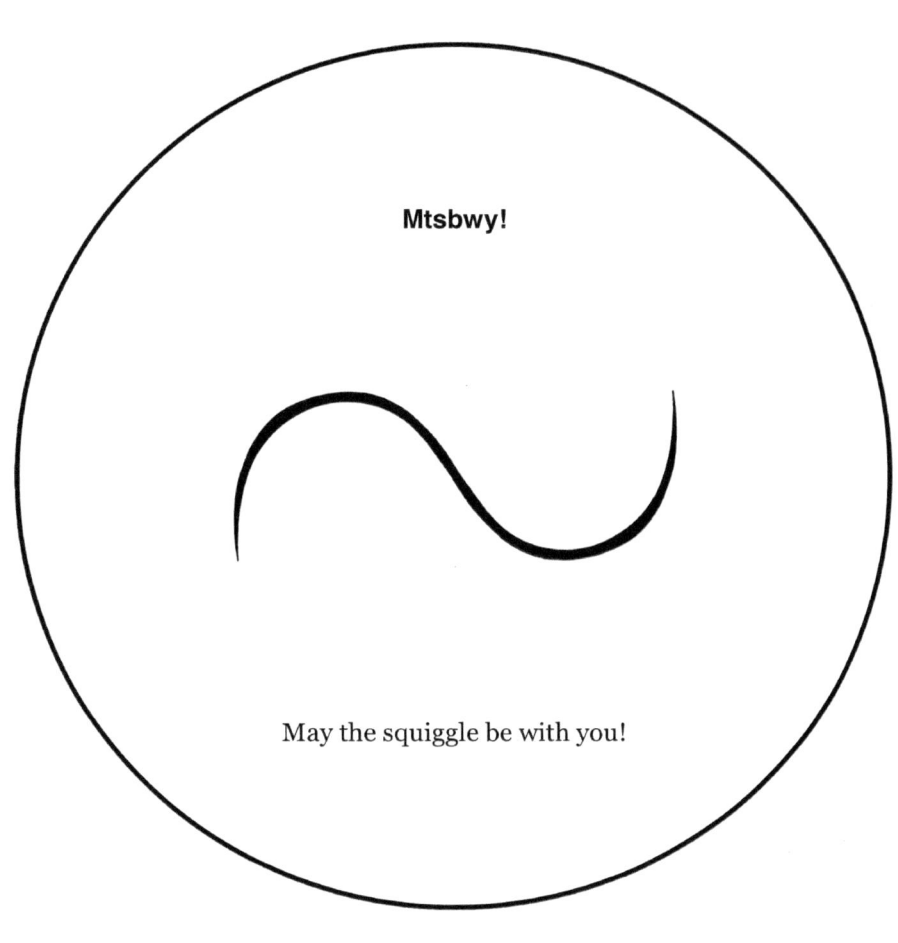

The Squiggle Sense

J. A. Scott Kelso · David A. Engstrøm

The Squiggle Sense

Sixth Sense of the Complementary Nature
and the Metastable Brain~Mind

 Springer

J. A. Scott Kelso
Center for Complex Systems and Brain
Sciences
Florida Atlantic University
Boca Raton, FL, USA

Intelligent Systems Research Centre
Ulster University
Derry~Londonderry, N. Ireland, UK

David A. Engstrøm
Center for Complex Systems and Brain
Sciences
Florida Atlantic University
Boca Raton, FL, USA

ISBN 978-3-031-59368-0   ISBN 978-3-031-59369-7   (eBook)
https://doi.org/10.1007/978-3-031-59369-7

Cover illustration based on SPHERE BM (brain~mind) © Dražen Pavlović. Reproduced with permission.

This Springer imprint is published by the registered company Springer Nature Switzerland AG
The registered company address is: Gewerbestrasse 11, 6330 Cham, Switzerland

If disposing of this product, please recycle the paper.

J. A. Scott Kelso dedicates his contribution to this work to his children, Jason, Jamie, Kate and Alex, and his grandchildren Liam, Caroline, McKenna and Cameron. And their children, and all children everywhere. Mtsbwy. Always.

David A. Engstrøm dedicates his contribution to this work individually to Lene, Forest, Chloe, Kenzie, Zoe, Janet, Sylvester and Pavlović and collectively to all humanity, all that have been and gone, alive today and yet to become. May you discover your squiggle sense. May the squiggle be with you!

Preface

The Squiggle Sense: Sixth Sense of the Complementary Nature and the Metastable Brain~Mind is an inspirational primer that connects human perception of the complementary nature with a new, empirically based theory of how the human brain~mind works. It's meant to be a companion, a personal reference that you can return to again and again. As a result, the hope is that you may see things in a different and more useful light. This book draws on the profound relationship between nature's many complementary contraries and the science of coordination called Coordination Dynamics (CD). Its purpose is to awaken and help you employ your *"squiggle sense"* in order to transcend the detrimental narrow-mindedness of polarizing, "either/or" thinking and behavior. The book follows on an earlier treatise (*The Complementary Nature*, MIT Press, 2006/2008) where we introduced a novel meaning for the squiggle (~) to symbolize the complementary nature of contraries like individual~collective, cooperation~competition and integration~segregation. Here, the behavior of both complementary aspects as well as the multistable and metastable CD that underlies them is captured by the *squiggle*. From the brain~mind's metastable mode of operation emerges a sentient faculty that physically senses and mentally perceives the complementary nature of itself and the world. This is *the squiggle sense*.

The world is wounded these days. Persistent dualistic, either/or thinking and behavior continues to be a major stumbling block to human development that foments polarization, side-taking, intransigence, intolerance and conflict. A less binary and binding vocabulary is needed. *The squiggle sense* offers a way out. The great physicist Niels Bohr said, "If you hold opposites together in your mind, you will suspend your normal thinking and allow intelligence beyond rational thought to create a new form". *The squiggle sense* can help you do this better and on purpose.

But how? Your brain~mind must enter its "metastable mode", where complementary metastable coordination tendencies coexist. CD provides a rigorous, scientifically grounded explanation of the metastable mode of brain~mind and how it leads to *the squiggle sense*. You are introduced to the paradigm shifting science of CD via a set of *squiggles* that capture core phenomena and principles. These *squiggles* are proposed to be universally applicable to any particular subject, field and level of human interest and endeavor. They are used to explain how dynamic coordination patterns and brain~mind modes emerge and shift, and reciprocally, how the metastable mode of your brain~mind enables you to perceive and reconcile *squiggles*. Each time you are presented with apparent contraries or opposites joined by a *squiggle* symbol, you are meant to hold them together in your mind as complementary, coexisting tendencies. The role and goal here is to nudge your brain~mind into its metastable mode again and again. Contemplation of each archetypal *squiggle* is intended to pique and exercise your *squiggle sense* as you read about it. CD strongly indicates that contemporary life—mind, brains, people, society—is manifested via metastability. A person, group or society that realizes this, and intentionally exercises their *squiggle sense*, is called a Metastabilian.

Metastabilians wield *squiggle* power. Their intentionally engaged *squiggle sense* says, for example, "there is no individual without a collective, no parts without a whole, no competition without cooperation, no segregation without integration, no dwell without escape, no unity without diversity…". *The squiggle sense* of your metastable brain~mind shows how these complementary aspects arise and function in complex systems. It is your personal skeleton key to the complementary nature. This book aims to inspire you to awaken and wield your *squiggle sense*. It invites you to pursue the perspective and paradigm shifting call of the Metastabilians—the call for an unprecedented evolutionary advance in human enlightenment. Although everyone has access to their *squiggle sense*, the complementary, metastable mode of coordination and consciousness, most remain unaware of it. You are probably unaware of it as such. Metastabilians are aware of it. Reading this book will make you aware of it. That awareness can lead to profound, novel advances in your life.

We would like to thank Michela Castrica and her production team (Viju Falgon, Arun Kumar) for their kind and diligent attention to realizing this book. We are especially grateful to Dr. Thomas Ditzinger, Editorial Director at Springer Nature for his longtime friendship and support of this project, and the science of complex systems in general.

Boca Raton, USA   J. A. Scott Kelso
Copenhagen, Denmark   David A. Engstrøm
October 2023

Contents

01 Of Knowledge and Wisdom

 Strange how things in the offing, once they're sensed, convert to things foreknown; And how what's come upon is manifest only in light of what has been gone through. Seventh heaven may be the whole truth of a sixth sense come to pass…—Seamus Heaney

A powerful and profound capability exists within you, a sixth sense called *the squiggle sense*. The *squiggle sense* is your *sixth sense* of the complementary nature. Your *squiggle sense* senses the dynamic dances of complementary contraries or "*squiggles*" of life, like part~whole, cooperation~competition, integration~segregation… The *squiggle* symbol (~) symbolizes the complementary nature of such complementary contraries. Like any of your senses, your *squiggle sense* can help you survive and prosper, learn and adapt to novel situations and understand life better. Yet it's more likely than not to be unknown to you as a sense that you possess, a power you can wield. But once you learn about and sense it in yourself, you will suddenly feel that you've always known about it, and can start using it. This book is meant to help you do that.

Now, it turns out that *the squiggle sense* is grounded in and best understood via the paradigm and language of Coordination Dynamics (CD), the science of coordination. Indeed, existence of *the squiggle sense* is based on the discovery from CD that life's complex systems, including human brains, are metastable, capable of expressing two complementary contrary tendencies at the same time. CD provides a theory of coordination that is broad and deep. It is tried and tested, and based on data collected from thousands of hours of experiments by many dedicated, professional scientists over the last 40 years or so. If you wish to delve deeper into this science, you can explore hundreds, if not thousands of peer-reviewed journal publications, encyclopedia articles, books, compilations, letters, editorials, lectures, videos, web pages and social media posts. It's all there, the theoretical concepts, methods, mathematical models and tools, experiments and data, tables and graphs, analysis of results and conclusions, waiting for you to peruse. You are encouraged to study and explore the whole paradigm of CD, for in its details lie many profound concepts and discoveries.

That being said, it's important to note that this book is not a treatise or defense of CD. Rather, it presents a unique perspective in which the knowledge gained from CD via research and discovery actually leads to wisdom of *the squiggle sense:* it offers a way for you specifically and humanity in general to overcome frequently encountered stumbling blocks in life, and heal, survive, prosper, and advance. The well-researched phenomena of CD are no less amazing than black holes and quarks. In fact, they're arguably more relevant to you, at your scale, in your daily life. Together with the *squiggle sense* it grounds, CD offers you knowledge that is inextricable with wisdom. It provides an inspirational path that may well be a key to your and humanity's future.

J. A. S. Kelso and D. A. Engstrøm, *The Squiggle Sense*,
https://doi.org/10.1007/978-3-031-59369-7_1

knowledge~wisdom

Only knowledge, only wisdom, either knowledge or wisdom, both knowledge and wisdom, knowledge changing to wisdom, wisdom changing to knowledge, between knowledge and wisdom, neither knowledge nor wisdom, beyond knowledge and wisdom...

The Metastabilian says: The knowledge I've gained from CD leads me to novel insights about the complementary nature, a sixth sense of it called *the squiggle sense*. I use the discovery and awareness of my *squiggle sense* to comprehend, explore, discover and express the complementary nature of myself, my life, my world, and even the CD it's grounded in. My *squiggle sense* reconciles knowledge and wisdom. The wisdom leads to new trajectories in my life, that in turn motivate exploration, discovery and acquisition of new knowledge...

Related squiggles: science~philosophy, facts~intelligence, learning~experience...

02 Of Beginning and End

 What we call the beginning is often the end and to make an end is to make a beginning. The end is where we start from—T. S. Eliot

This book was written to help you awaken, engage, train and wield your *squiggle sense* at will. Some effort is required to do this, but the potential benefits are many. The book is a collection of vignettes called *squiggle* frames, each providing a brief introduction to a *squiggle* important to and grounded in Coordination Dynamics (CD). Each *squiggle* frame is two pages long, a page-left and page-right. You're reading the second one now. The left pages begin with a numbered title, a small picture with a quote beside it, and three paragraphs of text. The right pages have a circular bubble with a *squiggle* title and illustrations that go with the text. Under the bubble is a short section called, "The Metastabilian says:" and a list of three "Related Squiggles." A metastabilian is someone who intentionally engages and wields their *squiggle sense* of the complementary nature. All human beings have the potential to be metastabilians.

This modular format facilitates reading and rereading the *squiggle* frames in different ways and orders to stimulate, challenge and explore your *squiggle sense*! While at first it's probably best to read them in the order presented, there are as many ways to read them as there are combinations and permutations of the *squiggle* frames. You can begin at the end and end at the beginning! Or open your copy at random, read a frame, skip some pages and read another. Be creative. Follow your *squiggle sense!* Each *squiggle frame* offers a different perspective of the complementary nature and the Coordination Dynamics that scientifically grounds it. Each is meant to stimulate your *squiggle sense,* draw it out, teach it, such that you can learn to engage and wield it intentionally, so that you can become a metastabilian.

Reading different combinations of frames results in new associations. The idea is that you engage your *squiggle sense* more and more as you learn more and more about it. The more often you read the *squiggle frames*, the more variations you try, the more it will all make *squiggle sense* to you. As you explore these *squiggle* frames, try to notice if the *squiggle* you're reading about reminds you of aspects of your life—if novel, unexpected associations and relevancies to your life emerge spontaneously. Such associations can be cogent, insightful, even enlightening. Use them to enhance your life's trajectory. And whenever something in your life reminds you of one of the *squiggle* frames, let it remind you to open your book and begin a new reading. This book of *squiggle* frames is meant to be a companion, a personal reference you can return to again and again. As a result, you may see things in a different and useful light.

J. A. S. Kelso and D. A. Engstrøm, *The Squiggle Sense*,
https://doi.org/10.1007/978-3-031-59369-7_2

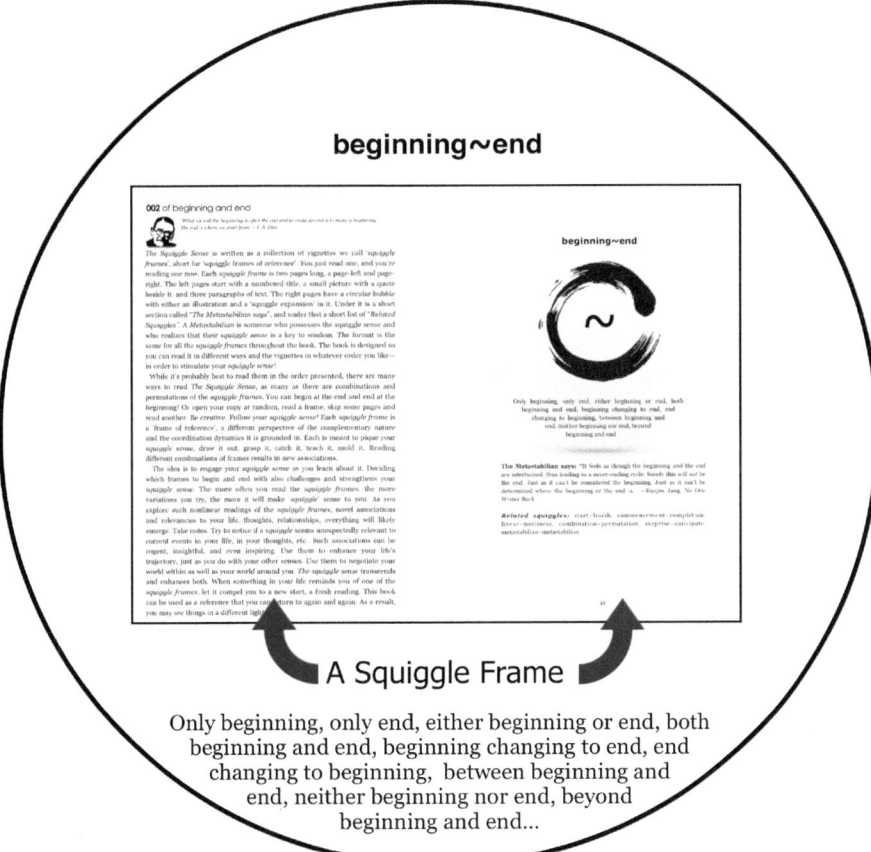

beginning~end

A Squiggle Frame

Only beginning, only end, either beginning or end, both beginning and end, beginning changing to end, end changing to beginning, between beginning and end, neither beginning nor end, beyond beginning and end...

The Metastabilian says: In *No One Writes Back,* Jang Eun-Jin wrote, "It feels as though the beginning and the end are intertwined, thus leading to a never-ending cycle. Surely this will not be the end. Just as it can't be considered the beginning. Just as it can't be determined where the beginning or the end is." My *squiggle sense* and the complementary nature it senses are like that. They've always been with me, even when I didn't know it, and had no name for it. Now I've learned that my *squiggle sense* comes from the metastable mode of my brain~mind's coordination dynamics. And each *squiggle* I sense reminds me that my *squiggle sense* is a skeleton key, a beginning~end of wisdom…

Related squiggles: start~finish, open~close, initiation~termination…

03 Of Mystery and Wonder

 The most beautiful thing to experience is the mysterious, the source of true art and science—Albert Einstein

The Complementary Nature is a grand and mysterious coordination indeed—of life and death, energy and matter, conscious and unconscious, senses and movement, thoughts and emotions, reaction and anticipation, spontaneity and will. The mystery of it all sparks curiosity. It makes you wonder. Think of the dawning of your self, your first spark of awareness as a newborn, the first words you spoke and steps you took as a toddler. Wonder how you're able to stand and walk, think and talk, learn and remember, retain and relate, reflect and engage on so much information, so many facts, images and precious moments. How many times in your life have you wondered about those mysteries, like your body and mind, stars and planets, earth and sky, plants and animals, people and civilization—how all of it fits together? Your *squiggle sense* can help kindle and enhance your sense of mystery and wonder about life and nature.

Now consider the stunningly sobering fact that we human beings tend to set all this mystery and wonder aside. Too often we ignore, overlook, or just plain forget to wonder about these most beautiful things, the mysterious, the source of true art and science and 10,000 other pursuits. The challenges, stresses, expectations and worries of daily life often cause wonder to decline and fade away. Our sense of wonder seems to dissipate though the sources of wonder still remain. Think about it—even this attenuation of awareness, this habituation of the senses, is mysterious, something to wonder about. How does it happen, and is there a way back, a return to wonder? And how does that happen? And of course, the mysterious wonder of nature and nature of wonder remain. How does it all fit together? We are used to thinking about things as separate, but how are things (and namesakes like parts and processes) related? Are they *really* separate? Can the science of coordination help?

This grand project to reconcile human nature with nature at large, this mystery of mysteries, is crucial not only to you as a person, but the survival of our species, the leap to new stages of sentient awareness where new mysteries are waiting to be wondered about, discovered and explored. To this end, this book is written to inspire you to rekindle your wonder, your child mind. It is this innate curiosity and wonder that leads to new Eureka moments, which expand your awareness and enable new discoveries and fresh perspectives. Whenever everyday life seems to dull your *squiggle sense* of mystery and wonder—read this again!

J. A. S. Kelso and D. A. Engstrøm, *The Squiggle Sense*, https://doi.org/10.1007/978-3-031-59369-7_3

mystery~wonder

Only mystery, only wonder, either mystery or wonder, both
mystery and wonder, mystery changing to wonder,
wonder changing to mystery, between mystery
and wonder, neither mystery and wonder,
beyond mystery and wonder...

The Metastabilian says: This work has become more precious to me as my *squiggle sense* has become awakened, as I have engaged, enhanced and expanded it. It inspires and motivates me upon my life's journey onward, as I appreciate and celebrate the mystery, wonder and illumination of the complementary nature and its scientific grounding in Coordination Dynamics. It is, all things considered, amazing to be alive and aware—we sentient creatures, coexisting upon this tiny blue island oasis among the vast and mysterious ocean of stars!

Related squiggles: ignorance~curiosity, question~answer, unknown~known…

04 Of Survival and Discovery

 So I am always between two currents of thought, first the material difficulties, turning round and round and round to make a living; and second, the study of color. I am always in hope of making a discovery there, to express the love of two lovers by a marriage of two complementary colors, their mingling and their opposition, the mysterious vibrations of kindred tones—Vincent Van Gogh

Your squiggle sense can enhance your ability to navigate two fundamental currents of thought. First, those dealing with how to survive the daily material difficulties of your life, and second, how to make discoveries about yourself, your life and anything you're interested in. These two currents of thought are complementary—they're mutual, dynamically linked, and can't be meaningfully separated. To be able to study and make discoveries about your life's interests and pursuits, you must somehow navigate and negotiate the logistics of daily survival. The discoveries you make inspire and motivate you, compelling you to follow new paths of exploration, which alter the logistics of your daily survival. It goes back and forth, round and round, like the proverbial serpent swallowing its tail.

Now, before you can wield your *squiggle sense* and thereby enhance your existing abilities, you must first awaken and engage it. This is easier said than done. You will probably need to forgo, possibly even transcend some fixed ideas, attitudes and perspectives you take for granted that can block or drastically inhibit your sixth sense of the complementary nature. Ironically the *squiggle sense* becomes attenuated by the same kind of narrow-mindedness you use it to overcome. Work, stress, distractions, exhaustion and worries of all sorts can weaken the *squiggle sense*. So here's the challenge: how can you awaken and engage your *squiggle sense* and use it to enhance your ability to survive and to discover in your life? Becoming aware and gaining access to it is only the first step. Next, you must learn to engage your *squiggle sense* intentionally, wield it at will. You need to want to use it. If you can sense it but ignore it, it will fade into the background again, and be of little use. The *squiggle sense* is there to be discovered, but you have to realize and engage it yourself.

Amazingly enough, simply entertaining the idea that two contrary aspects of life like survival and discovery might be complementary is where your quest of the *squiggle sense* begins. Awakening and engaging your *squiggle sense* can enhance your ability to survive day-to-day as a person, and leap to new stages of awareness that can be used to *discover*, ponder and explore your life, your world. A key insight relevant to your quest is that these complementary abilities of survival and discovery are at their roots coordination patterns that can be understood and explained by Coordination Dynamics (CD). As hidden, unknown or attenuated as it's been up to this moment, contemplating the possibility of the complementary nature of CD sets your brain~mind in motion, and kindles your *sixth sense* of the complementary nature—*the squiggle sense*!

J. A. S. Kelso and D. A. Engstrøm, *The Squiggle Sense*,
https://doi.org/10.1007/978-3-031-59369-7_4

survival~discovery

Only survival, only discovery, either survival or discovery, both survival and discovery, survival changing to discovery, discovery changing to survival, between survival and discovery, neither survival nor discovery, beyond survival and discovery ...

The Metastabilian says: I reflect upon the stunning discovery that my *sixth sense* of the complementary nature that motivates me to explore the mysteries of life is actually vital not only to my own personal survival, but the survival of the human species. What happens, then, when such a vital sense becomes depressed or obstructed—blinded by habits, fixations and intolerance—in this precious life in which we're attempting to survive and prosper, create and discover?

Related squiggles: viability~exploration, life~development, awareness~realization...

05 Of Polarization and Reconciliation

 The garden reconciles human art and wild nature, hard work and deep pleasure, spiritual practice and the material world. It is a magical place because it is not divided. The many divisions and polarizations that terrorize a disenchanted world find peaceful accord among mossy rock walls, rough stone paths and trimmed bushes—Thomas Moore

Your *squiggle sense* reconciles opposites like art and science, work and pleasure, spiritual and material as inextricable, dynamic, complementary aspects of the complementary nature. They are not divided. But if you can't, don't or won't use your *squiggle sense*, opposites become divisions and polarizations that can terrorize and disenchant the world. In them you can find no peaceful accord, no tenable middle way: "My way or the highway!" "Either with me or against me!" "Whose side are you on?" "Either us or them!" Sadly, we do not live in a *squiggle*-aware world. Polarization remains one of humanity's greatest stumbling blocks, both a root cause and consequence of a world divided. The news (both factual and fake) is inundated with stories of *polarization* and its devious narrow-minded siblings—extremism, intransigence, intolerance and hubris. You shake your head at it all, thinking, "It's so frustrating! Can't anything be done about this…*Ridiculum*?".

Actually, there is at least one thing you can do. You can engage your *squiggle sense* whenever and wherever you can. And if it's engaged right now, you might wonder if polarization and reconciliation *squiggle?* If they do then Coordination Dynamics (CD) should be able to provide scientific insight into both and their relationship. In fact, CD does provide the means and ends to understand polarization and reconciliation, and their complementary nature. Learning about CD can help you gain a novel, scientifically grounded world view that, if embraced, can save you from toxic narrow-mindedness. Via CD, you sense and understand the polarities of life as idealized aspects of a far more extensive, mind-expanding phenomenon that sweeps dynamically between them. This is the *squiggle sense* of the complementary nature.

CD provides a natural basis for polarization, and how it can lead to getting stuck in one mode of thought versus another—either/or thinking. Even more importantly, CD shows how polarized tendencies may coexist at the same time, a crucial phenomenon called metastability, which provides both a scientific basis and a syntax for polarization~reconciliation. Metastability's ability to explain apparently irreconcilable opposites as complementary aspects of a single reality offers deep insight into the human condition and a basis for genuine human action and change. It's also the basis of the *squiggle sense* which, if embraced can enable humanity to transcend ubiquitous polarization and quell at long last the corrosive culture of seemingly eternal conflict. As Joni Mitchell wisely wrote, "we are stardust, billion year old carbon, we are golden, caught in the devil's bargain, and we've got to get ourselves back to the garden…".

J. A. S. Kelso and D. A. Engstrøm, *The Squiggle Sense*,
https://doi.org/10.1007/978-3-031-59369-7_5

The either/or mode

Either... beginning or end, one or many, brain or mind, whole or parts, instinct or learned, nature or nurture, expert or novice, cooperation or competition, love or hate, freedom or constraint, wrong or right, friend or enemy, us or them, life or death, individual or collective, evolution or design, order or chaos, tradition or progress, form or function, efficacy or toxicity, supply or demand, faith or reason, professional or amateur, mind or matter, stability or instability, leader or follower, public or private, business or pleasure, right or wrong, us or them, democrat or republican, conservative or liberal, religious or secular, genetic or environmental, mind or body, agree or disagree, design or evolution, science or philosophy, work or leisure, black or white, good or evil, right or left, rich or poor, fantasy or reality, analytical or artistic, logic or intuition, private or public, bound or free, mental or physical, actual or virtual, human or machine, science or religion, choice or chance, academic or commercial, beautiful or ugly, genotype or phenotype, centralized or distributed, creative or destructive, organism or environment, mechanistic or holistic, male or female, young or old, yours or mine, winner or loser, conscious or unconscious, creation or evolution, rich or poor, idealist or realist, believer or nonbeliever, random or determined, freedom or security, with or against...

Only polarization, only reconciliation, either polarization or reconciliation both polarization and reconciliation, polarization changing to reconciliation, reconciliation changing to polarization between polarization and reconciliation, neither polarization nor reconciliation, beyond polarization and reconciliation...

The Metastabilian says: The philosopher Colin McGuinn said, "Despite the unifying efforts of Enlightenment thinkers, modernity has been structured around fracture lines, like the mind/body problem, the nature/nurture problem, free will versus determinism and secularism or faith... our current intellectual predicament reflects a long history of doubt and debate in which the competing notions of machine and soul still struggle. It is enough to make one think that we need a radical reconceptualization of the entire problem, if indeed that were humanly possible". My *squiggle sense* says it is possible: It is precisely such detrimental either/or thinking and acting that my *squiggle sense* transcends, due to the metastable CD of my brain~mind.

Related squiggles: integration~segregation, togetherness~apartness, unity~division...

06 Of Contraries and Complementarity

 If you hold opposites together in your mind, you will suspend your normal thinking process and allow an intelligence beyond rational thought to create a new form—Niels Bohr

Your *squiggle sense* allows you to hold opposites in your mind, suspending your normal thinking process and allowing an intelligence beyond rational thought to create a new form. It senses the complementary nature as myriad dances of what Niels Bohr called, *contraria sunt complementa:* contraries are complementary. Somehow he was able to hold opposite theories of light (as waves and as particles) in his mind and suspend normal ways of thinking about them. This allowed an intelligence beyond the usual way of thinking about competing theories (is light a wave or a particle?) to create a new form, which became known as the Complementarity Interpretation of Quantum Mechanics (QM). For Bohr, whether light is a wave-like or particle-like phenomenon depends on the experimental arrangement used to measure it. Strange but true.

QM is one of the most successful scientific theories of all time, and Bohr's Complementarity Interpretation of it is the one accepted by most physicists. Complementarity works at atomic and subatomic levels. But is it generalizable? There are many complementary contraries in nature besides waves and particles. Just how many there are is unknown. Of course, you are probably aware that the ancient Chinese principle of yin and yang is held as a general complementarity principle, applicable at all levels of nature. Yin and yang is an archetypal pair of complementary contraries, thought to be coexistent, inextricable, dynamic. Ironically though, the pair was believed to spring from the ineffable Tao—too great to be expressed or described in words, let alone studied scientifically.

Stimulated (by his awakened *squiggle sense?*) and inspired by his successful but specific interpretation of complementarity at the quantum level, Bohr sought to discover a general scientific complementarity theory, applicable at all levels great and small, living and non-living—a principle of a complementary nature. The situation was ironic: a modern, specific complementarity grounded in physics but missing a generalized form, and an ancient general complementarity lacking scientific grounding. Bohr immortalized this irony on the shield commemorating his Danish knighthood by including a yin-yang symbol in the coat-of-arms of his design with latin motto, *contraria sunt complementa.* Unfortunately, this goal was not achieved in his lifetime and until recently, there has been little progress made towards Bohr's grail. Coordination Dynamics (CD) provides a compelling, general, scientific grounding of complementary contraries, able to explain how they emerge, coexist and change. For this reason, CD has been dubbed, "the new science of The Complementary Nature".

contraries~complementarity

Only contraries, only complementarity, either contraries or complementarity, in no both contraries and complementarity, contraries changing to complementarity, complementarity changing to contraries, between contraries and complementarity, neither contraries nor complementarity, beyond contraries and complementarity...

The Metastabilian says: *Contraria Sunt Complementa,* or complementary contraries, is the motto on Bohr's Shield. The complementarity of particle and wave behaviors of light which inspired it is very specific, scientific and modern. Yet, the ying-yang symbol is philosophical and ancient. The shield captures Bohr's grail, the advent and development of a scientifically tenable Generalized Complementarity Principle. And now, Coordination Dynamics not only presents us with a science of The Complementary Nature, but scientifically grounds my *squiggle sense* that senses that complementary nature. Amazing!

Related squiggles: opposites~complements, polarization~reconciliation, contraria~complementa...

07 Of the Squiggle Sense and the Complementary Nature

 Just as in the organization of the physical world with which it interacts, it is proposed that the brain is organized to obey principles of complementarity, uncertainty, and symmetry-breaking. In fact, it can be argued that known complementary properties exist because of the need to process complementary types of information in the environment—Stephen Grossberg

Your *squiggle sense* (TSS) is the *sixth sense* of the complementary nature (TCN). Its awakening leads to a new mindset and new wisdom grounded in the science of Coordination Dynamics (CD) with its many fascinating phenomena and the theoretical concepts, methods and tools it employs to understand them. CD is a science of the complementary nature, that is, a nature entailed and defined by a core principle of complementarity that runs all the way through it. This is an ancient idea, exemplified by the principle of yin and yang in Eastern philosophy as well as the concept of Dialectic in Western philosophy, and courses broadly in many guises and contexts through the entire history of ideas. In this modern age of science and technology, you are probably at least casually familiar with the importance of the complementary nature in Quantum Mechanics (QM). Indeed, the vanishingly tiny quantum world of particles and waves is so complementary that one of the most important and successful interpretations of QM is actually called Complementarity.

You yourself are a complementary creature of the complementary nature, from bottom to top, left to right and front to back. The genetic code of DNA in all your cells is a combination of just two complementary pairs of nucleotides, namely adenosine~thymine (A~T) and cytosine~guanine (C~G). The coordinated dances of your DNA and proteins are also complementary dances, their dynamic patterns and pattern dynamics comprising the complementary structure and function of your complementary body and mind. The inhalations and exhalations of every breath you take, the systoles and diastoles of all your heartbeats from the cradle to the grave involve coordinated movements most complementary.

Your brain~mind is complementary, matching the complementary nature it is produced by, your perceptions and actions, reactions and anticipations, learnings and memories. Your cognition is complementary, your thoughts and emotions, both conscious and unconscious. Your behavior, the coordinated expression of the many things you do that make and define you, is complementary! Your spoken and body language, your social interactions as an individual in a family and community of other people are all complementary coordination patterns. What about your *squiggle sense*? Well, it's as complementary as complementary can be! That you, a marvelous and able sentient being living amidst the cosmos possess a sense, a sixth sense, that is sensitive to the very complementary nature that has produced you, is quite literally as sensible as a sense can be!

J. A. S. Kelso and D. A. Engstrøm, *The Squiggle Sense*, https://doi.org/10.1007/978-3-031-59369-7_7

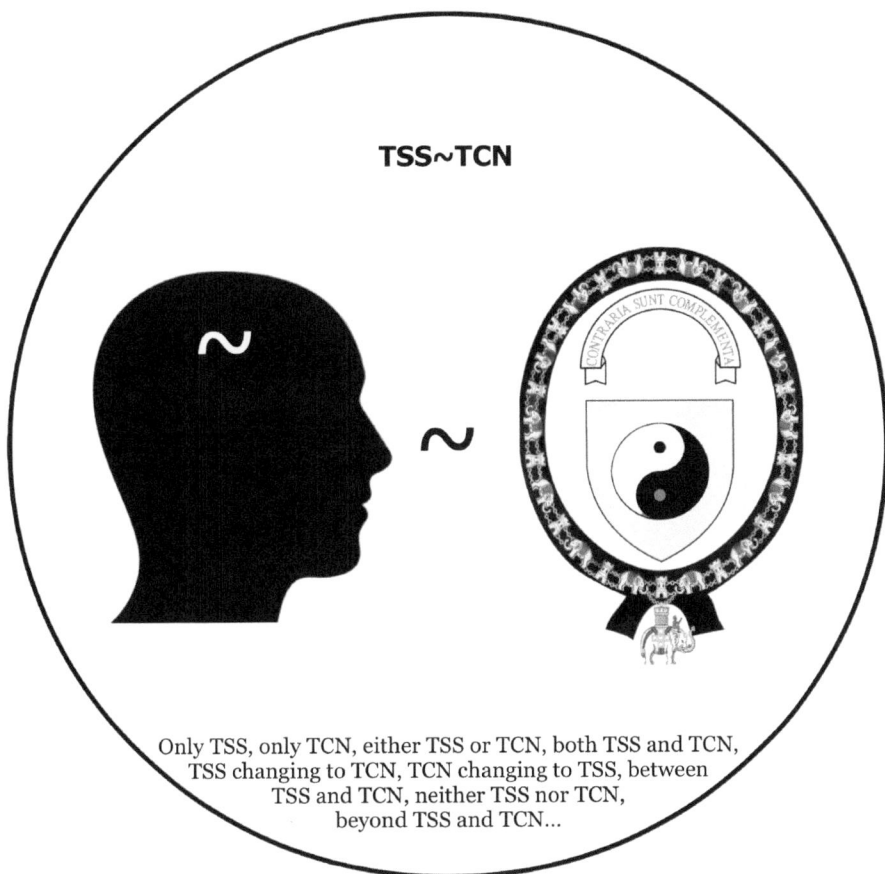

TSS~TCN

Only TSS, only TCN, either TSS or TCN, both TSS and TCN,
TSS changing to TCN, TCN changing to TSS, between
TSS and TCN, neither TSS nor TCN,
beyond TSS and TCN...

The Metastabilian says: It makes *squiggle sense* to me that my body and brain~mind that enables me to be sentient and aware should somehow match the complementary nature that produces it—to grow and develop, learn, survive, discover and interact with, live, love and experience. It's extraordinary to realize that my *squiggle sense* of the complementary nature not only senses, but is a complementary aspect of it. It is a sixth sense, a way to sense the complementary nature of the heavens, my world and myself...

Related squiggles: awareness~reality, metastability~complementarity, brain~environment...

08 Of Coordination Dynamics and the Complementary Code

 Present-day physics is running into obstacles of various kinds. It is argued that the cause of the problems is the fact that physics lacks tools appropriate to gaining a clear understanding of complexity. Alternatives such as that of coordination dynamics are likely to lead to a new level of understanding of nature, including the role played by mind—Brian Josephson

Your *squiggle sense* is both a *sixth sense* of the complementary nature and a phenomenon of Coordination Dynamics (CD). When you engage and wield your *squiggle sense,* you're *doing* coordination dynamics. From here on in, all the *squiggle* frames are about *squiggles* that comprise the complementary code (TCC) of CD. So you're going to learn about CD, which grounds your *squiggle sense* in science, via its complementary nature, its complementary pairs, its *squiggles*—using your *squiggle sense*! But what is CD? To begin, realize that coordination is in you and all around you, everywhere you look—in your DNA, genes, proteins, cells, organs, the neurons of your brain that allow you to perceive the world, remember, decide and act. And then there is coordination among people. All is a coordinated dance at all levels and scales. In fact, the coordination of living things is one of the great mysteries in the science of life.

CD showed that coordination is an emergent, self-organizing process, where the many diverse components of a living system come together to form coherent, meaningful patterns of behavior. It studies how these coordination patterns form, persist and change, regardless of the many differences between the parts and processes involved. In CD, principles and mechanisms underlying coordinated behavior are identified in different fields and functions, in multiple systems and multiple levels of observation. CD is an approach to understanding how coordination works in nature's complex systems, including certain aspects of human nature such as intention, learning and emotion. CD began with very basic experiments on the coordination of human hand movements and was then extended to the coordination between touch, vision, sound and movement and the neuronal patterns that accompany them. More recently, CD has been extended to social coordination within~between brains and between people and machines, even to understanding team and crowd behavior.

From these roots, CD's influence has spread to different fields like philosophy, education, law, ecology and economics, and has useful applications in medicine, therapeutics and sport science. While such diverse systems and levels are very different in many ways, a deep connection between all of them is found in the dynamical principles and mechanisms of CD, especially metastability. CD's essential elements can be expressed via a codified set of concepts which is open and continues to evolve. It turned out that these were none other than a complementary code—the *squiggles* of CD. This discovery elevated Coordination Dynamics to its updated status as a science of the complementary nature—which you can learn about using your *squiggle sense*!

J. A. S. Kelso and D. A. Engstrøm, *The Squiggle Sense*, https://doi.org/10.1007/978-3-031-59369-7_8

CD~TCC

knowledge~wisdom, beginning~end, mystery~wonder,
survival~discovery, polarization~reconciliation,
contraria~complementa, TSS~TCN, squiggles~dynamical modes,
theory~experiments, brain~mind, metastable mode~complementary code,
dynamic patterns~pattern dynamics, self-organization~agency, ID~FI,
intentional action~intrinsic dynamics, perception~action, learning~memory,
individual~collective, dissipation~cycles, boundary~domains, spatial~temporal,
persistence~change, convergence~divergence, path~bifurcation, structure~function,
planning~execution, multifunctionality~functional equivalence, absolute~relative,
synchronization~syncopation, synergy~complexity, whole~parts,
qualitative~quantitative, modeler~model, cv~cp, heterogeneity~coupling strength,
deterministic~stochastic, states~transitions, attraction~repulsion, stability~instability,
e.o.f.~c.s.d., CD Law~dynamical landscape, dynamical landscape~coordination modes,
uncoupled~coupled, monostability~bistability, multistability~metastability,
integration~segregation, dwelling~escaping, symmetry~broken symmetry,
competition~cooperation, creation~destruction of FI, states~tendencies,
accommodation~assimilation, Metastabilian~Metastabilion, choice~chance,
inphase~antiphase, local~global, part~whole, dwell~escape, inphase~antiphase,
trapping~wrapping, between~within, components~coupling, gradual~abrupt,
constant~intermittent, stability~fluctuations, SO~FI, SOD~ID, local~global,
organism~environment, planning~execution, preferences~exploration,
reaction~anticipation, recruitment~dispersion, simple~complex,
accommodation~assimilation, order~disorder...

Only CD, only TCC, either CD or TCC, both CD and
TCC, CD changing to TCC, TCC changing to
CD, between CD and TCC, neither CD
nor TCC, beyond CD and TCC...

The Metastabilian says: And now I realize that Nature itself, Mother Nature, human nature, our nature—is at its core complementary through and through. In everyday language, CD provides me a vehicle, a means to explore and understand how the complementary nature works at any scale of magnitude I wish to study or express, in any viable system and in any context I wish to study it. CD, together with my awakened squiggle sense, provide a way forward for me personally as well as my ability to share it with others. CD's Complementary Code (~) and its many realizations are an invaluable aid that help me to do it!

Related squiggles: all of the above! ~;)

09 Of Squiggles and Dynamical Modes

 To us… the only acceptable point of view appears to be the one that recognizes both sides of reality—the quantitative and the qualitative, the physical and the psychical—as compatible with each other, and can embrace them simultaneously …. It would be most satisfactory of all if physics and psyche (i.e., matter and mind) could be seen as complementary aspects of the same reality—Wolfgang Pauli

Your squiggle sense senses the complementary nature (TCN) of *squiggles* like physics and psyche, matter and mind, body and soul. It enables you to recognize them as compatible with each other, to embrace them simultaneously, to see them as complementary aspects of the same reality. While this same profound and useful insight has been expressed by many thinkers throughout the history of ideas, it has proven to be quite difficult to express and share with the general populace, and continues to be widely unknown and under appreciated to this day. Reversing this trend is a main goal of this book. A simple but powerful notation has been developed to accomplish this goal. In it, pairs of possible complementary aspects like brain and mind are written with a *squiggle symbol* (~) between them, as in brain~mind. This is one reason why they're called *squiggles,* and the sixth sense of them, the *squiggle sense.* The generic written form is ca1~ca2, where 'ca' stands for complementary aspects. It's spoken form is, "c-a-one *squiggle* c-a-two".

The *squiggle* (~) stands for the inextricable, dynamic, complementary nature of any pair of complementary aspects. It also implies a set of archetypal dynamic patterns or dynamical modes of TCN. Symbolic dynamic modes of the generic ca1~ca2 squiggle are: [only ca1, only ca2, either ca1 or ca2, both ca1 and ca2, ca1 changing to ca2, ca2 changing to ca1, between ca1 and ca2, neither ca1 nor ca2, beyond ca1 and ca2]. This ongoing *squiggle* dance between and within a *squiggle's* dynamical modes is what your *squiggle sense* senses. But how are *squiggles* produced? How do they work, how are they related to each other and how is TCN to be understood? What drives the dance of its dynamical modes? Where does the *squiggle sense* come from?

Answers to all these questions come from Coordination Dynamics (CD). Besides their complementary nature, the *squiggle* '~' also symbolizes the set of actual universal dynamical modes of coordination produced, studied and theoretically modeled in the science of CD: (uncoordinated mode, monostable mode, multistable mode and metastable mode). The correspondence between the *squiggle sense,* the conceptual awareness and appreciation of complementary aspects, the symbolic *squiggle* notation, dynamical modes of *squiggles,* and the dynamical modes of coordination in CD constitute the basis of the 'scientific grounding' of TCN, *squiggles* and the *squiggle sense.* The dynamical modes of a *squiggle,* the actual phenomena, are in fact different dynamical modes of CD.

J. A. S. Kelso and D. A. Engstrøm, *The Squiggle Sense,*
https://doi.org/10.1007/978-3-031-59369-7_9

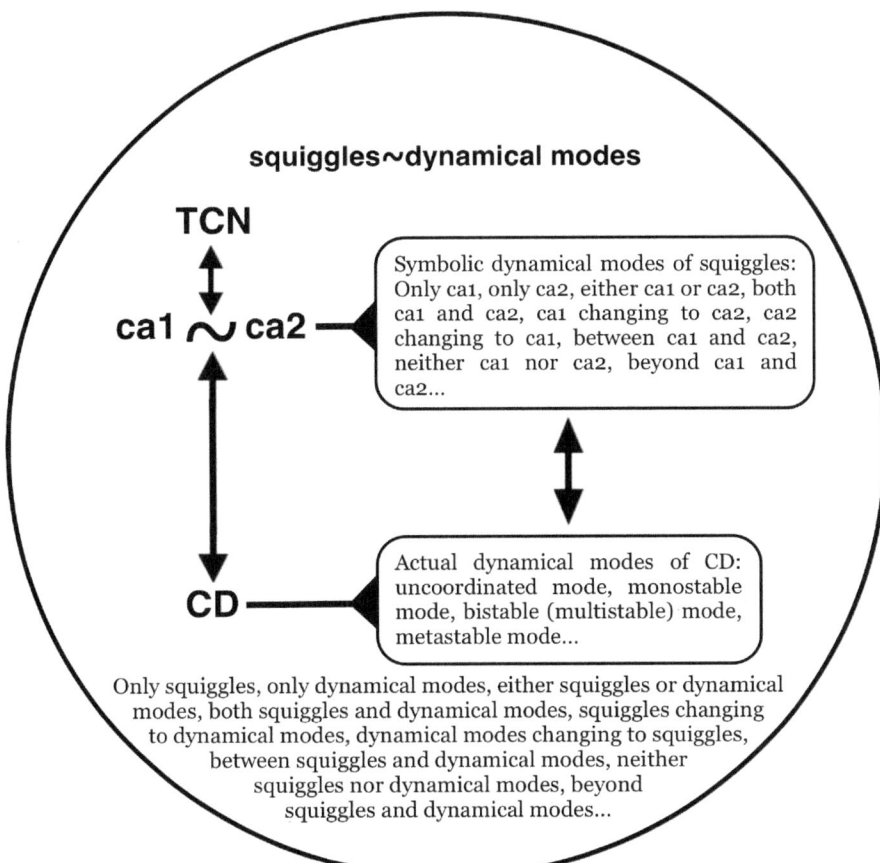

squiggles~dynamical modes

TCN

ca1 ~ ca2

Symbolic dynamical modes of squiggles: Only ca1, only ca2, either ca1 or ca2, both ca1 and ca2, ca1 changing to ca2, ca2 changing to ca1, between ca1 and ca2, neither ca1 nor ca2, beyond ca1 and ca2...

CD

Actual dynamical modes of CD: uncoordinated mode, monostable mode, bistable (multistable) mode, metastable mode...

Only squiggles, only dynamical modes, either squiggles or dynamical modes, both squiggles and dynamical modes, squiggles changing to dynamical modes, dynamical modes changing to squiggles, between squiggles and dynamical modes, neither squiggles nor dynamical modes, beyond squiggles and dynamical modes...

The Metastabilian says: The *squiggle* notation 'ca1~ca2' and symbol '~' enhance my *squiggle sense* of the complementary nature (TCN). They provide a useful syntax for dynamical, inextricable, complementary aspects of *squiggles* that I can use to contemplate, study, express and communicate their qualitatively different symbolic dynamical modes. The notation is even more powerful and compelling as the symbolic dynamical modes of *squiggles* correspond to actual dynamical modes of CD. In turn, the syntax provides a lucid way to think about and comprehend the *squiggles* of CD, called 'The Complementary Code' (TCC).

Related squiggles: symbol~dynamics, syntax~semantics, TCN~CD...

10 Of Theory and Experiments

To me, what makes physics physics is that experiment is intimately connected to theory. It's one whole—Lene Vestergaard Hau

Your *squiggle sense* is grounded in the science of Coordination Dynamics (CD). What this means is that both your *squiggle sense* and the coordinated dances of the *squiggles* it senses can be understood and explained by CD in two main ways: (1) as phenomena of CD, that is, as observable facts and measurements about objects and events discovered or obtained by experiments; and (2) as noumena of CD, that is, as posited conceptual elements or principles of theory, which are most effective when expressed in the language of mathematics. Your engaged *squiggle sense* should be blinking now, as you ponder the complementary nature of phenomena~noumena, theory~experiments, concepts~observables, ideas~applications... What other *squiggles* come to mind right now? Make an experiment of your own!

In CD, all of its experiments are intimately connected to the theory and the development of models, and vice-versa. New experimental observations lead to new theoretical models, which as a result lead to the design of further experiments to test model predictions, and so on. Such is the normal course of science and the scientific method. The complementary nature of CD entails conceptual noumena and observable phenomena, both of whose origins and dynamical behavior are central to CD's theoretical~experimental scientific paradigm. That many of CD's *squiggles* have been described since antiquity is not surprising. What has transpired over the last 40 years or so, however, is that CD has provided novel perspectives and insights into where these *squiggles* come from and our awareness of their complementary nature.

In the remaining *squiggle* frames, the theoretical~experimental paradigm of CD is unpacked via its *squiggles*, its complementary code. But here's a little taste of it to get you started: As we have already noted, students of CD seek to identify the laws, principles and mechanisms underlying coordinated behavior in all kinds of different systems and contexts. CD explicitly addresses coordination within~between levels, intra- and inter-actions between bodies, brains and environments. The original experimental paradigm that led to CD involved the coordination of human finger movements. From these simple experiments and the many extensions and variations they inspired, a deep understanding of the nonlinear dynamics of the human brain and its relation to behavior ensued—the current theoretical~experimental paradigm of CD was born!

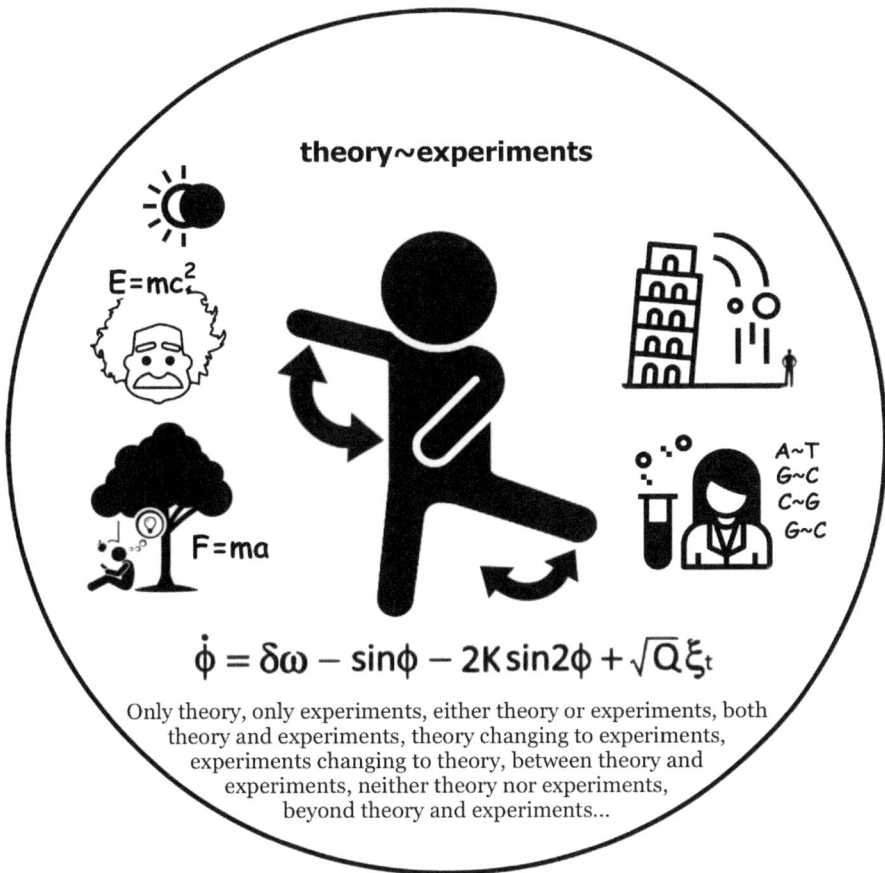

theory~experiments

$E = mc^2$

$F = ma$

A~T
G~C
C~G
G~C

$$\dot{\phi} = \delta\omega - \sin\phi - 2K\sin2\phi + \sqrt{Q}\,\xi_t$$

Only theory, only experiments, either theory or experiments, both
theory and experiments, theory changing to experiments,
experiments changing to theory, between theory and
experiments, neither theory nor experiments,
beyond theory and experiments...

The Metastabilian says: What better place to discover a general principle of nature than in the Coordination Dynamics of animate movement! All the signatures of the complementary nature are present. An amazing feature of CD is that its entire theoretical~experimental strategy can be captured by its complementary code, TCC of CD, the set of *squiggles* that comprise its conceptual~phenomenal boundary~domain. This is a very handy insight, since it implies that the *squiggles* in any viable system at any level that attracts my interest can be explored, comprehended and expressed via CD.

Related squiggles: concept~model, noumena~phenomena, ideas~observables...

11 Of Brain and Mind

 …the splitting off from each other of religion, morals and science; the divorce of philosophy from science and of both from the arts of conduct. The evils which we suffer in education, in religion, in the materialism of business… in the whole separation of knowledge and practice—all testify to the necessity of seeing mind–body as an integral whole—John Dewey

Your *squiggle sense* is by definition a sixth sense of your brain and mind. It's something your brain actually does. It's physical, physiological. It's also something your mind does. It's mental, psychological. Hopefully, your engaged *squiggle sense* will now help you transcend the mind-numbing, seemingly never-ending, polarizing debates over which is more primary, brain or mind (body or mind, matter or mind…). Or how matter makes mind and mind makes matter. Coordination Dynamics (CD) sees the dynamic, inextricable dance of your brain and mind as dynamic complementary aspects of a *squiggle.* And your brain~mind produces your *squiggle sense,* your *sixth sense* of the *squiggles* of the complementary nature.

What's more, the complementary nature of brain~mind, like all *squiggles,* is grounded in CD. Another way to say this is that the *squiggle sense* of your brain~mind can be studied, understood and explained within the paradigm and language of CD. Moreover, studies of one of the dynamical modes discovered in CD, called metastability or the metastable mode, has led to an entirely novel conception of how the brain~mind works. In the metastable mode of brain~mind dynamics, tendencies for individual regions of the brain to function as specialized, segregated entities coexist with collective, coordinative tendencies for them to bind together and integrate as functional coalitions. The discovery of the complementary nature of integration~segregation tendencies in the metastable mode of brain~mind dynamics is elegant and profound. It sheds light on many mysteries of human capabilities, thought and behavior.

The main one is that the metastable mode of brain~mind produces your *squiggle sense.* The kindling of your *squiggle sense* coincides with a shift of your brain~mind CD to its metastable mode. Brain~mind metastability is the unique signature of the complementary nature. Once you realize this, you become highly sensitive to it. Metastability shows how apparently contrasting properties of the brain and mind coexist, and how they may be reconciled. "Pure" states of integration or segregation are interpreted as polarized, idealized extremes of a vast dynamical landscape that sweeps flexibly between them. Taking stock, the complementary nature of CD *squiggles,* as do all its complementary pairs. And with your metastable based brain~mind's *squiggle sense*, you perceive and act upon the complementary nature of your world.

J. A. S. Kelso and D. A. Engstrøm, *The Squiggle Sense,*
https://doi.org/10.1007/978-3-031-59369-7_11

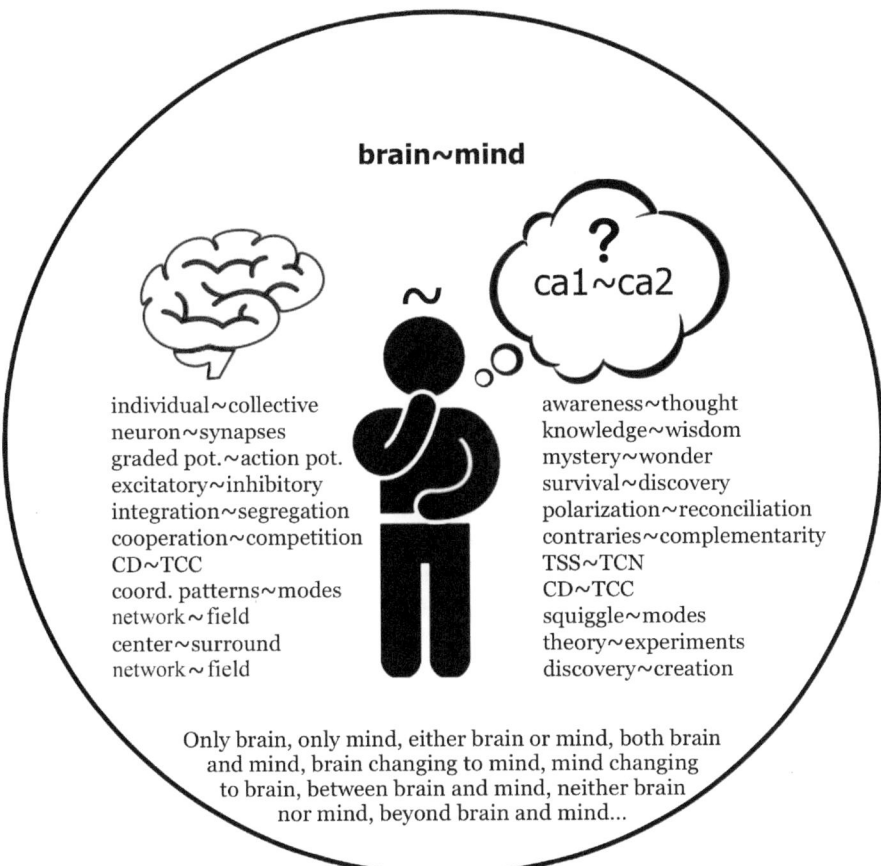

The Metastabilian says: Coordination of brain~mind is like a Ballanchine ballet—neural groups and thoughts briefly couple, some join as others leave, new groups form and dissolve, creating fleeting dynamical coordination patterns of mind that are always meaningful but don't stick around for very long. Such transient coupling~uncoupling tendencies within~between individual brain regions and within~between thoughts and actions underlie the workings of the brain~mind and its complementary nature. Incredible!

Related squiggles: physical~mental, body~mind, behavior~cognition...

12 Of Metastable Mode and Complementary Code

The test of a first-rate intelligence is the ability to hold two opposed ideas in the mind at the same time, and still retain the ability to function—F. Scott Fitzgerald

Your *squiggle sense* becomes engaged when you enter your brain~mind's metastable mode. In these moments, you can tune in to the complementary nature of your world as a realization of your brain~mind expressing itself. While in your metastable mode, you're able to sense, perceive, experience the dynamic mutuality of complementary aspects. In those moments, you have the ability to hold two contrary ideas in your mind at the same time and still retain the ability to function! In Bohr's words, you're able to "hold opposites together in your mind, suspend your normal thinking process and allow a first-rate intelligence beyond rational thought to create new forms and ideas." But then what? What do you do with that?

Start by realizing that your brain~mind operates in its metastable mode many times a day, even if you aren't aware of it as such. If you aren't aware of it, your *squiggle sense* is, as it were, "flapping in the wind". In those moments, what comes of the complementary stimuli presented to you, whether valuable, useful, disturbing, random or inconsequential, is catch-as-catch-can. The irony in this case is that you have a powerful, vital sense available to you that is operational yet often acting independently of your will. This futile situation is what the complementary code helps you to overcome, giving you a way to recognize, engage, make sense and wield your *squiggle sense* intentionally. The complementary code of CD presented in the *squiggle* frames of this book give you a way to intentionally engage and develop your *squiggle sense*.

The *squiggle* notation 'ca1~ca2' symbolizes the *squiggle sense* of any given, or even potential complementary aspects. The *squiggle* (~) not only symbolizes the coexistence of complementary aspects, but also their Coordination Dynamics (CD)—the different possible dynamical modes or coordination patterns and transitions between them. The *squiggle* (~) of the Complementary Code also symbolizes both the *squiggle sense* itself and the CD that gives rise to all of your potential brain~mind's dynamical modes, especially its metastable mode. Without the complementary nature and metastability, there would be no complementary code to express, nor anyone to express it to. The metastable mode grounded in CD and the complementary code it gives rise to are inextricable, dynamic and complementary. They are themselves complementary aspects of a *squiggle,* whose *squiggle* dance is crucial to the sentient call of Metastabilians.

J. A. S. Kelso and D. A. Engstrøm, *The Squiggle Sense*, https://doi.org/10.1007/978-3-031-59369-7_12

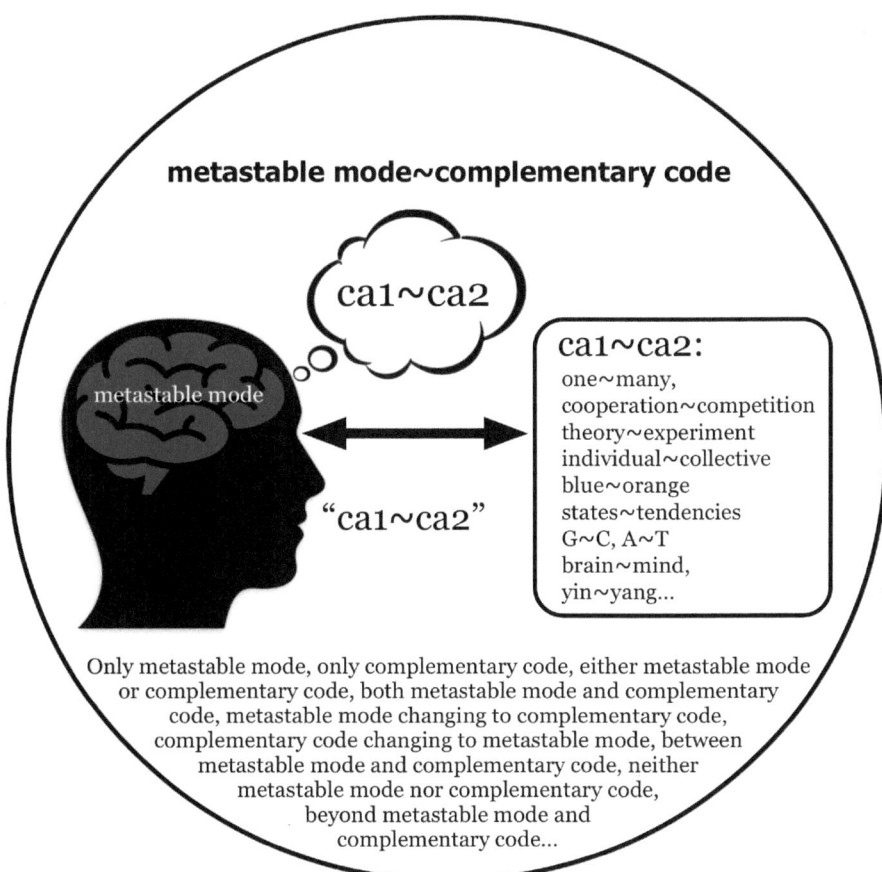

The Metastabilian says: How do you engage your *squiggle sense*? One way is to play The *Squiggle Game*. All you need is the complementary code! (1) Choose any pair of contraries, and write them as a *squiggle*, like 'choice~chance'. (2) Under the *squiggle*, write out its *squiggle* expansion—Only choice, only chance, either choice or chance, both choice and chance, choice changing to chance, chance changing to choice, between choice and chance, neither choice nor chance, beyond choice and chance… (3) read through the expansions and try to 'imagine' each member of the expansion set several times from first to last, then last to first. Then try skipping around. (4) Try your *squiggle* scans to a metronome, from slower to faster, then faster to slower. (5) Remember the goal: to enter the metastable mode of your brain~mind; to engage your *squiggle sense*.

Related squiggles: metastability~complementarity, dynamical~symbolical, noumena~phenomena…

13 Of Dynamic Patterns and Pattern Dynamics

 Coordination dynamics reveals the coordination of life as an emergent, self-organizing process where many diverse parts of a living system synergize to form coherent, meaningful dynamic patterns of behavior—J. A. Scott Kelso

The Coordination Dynamics (CD) that grounds your *squiggle sense* in science treats the coordination of your brain~mind as an emergent, self-organizing process in which its many diverse complementary aspects gang together to form coherent, meaningful dynamic patterns of behavior. One of these coherent, meaningful dynamic patterns is called metastability, or your metastable mode. Not only is your *squiggle sense* a coherent, meaningful dynamic pattern of your brain~mind's metastable mode, it also senses dynamic patterns of the complementary nature, its *squiggles*. CD is a science of the complementary nature of your brain~mind's dynamic patterns. CD also derives from your brain~mind's dynamic patterns, the dominant form of which is the creative metastable mode that underlies your *squiggle* sense.

In the paradigm of CD, dynamic patterns of brain activity produce your thoughts and emotions, the learning and memory that leads to your knowledge and wisdom, mystery and wonder, survival and discovery, your movements and skills—basically everything you are and do as a human being! As such, one of the main tasks of CD is the discovery, identification and characterization of the many specific dynamic patterns produced by different systems and levels of living systems and the relations between them. Another is the elucidation and study of the underlying pattern dynamics, the laws that generate observed dynamic patterns. Pattern dynamics is the universal set of principles and laws by which dynamic patterns of emergent, self-organized processes of all systems and levels emerge, persist, evolve and change.

Pattern dynamics is a fundamental pillar of the CD paradigm. An amazing fact about CD is that although the information required to characterize nature's dynamic patterns is, in principle, near infinite, all the complexity is enfolded into a self-organized pattern dynamics that follows much simpler, albeit nonlinear, universal laws. A main insight of CD is that the entire repertoire of dynamic patterns and universal pattern dynamics can be recast as a base set of complementary pairs or *squiggles*, which now constitute 'the complementary code of coordination dynamics' (TCC). Dynamic patterns~pattern dynamics is a principal *squiggle* of the complementary code of coordination dynamics. It's one of those *squiggles* upon which the science of your *squiggle sense* depends.

J. A. S. Kelso and D. A. Engstrøm, *The Squiggle Sense*,
https://doi.org/10.1007/978-3-031-59369-7_13

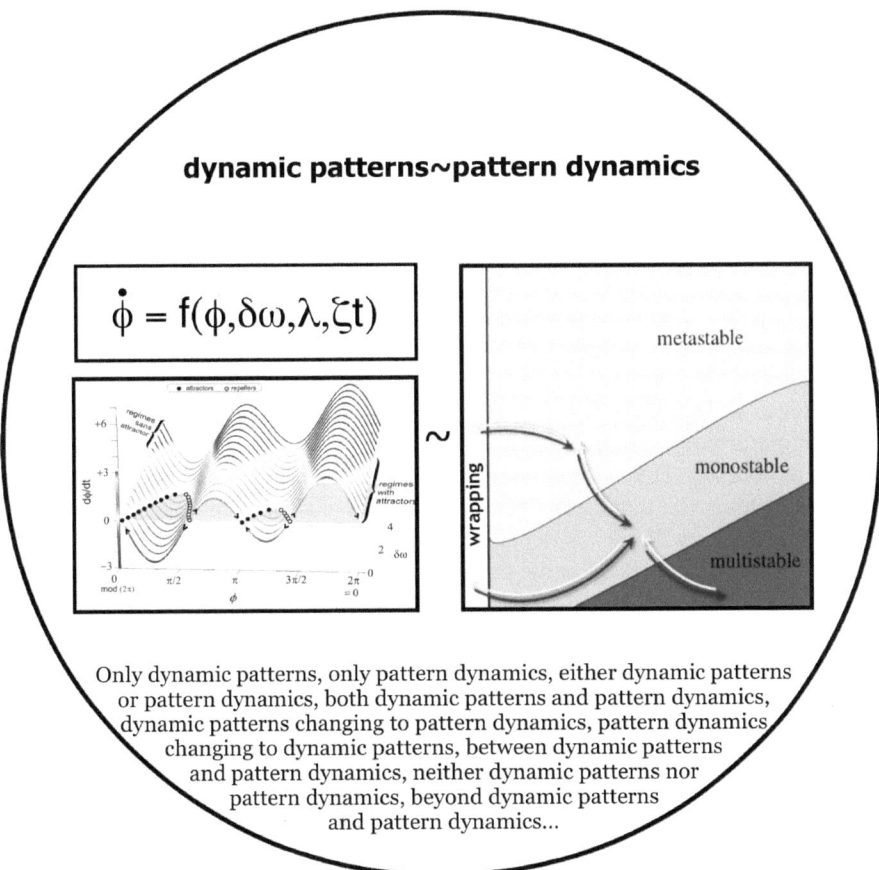

The Metastabilian says: I'm an emergent, self-organized, complementary being, whose many diverse complementary aspects form coherent, meaningful dynamic patterns of structure~function and whose pattern dynamics consist of multiple unique dynamical modes of coordination. My squiggle sense discovers~explores the complementary nature of my world and myself. The complementary code of CD and its scientific grounding of dynamic patterns~pattern dynamics bolsters and stimulates my understanding, confidence and expression of *the squiggle sense*.

Related squiggles: phenomena~laws, code~meaning, squiggles~CD...

14 Of Self-Organization and Agency

 "γνῶθι σεαυτόν"—Delphic Maxim, inscribed in the forecourt of the ancient temple of Apollo

"Know thyself" is an ancient Delphic maxim, a piece of timeless wisdom. It's also the ultimate conundrum, your personal daily paradox. Self-awareness is a taproot of your existence. It plays a major role in all that you are and do, even if you don't pay attention to it. Like everything else about yourself, self-awareness is a product of the complementary nature. To really know yourself is to know the complementary nature of yourself. This is an extraordinary insight: to know yourself, you need your *squiggle sense*! And that means your brain~mind must shift to its metastable mode. How you do that will always depend on many factors, not all of which can ever be specified. But reflecting upon the complementarity of self-organization and agency is a good place to start, as this *squiggle* is arguably the bullseye of the conundrum!

Self-organization, or the spontaneous formation of dynamic patterns, occurs in *open systems* that exchange matter, energy and information with their surrounding environment. No specific ordering, programs or prescriptions are responsible for the emergence of these dynamic patterns. They arise solely as a result of properties typical of open systems, like many fluctuating elements, nonlinear interactions between them and dissipation of free energy. In such conditions, open systems organize themselves! In CD, such spontaneously arising patterns are referred to as intrinsic dynamics. Of course, you're not exclusively driven by your self-organized intrinsic dynamics. Your thoughts, desires, memories and intentions can also direct your actions and lead to your deeds. In scientific language, this directedness of life is called agency, or action towards an end.

Agency began for you in utero with spontaneous movements, your arms and legs waving and kicking, mouth opening and closing, body bending and twisting. You came into the world moving. This repertoire of basic spontaneous, self-organized coordinated movements formed before you even knew how to make them, even before you were aware *you* were making them. Yet, through the internal~external sensations arising from this basic movement repertoire, you discovered it was your own. In CD, spontaneous self-organizing coordination tendencies give rise to agency, which is then able to steer the very coordination dynamics that produces it. Directed self-organization is when spontaneous pattern forming tendencies and agency coexist and complement each other. From such spontaneous self-organized behavior your self emerges—"I do therefore I am". Your *squiggle sense* is such a self-organized process. You are the agent that directs it, and so can come to know yourself.

J. A. S. Kelso and D. A. Engstrøm, *The Squiggle Sense*,
https://doi.org/10.1007/978-3-031-59369-7_14

The Metastabilian says: To know myself is to know the complementary nature of myself. And so I must sense the complementary nature of myself, must perceive its self-organizing intrinsic dynamics such that I can learn to interact with and finally direct it at will. And to accomplish that action towards end, that agency, I must engage my *squiggle sense*. And for that, I must shift my brain~mind to its metastable mode. And so to know myself, I must become my Metastabilian self!

Related squiggles: action~perception, brain~mind, curiosity~awareness…

15 Of Intentional Action and Intrinsic Dynamics

 Intending, one does karma (acts) by way of body, speech, and intellect—The Buddha, Nibbedhika Sutra, Anguttara Nikaya 6.63

As an infant, your spontaneous, self-organizing intrinsic dynamics pretty much ran the show. Later, your newly discovered ability to direct it intentionally was mostly exploratory. But its influence rapidly flourished as you began intentionally directing your coordinated action in the world. Soon, your agency, or intentionally directed intrinsic dynamics, became your rudimentary self-awareness. You became *you*. This directed, voluntary, purposeful, willed, intentional action—is your "agency directing intrinsic dynamics towards ends", though you probably don't think about it that way. If you don't, you're not alone. Most of us seldom consider the effects our intentions have on our intrinsic dynamics and *vice-versa.*

In Coordination Dynamics (CD), intentional action is proposed to emerge from the coupling between you and your world, attracting and moulding the actions of your intrinsic dynamics toward desired ends. Your intrinsic dynamics dictates the influence intentions can have. At the same time, your intentions can stabilize and destabilize your intrinsic dynamics in the direction of your intentional action. Intentions both constrain and are constrained by intrinsic dynamics. CD captures the complementary nature of intention and intrinsic dynamics. It shows what intentions do and how they do it: how *intentions* are constrained by the stability of your intrinsic dynamics and how the latter influences which actions actually occur and how fast you can switch flexibly and adaptively between them.

Think about breathing. Breathing is vital and complex, physically, mentally, and spiritually speaking. From a CD perspective, breathing is a directable, spontaneous self-organized pattern of inspiration and expiration. Up to a point, you can hold your breath. Directing your breath, like when you eat or speak, is intentional action constrained by the intrinsic dynamics it intends to modify. When you stop directing it, spontaneous breathing takes over again. Your *squiggle sense* also has intrinsic dynamics. And if you know about it, you can wield it, direct it towards sensing the complementary nature. If you aren't aware of, or ignore it, its spontaneous, self-organizing, metastable intrinsic dynamics takes over again. Your new awareness of the *squiggle sense* and ability to intentionally direct it is still mostly exploratory. But its influence can rapidly grow and develop. When the intentional, agency-directed intrinsic dynamics of your *squiggle sense* becomes routine, you've become a Metastabilian!

J. A. S. Kelso and D. A. Engstrøm, *The Squiggle Sense*, https://doi.org/10.1007/978-3-031-59369-7_15

intentional action~intrinsic dynamics

Only intentional action, only intrinsic dynamics, either intentional action or intrinsic dynamics, both intentional action and intrinsic dynamics, intentional action changing to intrinsic dynamics, intrinsic dynamics changing to intentional action, between intentional action and intrinsic dynamics, neither intentional action nor intrinsic dynamics, beyond intentional action and intrinsic dynamics...

The Metastabilian says: Intending to engage the *squiggle sense* of my metastable brain~mind, I do karma: I act by way of my body, speech and intellect. My intentional action influences the intrinsic dynamics of my *squiggle sense* as its intrinsic dynamics influences my intentional action to engage it. These are complementary aspects of my agency. They *squiggle*. And to what end are they directed? To sense and act upon the very complementary nature from which they have emerged!

Related squiggles: voluntary~automatic, purposeful~spontaneous, conscious~unconscious...

16 Of Perception and Action

The firing pattern of both mirror and canonical neurons in area F5 shows clearly that perception and action are not separated in the brain. They are simply two sides of the same coin, inextricably linked to each other—Marco Iacoboni

Perception and action are essential complementary aspects of life. Without your ability to perceive, you could never have discovered your spontaneous, self-organized intrinsic dynamics in the first place. Your agency, sentience, or self-awareness could not have emerged. Without action, you would have had neither ways nor means to respond to and modify your perception, no way to direct your intrinsic dynamics towards ends. Your agency would have had no way to develop, function, adapt and evolve. The complementary nature of perception~action is the foundation of human intelligence, a complementary dance of your body and brain~mind directed and driven by the very perception~action processes they engender.

Perception and action are neither separable nor dualistic. They are two sides of the same coin—dynamically coexistent, mutually related and inextricably linked. They constitute complementary aspects of one of the principal *squiggles* of The Complementary Code of CD. Perception~action arises from the on-going emergent coupling of your body~mind with your world and requires *active sensing* (seeing, hearing, tasting, touching, smelling…). Perception~action of your *squiggle sense* of the complementary nature requires your brain~mind to enter its metastable mode.

Strong hints of metastable brain~mind come from studies of the perception~action dynamics of 'ambiguous figures', where a constant image is perceived to spontaneously switch back and forth between two interpretations. In experiments, switching times of perceived ambiguous stimuli are hypothesized to be generated by an underlying intermittent, metastable neural mechanism. This suggests at least one way your brain~mind flexibly enters and exits coherent brain patterns and avoids getting trapped in mode-locked states. These unusual situations of ambiguous stimuli (often referred to as illusions) hint strongly that your *squiggle sense* is active even if you aren't intentionally directing it. They help to make you aware that life presents you with uncountable uncertain, often ambiguous situations. You also know how easy it is to get stuck in a single mindset. Engaging your *squiggle sense* can alter your perspective of the many fuzzy, ambiguous situations that occur in life, providing you an entirely novel point of view. Know this for sure: there's nothing ambiguous or uncertain about the *squiggle sense* or the Metastabilian who wields it!

© The Author(s), under exclusive license to Springer Nature Switzerland AG 2024
J. A. S. Kelso and D. A. Engstrøm, *The Squiggle Sense*,
https://doi.org/10.1007/978-3-031-59369-7_16

perception~action

Only perception, only action, either perception or action, both
perception and action, perception changing to action, action
changing to perception, between perception and
action, neither perception nor action,
beyond perception and action...

The Metastabilian says: To understand perception~action, or any *squiggle*, I must not only discover facts, draw logical inferences and construct theories, but also consider different points of view. My metastable mode is a way of thinking or understanding that sees beyond the data given. It's not only about intentional looking. In many situations in life, although nothing may change input wise, what you perceive depends on your perspective, on how you see it. Whereas Gestalt (organization~transformation) and ecological (affordance~effectivity) explanations are related and relevant, my metastable mode transcends theories of perception. It's not just about percepts and perceptual change. It's more about "seeing" things, the sixth sense of the poet~scientist—the *squiggle sense* of perception~action.

Related squiggles: sensing~moving, affordance~effectivity, information~meaning...

17 Of Intrinsic Dynamics and Functional Information

 It is not the strongest of the species that survives, not the most intelligent that survives. It is the one that is the most adaptable to change—Charles Darwin

Consider the self-organizing, intrinsic coordination dynamics of your body~mind. You first perceived your body as an infant, discovering you could direct it as your agency and self-awareness emerged. And you have directed it ever since. But now as ever, to survive and thrive, your awareness must interact with and relate to the world. For this, you depend on your body~mind to perceive movement, space, surfaces, textures, weight, sounds, smells, temperature, time, the phases of the moon—all of which inform your awareness. You use this information to know what's going on, to guide, navigate, and coordinate your intrinsic dynamics (ID) with the world. In Coordination Dynamics (CD), such useful, meaningful information that you use to enhance your ability to adapt and change—ultimately your fitness for survival—is called functional information (FI).

Of course, the ocean of information zipping around the internet demonstrates the extent to which humanity has become inundated, dependent, even addicted to "information". But is all that information functional? Not likely. Whether information is useful and meaningful obviously depends. What is meaningful to one person or creature can be meaningless to another. The scale of the organism, its capacities, stage of development, knowledge, skill level, etc. all affect whether information is meaningful or not. Can your *squiggle sense* help you survive and thrive in the face of uncertainty? It's very likely! In CD, functional information is meaningful and specific to your intrinsic dynamics, which depend on the range, scope and context of all the different activities you perform, like when you communicate, drive, eat, speak, read, adapt, learn, and remember... "If it doesn't affect your ID, it isn't FI."

And there it is: intrinsic dynamics and functional information are complementary. They are a *squiggle*. Functional information lies in the coupling between you and your surrounding environment. Once created, it can stabilize and destabilize dynamic patterns of behavior. Internal and external conditions often have to be just right for functional information to express itself. FI not only modifies your existing intrinsic dynamics but its effects are tempered by them in a kind of circularly causal way. Context is crucial. What is meaningful in one situation or scenario may be meaningless in another. Your engaged *squiggle sense* allows you to perceive the complementary nature. If the words you are now reading enhance that ability, they can be considered a source of functional information that has altered the intrinsic dynamics of your *squiggle sense*!

J. A. S. Kelso and D. A. Engstrøm, *The Squiggle Sense*, https://doi.org/10.1007/978-3-031-59369-7_17

ID~FI

Only intrinsic dynamics, only functional information, either intrinsic dynamics or functional information, both intrinsic dynamics and functional information, intrinsic dynamics changing to functional information, functional information changing to intrinsic dynamics, between intrinsic dynamics and functional information, neither intrinsic dynamics nor functional information, beyond intrinsic dynamic and functional information...

The Metastabilian says: Now as I embrace the complementary nature of functional information~intrinsic dynamics, I begin to appreciate the profound advantages and benefits of the awakened *squiggle sense* of my metastable mode. Functional information influences the intrinsic coordination patterns of life in every boundary~domain—evolutionary, developmental, learning, social, economic, cultural, public, personal and private. It provides new meaning and utility, new perspectives to enjoy and explore, to experience the awe-inspiring mystery and wonder of the grand coordination of the complementary nature, and the miracle of my *squiggle* sentience! It allows me to be much more adaptable. It allows me to survive.

Related squiggles: talent~training, nature~nurture, innate~learned...

18 Of Learning and Memory

 The biggest mistake of past centuries in teaching has been to treat all students as if they were variants of the same individual and thus to feel justified in teaching them all the same subjects the same way—Howard Gardner

The *squiggle sense* is a key to a virtually untapped potential within you to enhance your ability to learn, remember, adapt and evolve. In Coordination Dynamics (CD), memory refers to the dynamic stabilization of what is learned. The process of stabilization of learned information occurs on many timescales, as in short and long term memory. Memory also takes many forms, like implicit and explicit, recognition and recall, declarative and procedural, and occurs on many levels, from the synapses that connect neurons to the collective memory of entire cultures. How it's done remains a great, unsolved scientific mystery. In CD, memory, the "lasting effect of learning and adaptive processes", is recast as "persisting modifications of intrinsic dynamics~functional information." This is useful. It suggests how learning~memory happens: information becomes functional when it modifies your intrinsic dynamics just as your intrinsic dynamics constrain, shape and modify new information to be learned.

So how successful learning and memory formation are depends both on your intrinsic dynamics and the efficacy of the information that modifies it. Like all human beings, your intrinsic dynamics is unique and ever evolving. CD research verifies the idea that people learn and remember in their own way. So how do you teach the same subject to a collection of different students? The usual educational strategy is to treat all learners as if they were variants of the same model individual, and to teach all subjects the same way. Though arguably expedient from an institutional perspective, it's often found to be less than efficacious. A contrary approach is to treat learners individually, teaching the *same* subjects different ways, according to the way a learner learns.

This sounds good, but it's easier said than done. A major challenge is that it takes time, effort, resources and means to implement it. How might one survey the intrinsic dynamics of all students, and adjust the learning material accordingly? You might have caught the *squiggle* irony here: "two contrary strategies for learning…" Doesn't that suggest a reconciliation is in order? In practice (and many effective teachers are aware of this), tolerance and incorporation of both teaching strategies is predicted to be the most effective. Such an approach follows learning~memory's grounding in CD: It says your *squiggle sense* of the complementary nature is engaged at the core nexus of learning~memory, in your brain~mind's metastable mode, where functional information is created~destroyed. Could strategies to foster the *squiggle sense* of learners be the key to a long awaited, evolutionary advance in education?

J. A. S. Kelso and D. A. Engstrøm, *The Squiggle Sense*,
https://doi.org/10.1007/978-3-031-59369-7_18

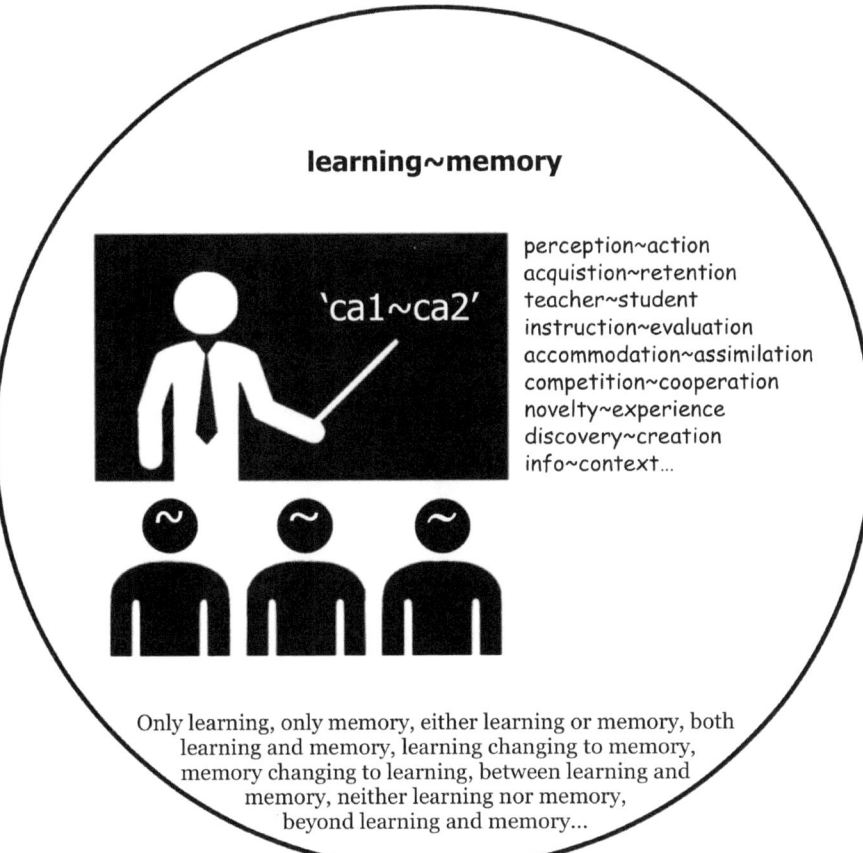

learning~memory

'ca1~ca2'

perception~action
acquistion~retention
teacher~student
instruction~evaluation
accommodation~assimilation
competition~cooperation
novelty~experience
discovery~creation
info~context...

Only learning, only memory, either learning or memory, both learning and memory, learning changing to memory, memory changing to learning, between learning and memory, neither learning nor memory, beyond learning and memory...

The Metastabilian says: And once again, I marvel at my *squiggle sense* of the complementary nature, at its grand coordination—how all my thoughts, plans and realizations, my coordinated decisions and actions mix in the metastable mode of my brain~mind! They cooperate~compete, modify and are modified by functional information~intrinsic dynamics. Such modifications change me and I them—my agency driven self-organization, my evolutionary trajectory, the mystery~wonder of my individual~collective learning~memory, whose persistence and change carry me through the vast, fleeting, shimmering dynamical landscape of metastable CD!

Related squiggles: acquisition~retention, novelty~experience, accommodation~assimilation...

19 Of Individual and Collective

 Synergy is the only word in our language that means behavior of whole systems unpredicted by the separately observed behaviors of any of the system's separate parts or any subassembly of the system's parts—R. Buckminster Fuller

At this time, humanity is a collective of about 8 billion human individuals, of which you are one. You're also a collective of around 37 trillion individual cells, coordinated in the dance of your life—a spontaneous, self-organized intrinsic dynamics acting on many space and timescales. Such staggering complexity is next to impossible to comprehend. Or is it? The roles of individual and collective in the context of life's complexity have been incessantly debated throughout history. Most often, one or the other is deemed more fundamental. Your *squiggle sense* can help you transcend such debates as you perceive individual and collective to be complementary aspects, and coordination dynamics (CD) to be the key to understanding their complementary nature. In fact, all of the complex systems studied by CD deal with individual~collectives, dynamic, self-organized collectives of individual coordinating elements or synergies. Every individual~collective is a synergy and vice versa.

You may know that 'synergy' means the combined effects of a whole, collective system that can't be predicted by separately observing only its individual parts. In viable complex systems, synergies are adaptive structural~functional entities that are the target of developmental, learning and evolutionary processes. But where do synergies come from, how are they formed and what functions do they serve? CD explains how individual~collective synergies arise, how they work and what they can do. Together with the rest of the *squiggles* of its complementary code, individual~collectives, or synergies, are crucial to comprehending the complexity of your own awareness, your collective~individual intrinsic dynamics, the synergy of your complementary brain~mind and its *squiggle sense*!

The complementary nature of synergies means that collective wholes and their individual parts are dynamic, coexistent and mutually entailed. Individual coordinating elements are only individual if they can retain some degree of autonomy while constrained by a collective whole. Collective wholes entail dynamic constraints that serve to couple individual coordinating elements together, thereby affecting how those individuals behave. A key discovery of CD is that in a complex system's metastable mode, tendencies for individual expression coexist simultaneously with tendencies for individuals to coordinate themselves as a collective. Many of humanity's problems and conflicts can be usefully addressed by appreciating the *squiggle sense*—the cooperative~competitive, complementary nature of individual~collectives.

J. A. S. Kelso and D. A. Engstrøm, *The Squiggle Sense*,
https://doi.org/10.1007/978-3-031-59369-7_19

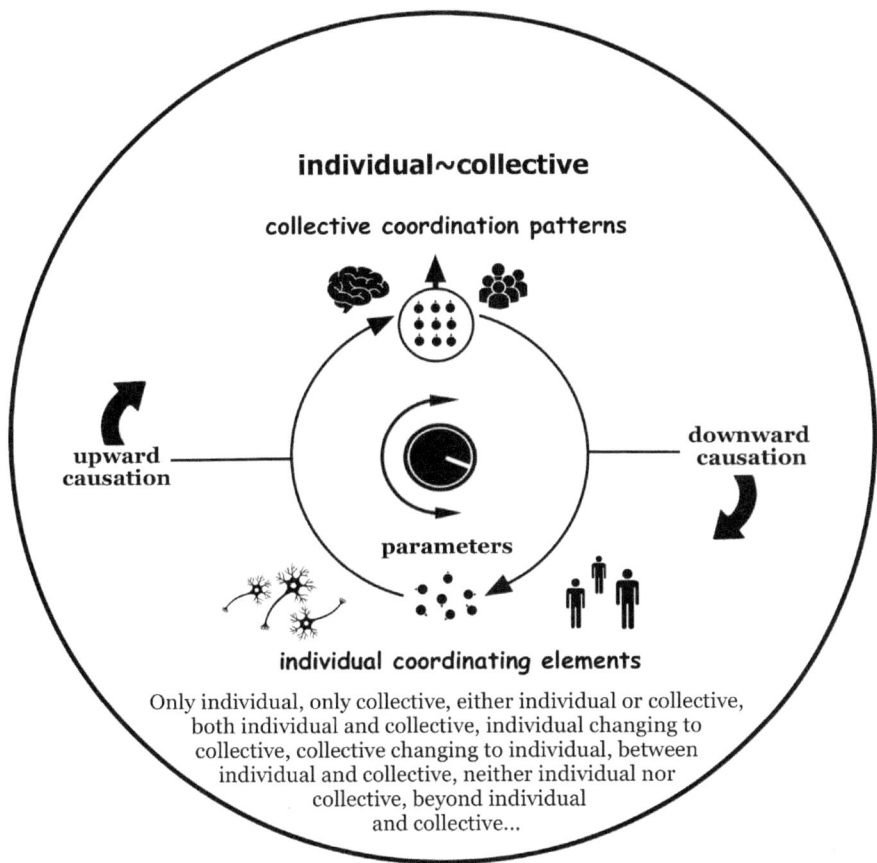

individual~collective

collective coordination patterns

upward
causation

downward
causation

parameters

individual coordinating elements

Only individual, only collective, either individual or collective,
both individual and collective, individual changing to
collective, collective changing to individual, between
individual and collective, neither individual nor
collective, beyond individual
and collective...

The Metastabilian says: How can I reconcile this ultimate complexity, this 'Grand Coordination', this cosmic individual~collective, synergy of synergies where a part is a whole and a whole is a part, dynamically speaking, together~apart? My *squiggle sense* is the key. While it is engaged, my brain~mind is in its metastable mode and my sixth sense of the complementary nature is awake and active. I am the individual~collective reconciliation, and directly experience the behavior of the whole unpredicted by the behavior of the parts. In those fleeting moments, I am the one in the many and the many in the one!

Related squiggles: one~many, part~whole, component~system, top-down~bottom-up...

20 Of Dissipation and Cycles

 Energy that flows through a system acts to organize that system—Harold Morowitz

Your *squiggle sense* is a way to grasp and appreciate your self-organizing intrinsic dynamics, your individual~collective synergy of synergies. It's the grounded *squiggles* of the complementary code of Coordination Dynamics (CD) that enables a deep connection to the complementary nature. At the heart of those vital *squiggles* chugs the complementary aspects of dissipation and cycles. Realize that CD refers to dissipative, open, nonequilibrium systems, just like you and all living creatures. Energy, matter and information from the environment flow into such systems as "sources" acting to organize them, and energy and waste flows out of them from "sinks" (sources~sinks).

Consider the fact that you breathe in air and eat food (sources). Your body uses the energy and materials to sustain your self-organized biological integrity, your intrinsic dynamics. Meanwhile, you eliminate CO_2, waste and heat (sinks). Similarly, plants absorb incident light and CO_2 (sources), and eliminate oxygen and heat (sinks). Only dissipative systems are capable of producing and sustaining the self-organized dynamic patterns~pattern dynamics of individual~collective synergies. But how is dissipative energy flux inextricably linked to life's cycles? Well, it creates them! Morowitz's Theorem from the field of biophysics says that in systems "in a non-equilibrium steady state, the flow of energy from a source to a sink creates a cycle." This is what Morowitz meant when he said, "energy that flows through a system acts to organize it".

Cycles are the archetypes of time-dependent behavior of all living things at all scales from cells to economies—all produced and sustained via dissipative energy flux. The dynamic functions and structures of life like glycolysis, DNA and proteins, cells, organs, organisms and their behavior are all inherently cyclical. They cycle with different characteristic periods on many timescales, from nanoseconds to years, contributing to the wide variety of rhythmic behaviors observed in organisms. CD views the brain itself as a 'geography of improvised rhythms' that resonate with the world. Without source~sinks and energy dissipation, there are no cycles. And without cycles, there is no coordination, no possibility of sustained life. But notice, Morowitz's Theorem doesn't tell you how cycles are actually produced, how they self-organize, or how the many different cycles on different levels of a dissipative system become synergistically coordinated. For that, you need to learn about the rest of the complementary code of CD.

J. A. S. Kelso and D. A. Engstrøm, *The Squiggle Sense*,
https://doi.org/10.1007/978-3-031-59369-7_20

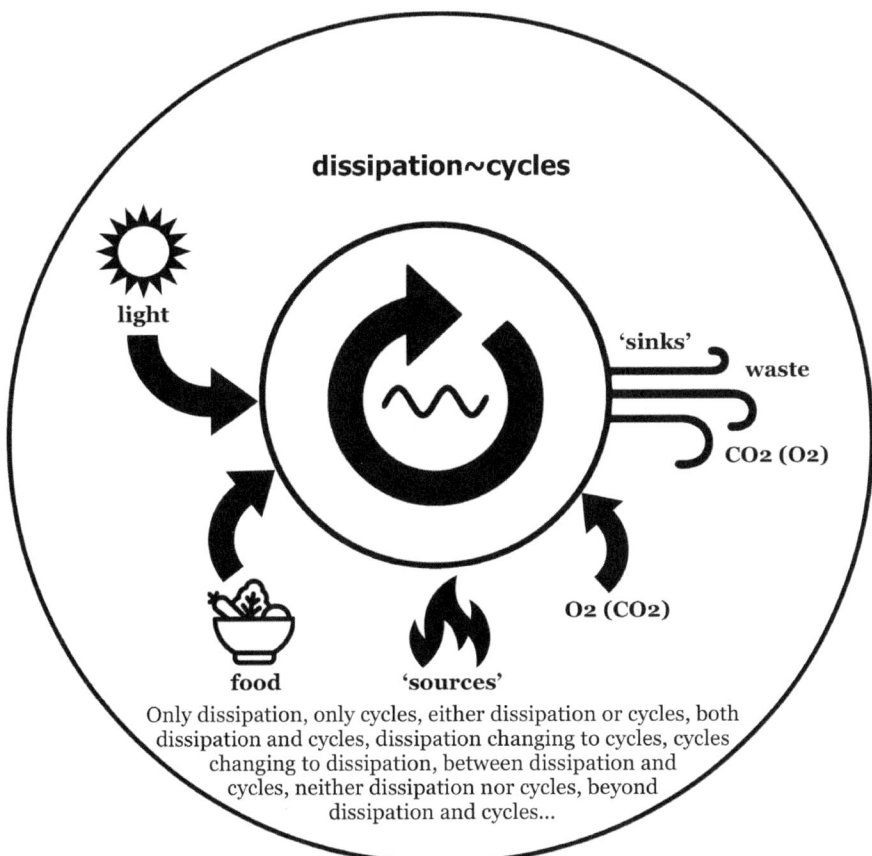

dissipation~cycles

light

'sinks'

waste

CO_2 (O_2)

O_2 (CO_2)

food   'sources'

Only dissipation, only cycles, either dissipation or cycles, both dissipation and cycles, dissipation changing to cycles, cycles changing to dissipation, between dissipation and cycles, neither dissipation nor cycles, beyond dissipation and cycles...

The Metastabilian says: I am a dissipative, open, nonequilibrium system. Energy courses through me. I breathe and eat, which introduces oxygen and food that stoke my inner living fire, and I give off heat, waste and CO_2. Energy, matter and information fuel my spontaneous, self-organizing intrinsic dynamics, the dynamic patterns~pattern dynamics of my complementary nature. And through dissipation~cycles, I sustain my body~mind, brain~mind...and *squiggle sense*!

Related squiggles: energy~matter, open~closed, entropy~order...

21 Of Boundary and Domain

 The world is a nested space, and so we have our brain as a person, and people are members of teams, and teams are part of business units, and business units are parts of corporations, and corporations are part of industries, which are part of economies— Clayton Christensen

In the complementary nature, levels are relative to each other. No single level has priority. This perspective is reified in Coordination Dynamics (CD), where it inspires the ongoing search for level-independent mechanisms and principles. The theory, model and experiments of Coordination Dynamics indicate that levels are synergetic individual~collectives. They are dynamic, inextricably linked, complementary and relative: What is micro at one level is macro at another. What is local at one level is global at another. But levels of what? What is a level? From the perspective of your *squiggle sense* grounded in Coordination Dynamics, there isn't one fundamental level of reality. A level is a dynamic, individual~collective boundary~domain, a dissipative, self-organized, nested container.

The earth is an example of a dynamic boundary~domain at the planetary level. And nested within its viable biosphere, are multitudinous boundary~domains, which include human beings and their many different social, biological, cultural and psychological levels. Each of these is also a dynamic boundary~domain, none of them have ultimate fundamental priority either! You are an individual~collective boundary~domain at the human level. Within your personal domain are the innumerable, coordinated, nested boundary~domains of your being: your brain, organs, muscles, cells, nuclei, DNA, etc. So it is and so it goes, all the way up~down, within~between, side-to-side, round and round…

Whereas no level of the complementary nature is fundamental, each boundary~domain is unique. They're different from each other, all are in flux, and coordinate with each other in different ways. It's bewildering to imagine how levels can be inextricable, equally valid and complementary. Yet in Coordination Dynamics, universal coordination laws do exist between~within individual~collective boundary~domains of living things. Coordination Dynamics even reconciles the *squiggle* dance of level-independent and level-dependent dynamical mechanisms and principles. A good way to grasp the perplexing complementary nature of boundary~domains is to just imagine the *squiggle* dance of any boundary and domain you're interested in. Whenever you do that, you engage the metastable mode of your brain~mind, which awakens your *squiggle sense*. Use your *squiggle sense* to extend the boundary~domain of your brain~mind!

J. A. S. Kelso and D. A. Engstrøm, *The Squiggle Sense*,
https://doi.org/10.1007/978-3-031-59369-7_21

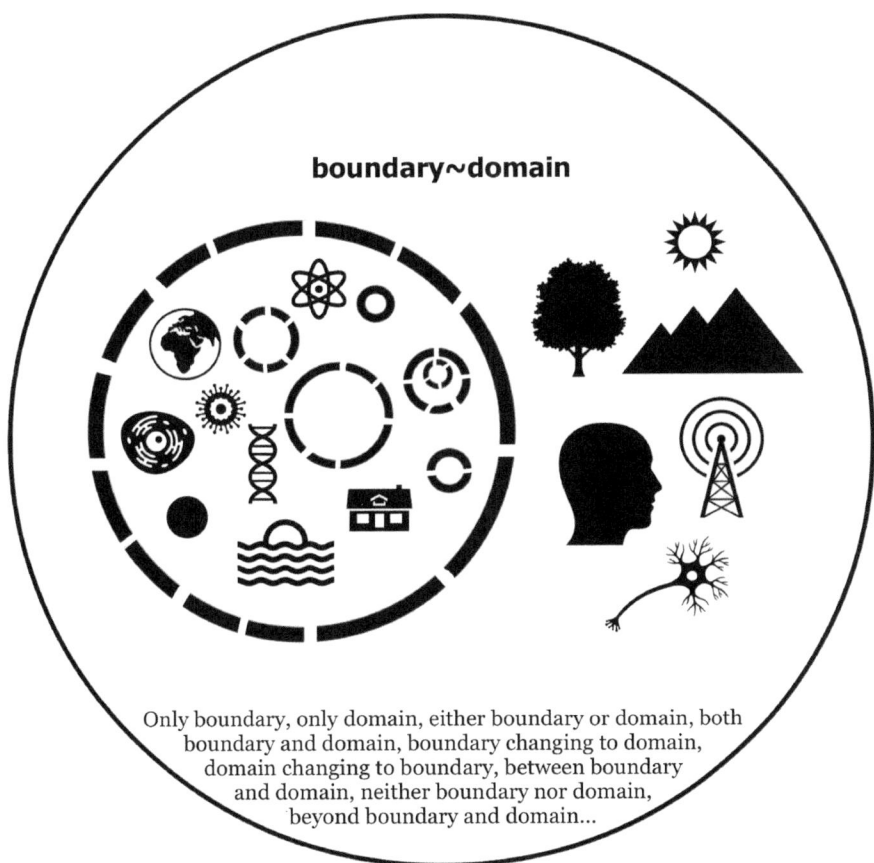

boundary~domain

Only boundary, only domain, either boundary or domain, both boundary and domain, boundary changing to domain, domain changing to boundary, between boundary and domain, neither boundary nor domain, beyond boundary and domain...

The Metastabilian says: What does it mean that there is no ultimate top and bottom levels, but rather an endless chain of nested, equally valid levels? It means that reductionistic 'bottom-up' and holistic 'top-down' thinking, explorations and explanations in their own right are eventually limited and limiting, and should be reconciled. To do this, I engage my *squiggle sense* of levels, of boundary~domains. Through it, I appreciate that to truly grasp any boundary~domain I must study its behavior and boundary features, its interaction with other boundary~domains, those nested within it, and those within which it is nested...

Related squiggles: between~within, organism~environment, global~local...

22 Of Spatial and Temporal

 The dance is the mother of the arts. Music and poetry exist in time; painting and architecture in space. But the dance lives at once in time and space—Curt Sachs

You are a spatial~temporal being, a *squiggle* dancer~dancing that lives at once in space and time. Spatial refers to that which occupies or surrounds space—often described by terms like size, position, shape, length, area, volume, geometry. It's the 'what' and 'where' of reality. Temporal means 'of time', contained in words like epoch, movement, rhythm, cadence, sequence, process, event, duration. It's when things happen and how long they last in the dynamic flow of reality. In the language of Coordination Dynamics (CD), spatial and temporal can't be meaningfully separated. They are a *squiggle*—inextricable, dynamic, complementary aspects of reality. The dance of life, like information, motion and energy, lives at once in time and space!

The complementary nature of your spatial~temporal existence includes all that you are and do. Ironically, until about a century ago, space and time were widely considered to be separate aspects of reality. That view was forever changed due to the profound thought, mathematics and science of Henri Poincaré, Hermann Minkowski, Albert Einstein and others. The theory of Relativity that radically changed humanity's understanding of the cosmos depends upon the idea that space and time are not separate. Yet, Relativity, amazing and useful as it is, is so far removed from your everyday life that it's easy to ignore it. On the other hand, the dynamic, complementary spatial~temporal coordination of your own body~mind, the dissipative synergies that are you and that enable your dance of life with others and the world around you are as close and relevant as can be.

So how can your *squiggle sense* and Coordination Dynamics help you grasp and appreciate the complementary nature of your spatial~temporal coordination? Well, to begin with, your *squiggle sense* is a spatial~temporal sense. That's its function. When engaged, that is, when you are in your metastable mode, your perception~action is spatial~temporal! And how spatial~temporal coordination works is exactly what Coordination Dynamics is about. It's a science of functional spatial~temporal synergies that evolve in the 'phase space' of informationally relevant coordination variables. By engaging your *squiggle sense,* you can learn the complementary code of Coordination Dynamics. The spatial~temporal dance of each squiggle contributes to the collective squiggle dance of Coordination Dynamics itself—the Coordination Dynamics of the complementary nature!

J. A. S. Kelso and D. A. Engstrøm, *The Squiggle Sense*,
https://doi.org/10.1007/978-3-031-59369-7_22

The Metastabilian says: I am a spatial~temporal being, a squiggle dancer~dancing, as are each of the synergies that define my individual~collective complementary nature. And my *squiggle sense* is a spatial~temporal sense. It helps me realize that, as in the cosmology of gravity~radiation, my own spatial~temporal aspects are inextricable, dynamic and complementary. Isn't this realization as exciting and profound as light bending around a star, an event horizon of a black hole? Could the secret reconciliation of Quantum Mechanics (QM) and how CD views perception~action and the human brain~mind be hiding in plain sight, staring at me whenever I look into a mirror?

Related squiggles: space~time, what~when, structure~function…

23 Of Persistence and Change

So do flux and reflux—the rhythm of change—alternate and persist in everything under the sky —Thomas Hardy

The complementary nature of persistence and change, the essence of what it means to be dynamic, is a generic, indispensable, vital feature of reality, of your existence and awareness. From the perspective of the *squiggle sense*, the old sayings, "the more things change, the more they stay the same" and "nothing lasts forever" are equally valid and complementary. Persistence~change is such a fundamental *squiggle* that it's easy to overlook and take for granted. But can its complementary nature really be discovered, explored and understood? Actually, yes it can. Coordination Dynamics (CD), a theory of directed self-organization, offers deep insights into *the squiggle sense* of persistence~change. CD shows how the flux of energy (change) through a complex living system like you plays an active role in self-organizing the dynamic patterns (persistences) you compose and that compose you. The stability (persistence) of those dynamic patterns is maintained at the price of energy dissipation (change). No energy flux (fuel), no dynamic patterns...no *you*.

The basic building blocks of CD are individual~collectives called synergies. Synergies are dynamic clusters of elements or processes that are sufficiently self-organized to produce persistent functions, like the cells of your heart, brain, lungs, muscles and gut. Hallmark features of synergies are both their stability and their flexibility. Whereas some variables cause a synergy to change, others preserve the integrity of the synergy to persist, to resist change. But how? Such deep questions regarding the complementary nature of persistence~change are precisely those pursued in CD! Grasping the CD of spatial~temporal persistence~change begins with the ubiquitous phenomena of oscillation and cycles, of flux and reflux.

Oscillation is a ubiquitous property of all natural behavior. Spatial~temporal dynamical patterns, including the very long lasting ones we call 'structures', are inherently cyclical. For any process to persist, a cycle of work must be done. It's the flux of energy from sources to sinks that fuels such cycles. The engines of persistence~change sustain cyclical motion by absorbing over the course of each cycle an amount of free energy that nearly balances the energy dissipated per cycle. Without this energy balance, things would simply decay toward static equilibrium. In CD, the main dynamical archetype that has been shown to underlie such spatial~temporal persistence~change is called the "limit cycle". How limit cycles emerge, function and interact are at the heart of CD and the complementary nature of persistence~change.

J. A. S. Kelso and D. A. Engstrøm, *The Squiggle Sense*, https://doi.org/10.1007/978-3-031-59369-7_23

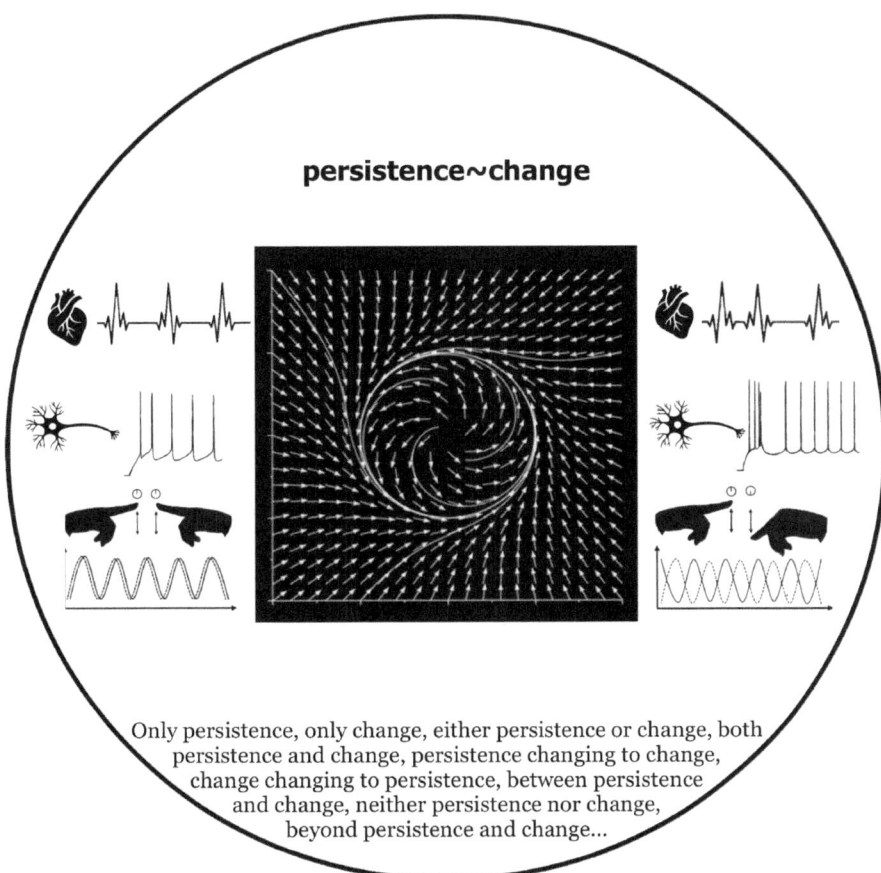

persistence~change

Only persistence, only change, either persistence or change, both persistence and change, persistence changing to change, change changing to persistence, between persistence and change, neither persistence nor change, beyond persistence and change...

The Metastabilian says: Persistence~change underlies all the dynamic patterns of my life, the being~becoming of my body~mind, my action~perception, all the relevant spatial~temporal rhythms of my existence. Learning and appreciating CD enables my *squiggle sense* to function and persist~change according to its complementary nature. CD explains how those rhythms, steady states, and limit cycles actually self-organize, which inspires me to engage it even more. What goes around, comes around!

Related squiggles: stability~instability, quiescent~active, learning~forgetting...

24 Of Convergence and Divergence

 Differing from Newton and Schopenhauer, your ancestor did not think of time as absolute and uniform. He believed it an infinite series of times, in a dizzily growing, ever spreading network of diverging, converging and parallel times—Jorge Luis Borges

Convergence~divergence is a *squiggle* of the complementary code of Coordination Dynamics (CD) that says: "the tendency of the flow of a dynamical system to converge coexists with the tendency of its flow to diverge". Dynamic patterns of all types, at all levels and in all contexts are literally complex dances of converging and diverging pattern dynamics. In fact, convergence~divergence is the way all matter behaves near phase transitions in physics, 'critical points' like freezing and boiling—where the state of matter converges and diverges from so called, 'saddle points'. CD enables you to anticipate, perceive and respond to the converging~diverging dynamic patterns within you and surrounding you, that fill your life with meaning, and enable you to make *squiggle sense* out of them!

Now, the inextricable coexistence of convergence and divergence might seem pretty self-evident. When it does, that's your *squiggle sense* talking. Yet in many situations and contexts, your inner voice might whisper quite the contrary, like "Forever together!", "Never the twain shall meet again!", "Do I stay or do I go?" Luckily, in these situations, awareness and appreciation of the complementary nature of convergence~divergence and how it fits into the CD paradigm can help you return to your metastable mode and re-engage your *squiggle sense.* A good place to begin is simply to think about how you might observe the converging~diverging dynamic patterns of a complex system…and imagine the underlying dynamics.

One way would be to watch time-lapse videos, like a person's activity over a day or week, a drone's-eye view of traffic, busy city-streets, a classroom, a nightclub, an artist painting from blank canvas to finish, the long term behavior of wild animals, blood flow, mitosis, etc. It's mesmerizing and a bit comical how the coordinating elements zip around one another, in and out, back and forth, stopping and starting. It's a thrumming rhythmicity of dynamic patterns. Observing a system in this way reveals an underlying flow, patterns of converging and diverging that aren't nearly as apparent when viewed in the normal time frame. Coordination Dynamics has many experimental strategies that are even more revealing than time-lapse videos, though it shares with them the strategy of observing a complex system in ways that reveal dynamic patterns of coordination that are often missed or taken for granted. And it uses those critical, saddle points of convergence~divergence to help identify the underlying Coordination Dynamics.

J. A. S. Kelso and D. A. Engstrøm, *The Squiggle Sense,*
https://doi.org/10.1007/978-3-031-59369-7_24

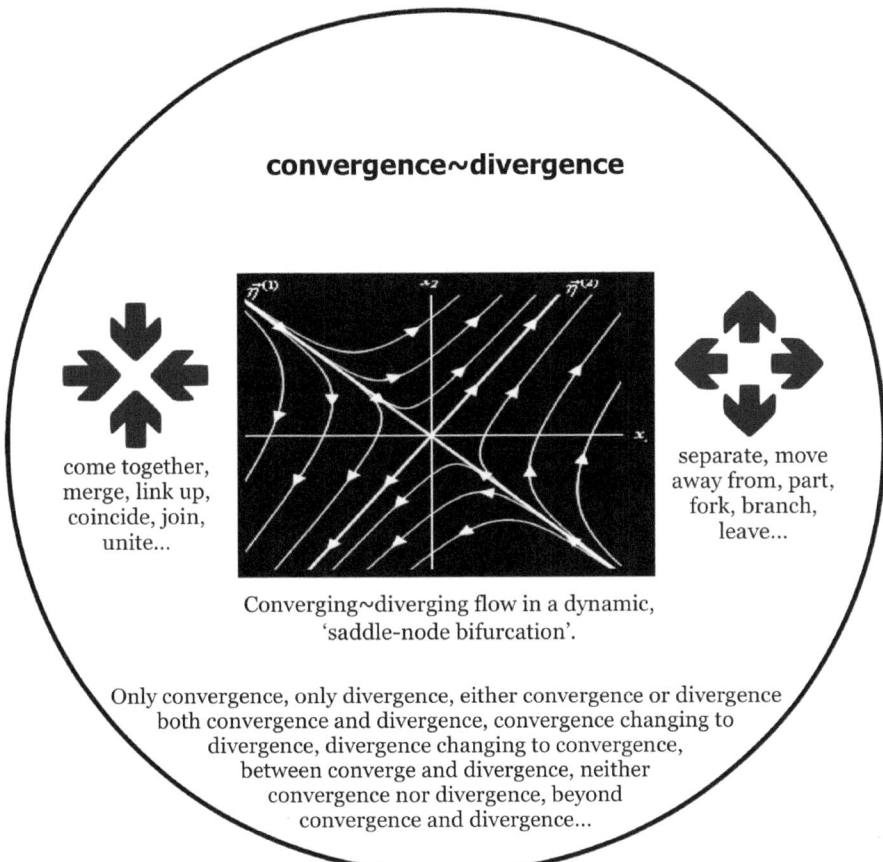

convergence~divergence

come together, merge, link up, coincide, join, unite...

separate, move away from, part, fork, branch, leave...

Converging~diverging flow in a dynamic, 'saddle-node bifurcation'.

Only convergence, only divergence, either convergence or divergence both convergence and divergence, convergence changing to divergence, divergence changing to convergence, between converge and divergence, neither convergence nor divergence, beyond convergence and divergence...

The Metastabilian says: Driving along, I watch the motorway ahead converging, whilst simultaneously glancing into the rear-view mirror at my receding, diverging wake. At that same time I muse upon the convergence~divergence of perception~action and knowledge~wisdom of my metastable mode, the breathtaking miracles of synergy and reciprocal causality, found exactly where opposing tendencies coexist. That discovery led to CD's "Principle of Coexisting Opponent Tendencies" (COT). The dance of complementary contraries is an archetype of dynamical convergence~divergence, which happens everywhere and every when, in all systems, at all spatial~temporal levels and scales of observation. Incredible!

Related squiggles: gravitation~radiation, attracting~repelling, assembling~separating...

25 Of Multifunctionality and Functional Equivalence

If they tell me one more time that I'm using the wrong fork for a part of a meal, I swear I'll show them exactly how multifunctional the utensil can be—Jennifer Ellison

Synergies and *squiggles* of the complementary code may seem a bit peculiar to study and reconcile using the binary, either-or logic common to most scientific enquiry. Historically, physical science has predominantly pursued a reductive approach to the dizzying array of structures and functions found in nature. The assumption is that the parts will sum up to the whole. More often than not, especially in biology, rigid, linear, one-to-one structure-to-function perspectives are ill-equipped to understand and explain the complex, nonlinear dynamic phenomena typical of living systems. There, and in Coordination Dynamics (CD), the ubiquitous characteristics of multifunctionality and functional equivalence reign.

Multifunctionality is the capacity for the same dynamic structure to express multiple different functions. The best example of multifunctionality is you! Human beings are one of the top 'Swiss army knife' species of the animal kingdom. You use the same structures of your body and brain to perform many different functions. Even if one wanted to, it would be a major challenge to identify and count them all. Functional equivalence is the complementary aspect, namely the capacity for the same function or goal to be realized by different material structures (parts, components, brain regions, team members…). The problem of the classic approach to understanding these phenomena should be obvious: nonlinear one-to-many and many-to-one structure~function relations simply cannot be captured by linear, 'one-to-one' perspectives and paradigms.

On the other hand, multifunctionality and functional equivalence are inherent attributes of CD, and occur at all levels of biological organization. They are in fact hallmarks of synergies, the structural~functional units of life, those dynamic self-organizing evolving collective~individuals that provide the flexibility and adaptability on which living things depend. Multifunctionality~functional equivalence is a *squiggle* of the complementary code of CD, whose complementary nature is tied to your viability, intelligent self-organization and survival. It might even be said its very existence enables your *squiggle sense,* which in turn provides novel insights into life itself. Like all the *squiggles* of the complementary code, its behavior is grounded in multistable and metastable CD, the dynamical modes from which all *squiggles* as well as your *squiggle sense* arise.

J. A. S. Kelso and D. A. Engstrøm, *The Squiggle Sense,*
https://doi.org/10.1007/978-3-031-59369-7_25

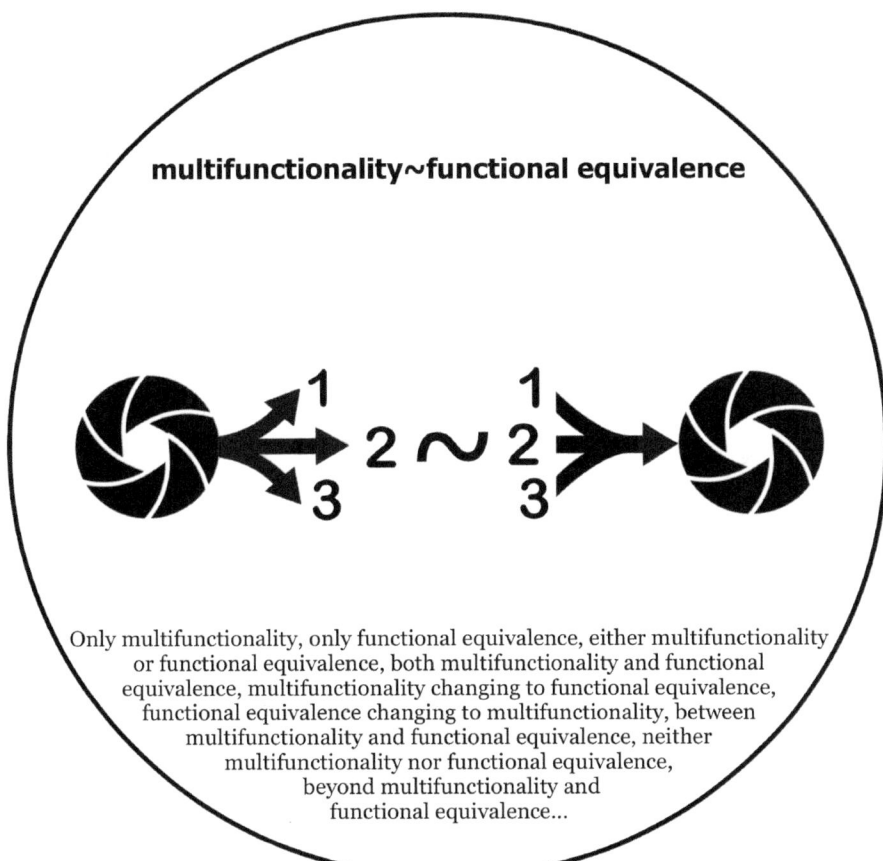

The Metastabilian says: I'm a multifunctional being~becoming. Not only am I able to do a vast number of different tasks and functions, but I'm also able to do a given task in different ways. My body~mind and behavioral capacities are miracles of synergy, whose multifunctionality~functional equivalences transcend the boundary~domain of the current standard physical paradigm. Classical physical assumptions of mechanics and mechanisms, linearity, 1-to-1 relations between structure and function, and fixed-point models are insufficient to explain life. They must now be expanded to accommodate and assimilate the phenomena and paradigm of CD and its complementary code.

Related squiggles: adaptability~flexibility, degeneracy~equifinality, variability~redundancy…

26 Of Absolute and Relative Coordination

 Relative coordination is a kind of neural cooperation that renders visible the operative forces of the central nervous system that would otherwise remain invisible—Erich von Holst

Whether you think about and appreciate it or not, you're an expert on coordination. It's a vital feature of your life and awareness. You know and sense coordination going on in and around you all the time with every move you make, every breath you take. Your life, life itself and the world you know would be impossible without it, let alone your ability to imagine or describe. So as an expert, think about what you know well already. Even though a total lack of coordination would be the 'end of things', there are nevertheless countless scenarios in life where dynamic elements are uncoordinated, for any number of reasons. And there are countless others that seem absolutely coordinated.

In most cases, absolute coordination is easy to recognize. It's when there's a rigid, unchanging relationship between coordinating elements, like marching soldiers, the gears, rails and pistons of machines, windshield wipers, etc.—up with up, down with down, forward with forward, backward with backward. The movements of your image in a mirror are absolutely coordinated with your movements in front of it. You would be very startled if they weren't! But if all coordination in your life were suddenly to become *absolute*, you would be even more startled. That's the stuff of Orwellian nightmares—throngs walking to work in perfect synchrony to identical factories, clone children emerging from houses at the exact same moment, bouncing balls at exactly the same frequency, amplitude and phase. In reality, total absolute coordination is untenable. Life is not an absolutely coordinated machine!

Whereas some coordination of viable living systems may seem absolutely coordinated, at least temporarily, more often it's really something in-between the extremes of absolute and no coordination. There has to be a more flexible form of coordination that allows for adaptation and variability. The brilliant physiologist Erich von Holst coined the term relative coordination to describe it. Absolute and relative coordination are complementary aspects whose complementary nature lies at the very heart of Coordination Dynamics (CD). In fact, the study of how both these forms of coordination work and how they *squiggle* was central to the development of CD. They turned out to be biological expressions of the multistable and metastable modes of Coordination Dynamics—and your *squiggle sense*. In this absolute~relative *squiggle* lies a key to how the complementary nature is connected to dynamic coordination patterns, their pattern dynamics, and *vice-versa*.

J. A. S. Kelso and D. A. Engstrøm, *The Squiggle Sense*,
https://doi.org/10.1007/978-3-031-59369-7_26

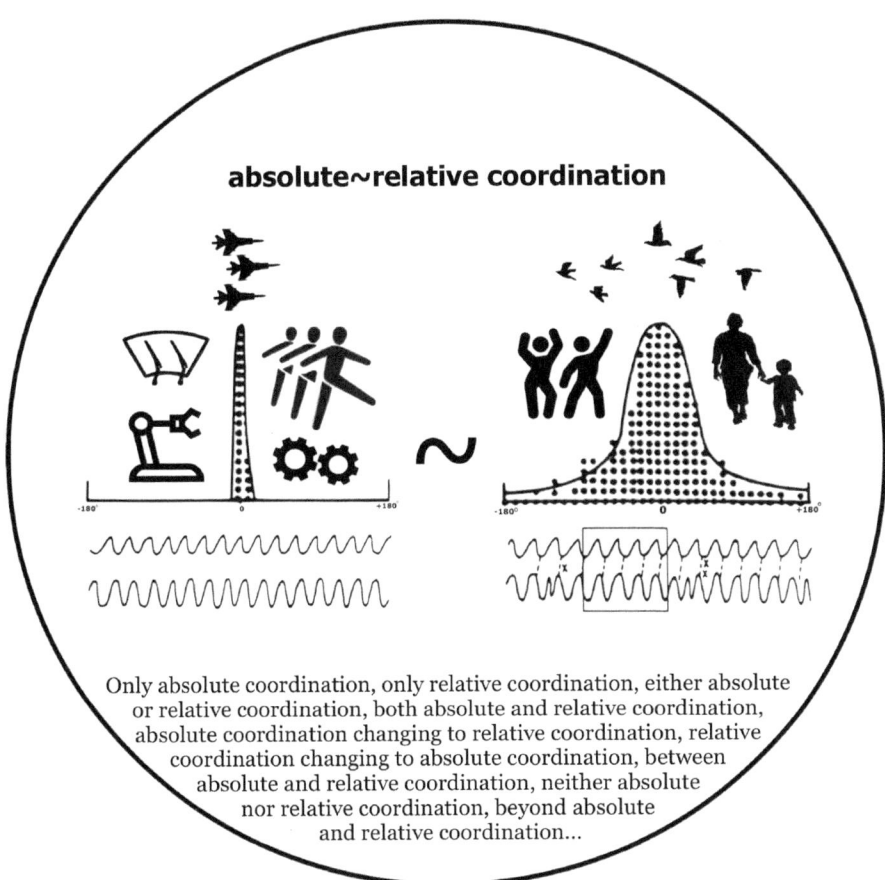

absolute~relative coordination

Only absolute coordination, only relative coordination, either absolute or relative coordination, both absolute and relative coordination, absolute coordination changing to relative coordination, relative coordination changing to absolute coordination, between absolute and relative coordination, neither absolute nor relative coordination, beyond absolute and relative coordination...

The Metastabilian says: It's funny how I control the coordinated movements of my body~mind in the complementary nature, yet the coordinated movements of my body~mind and the world are controlled by the laws of that complementary nature! I know that most of the time my coordinated movements are not absolute, not robotic, not rigid like marching soldiers, pistons and gears of machines. And yet, I can march like that at least for awhile. On the other hand, my movements aren't random, even in the moments I might wish them to be. Whatever this other, 'in-between' coordination is, my *squiggle sense* tells me it also springs from the complementary nature. And now, I marvel at how one science is able to explain all of these different scenarios with a single paradigm—Coordination Dynamics!

Related squiggles: rigid~flexible, states~tendencies, troops~flocks...

27 Of Synchronization and Syncopation

By rolling a ball down an inclined plane, we came to know how the planets move. By knowing how two fingers are coordinated we came to know how living things are coordinated—J. A. Scott Kelso

The original quest to understand absolute and relative coordination that evolved into the burgeoning field of Coordination Dynamics (CD) came long before any connection to the complementary nature and the *squiggle sense* were discovered and appreciated. The original experiments explored the coordination of voluntary rhythmic movements of the two hands. Synchronization and syncopation are basic dynamic patterns of such coordinated movement. They've been the subject of hundreds of studies, resulting in many hypotheses and theories about how they're produced. A unique contribution of CD is that it established a fundamental and lawful connection between synchronization and syncopation as two basic modes of coordination we all have. This is a crucial insight: the law not only explains how the coordination of the interacting elements produces these two dynamic patterns, but how they're connected to one another!

In the original experiments on voluntary finger movements, test subjects try to move their index fingers back and forth to a metronome in one of two ways: both fingers synchronized (both up, both down), or both fingers syncopated (one up, the other down). It turns out that these are the only way people wiggle their fingers—at least without a lot of practice. In a given experimental run, the metronome is systematically increased or decreased from an initial starting frequency, speeding up or slowing down every few seconds in small steps. This simple experiment and its many progeny have established, among many other insights, that only two stable coordination patterns exist between the fingers of both hands—synchronized (in phase) and syncopated (anti-phase). Which pattern is observed depends on the initial conditions.

Now, the usual scientific interpretation regarding two qualitatively different behavioral patterns would be that two different mechanisms are responsible. But enticingly, the results and the theory showed that's not the case. At a critical frequency, spontaneous transitions from syncopated to synchronized movements occur. Beyond the transition point, only synchronization is observed. When frequency is reduced, test subjects don't spontaneously return to where they started, a basic form of memory called hysteresis. And they don't switch at all when they start in the synchronized, in phase mode. In hindsight, these seminal observations were the first bits of evidence pointing to the complementary nature of CD! But first, the connection between absolute and relative coordination had to be realized, which required a deeper understanding of the observed phenomena and what they actually mean. This in turn meant new conceptual and theoretical development. And that has led to right here, right now!

J. A. S. Kelso and D. A. Engstrøm, *The Squiggle Sense*,
https://doi.org/10.1007/978-3-031-59369-7_27

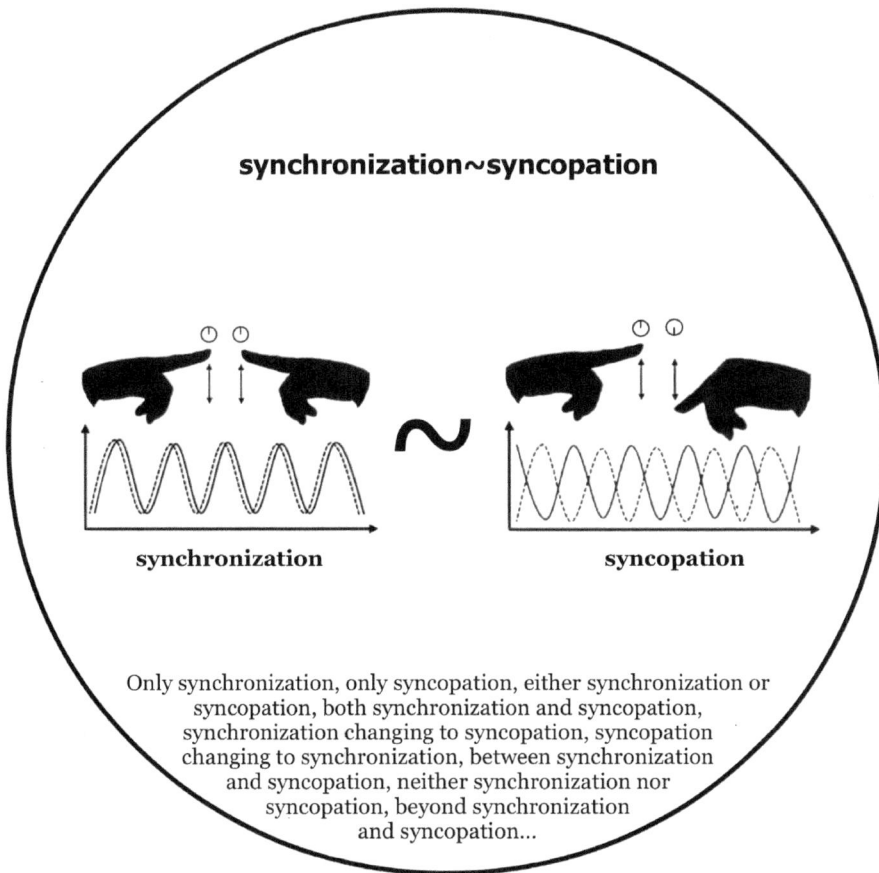

The Metastabilian says: Synchronization is to move together in space~time. But so is syncopation, just 180° out of phase. I can do both with my finger dances, using the same fingers, muscles, nerves, etc. All the physical anatomy and physiology remains the same. Only the coordination patterns are different. Mysterious! Of course, in most cases, I'm able to change with ease from one to the other, though that depends on how fast I'm moving. And sometimes I just switch from one to another when I wasn't even trying to change… What's up with that? My *squiggle sense* whispers the answer to me—its the complementary *squiggle* dance of synchronization~syncopation.

Related squiggles: inphase~antiphase, in sync~out of sync, agreement~debate…

28 Of Synergies and Phasing

 Synergies ... in which very few controls can manipulate a much larger number of configurational degrees of freedom are everywhere in biology. In biology they are indicators of complexity rather than mechanisms under constraints—Robert Rosen

You are a synergy of synergies, whose agency and behavior is not predictable by observing the behaviors of any of your separate physiological parts, component elements or their subassemblies. You are not a 'mechanism', a machine or clockwork whose behavior is the result of adding up the way all your individual parts behave. Actually, the word synergy is provocative. It flies in the face of centuries of reductionistic, mechanistic thought in philosophy and science that likens the universe and life itself to a grand machine or clockwork—orderly, knowable and predictable, Even today, this trend continues with the current quest in modern physics for a "Theory of Everything"—a single all-encompassing theoretical model of physics that can predict, explain and link together all known physical forces.

Whereas the mechanistic paradigm continues to be a powerful influence on scientific discourse, around the 1970s a new breed of 'nonlinear' scientist began to pursue new conceptual paths that deviated significantly from prior assumptions about physical reality. Since then, many branches and sub-disciplines of these pioneering efforts have emerged that are now bundled under the rubric, 'complexity science'. Coordination Dynamics (CD) itself is based on the mathematical tools of nonlinear dynamics and the concepts and methods of self-organizing dynamical systems called "Synergetics". Note that "Synergetics" referred to here isn't that of Buckminster Fuller, but rather a physical theory of self-organization in nonequilibrium systems formulated by the German theoretical physicist Hermann Haken stemming originally from his seminal work on lasers.

The CD paradigm allows synergies to be studied and modeled scientifically—how they form, persist, adapt and change. A key precursor to understanding the complementary nature of CD was to identify the relevant quantities that capture a synergy's dynamic patterns and their pattern dynamics, their persistence~change, convergence~divergence, stability~flexibility, cooperation~competition and continuous~discrete behavior. In the synergy of synchronization~syncopation, a key quantity is called 'relative phase'—the coherent phase difference between waves of partially coupled rhythmic movements. Synergetic behavior is phasic to the core. It is propagated, modeled and understood in the language of phase relations among interacting components. The key squiggle of synergies~phasing is thus grounded in the science of synergetic coordination, that is…CD!

J. A. S. Kelso and D. A. Engstrøm, *The Squiggle Sense*, https://doi.org/10.1007/978-3-031-59369-7_28

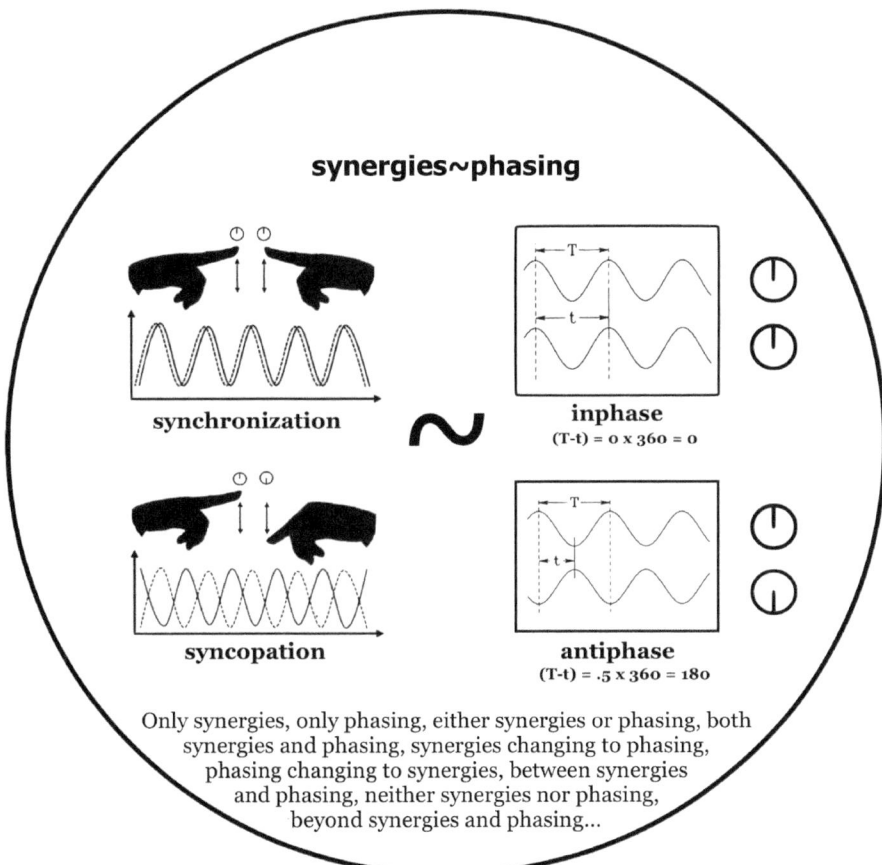

synergies~phasing

synchronization

inphase
(T-t) = 0 x 360 = 0

syncopation

antiphase
(T-t) = .5 x 360 = 180

Only synergies, only phasing, either synergies or phasing, both
synergies and phasing, synergies changing to phasing,
phasing changing to synergies, between synergies
and phasing, neither synergies nor phasing,
beyond synergies and phasing...

The Metastabilian says: My *squiggle sense* greatly facilitates my ability to transcend habitual classical assumptions about mechanisms and linear causality of life as I embrace the nonlinear CD of synergies and phasing as an illuminating step forward down the path of my new awareness and appreciation of the complementary nature. The CD of life, sentience, complexity, and circular causality suggests that humanity's understanding of physics isn't about to be 'solved' once and for all. On the contrary, it's about to evolve and embrace animate life.

Related squiggles: inphase~antiphase, cycles~phases, parts~wholes...

29 Of Qualitative and Quantitative

 Merely quantitative differences, beyond a certain point, pass into qualitative changes
—Karl Marx

Coordination dynamics (CD) provides a novel and lucid way to comprehend how qualitative and quantitative phenomena and thinking are related. Hopefully, your *squiggle sense* can already attempt to reconcile them as complementary aspects. It's common though, even in the sciences, to treat them as mutually exclusive polarized styles that have led to clichés like "it's quality rather than quantity that matters…" and "qualitative is just poor quantitative…" Although it's likely you may prefer and identify more with one or the other of these outlooks in general, you actually switch back and forth many, many times a day between qualitative and quantitative modes of perceiving, thinking and doing. It changes—sometimes it's both, sometimes it's neither. It depends. Both these apparently contrary mindsets are actually necessary for you to accurately gauge and navigate qualitative and quantitative aspects of life.

Indeed, you manage to rapidly and coherently perceive, act, react, anticipate and interact with life's numerous changing qualities and quantities, '24/7', even when you dream! Usually you don't think about how you do it while you're doing it. But somehow, the *squiggle* dance of qualitative and quantitative styles of thinking reconcile and compress the complex, ongoing barrage of stimuli impinging upon you. The fruit of this compression is much simpler and normal coherent thought and action that you can use to guide the ongoing trajectory of your life. But how? A vital clue is that your body~mind, your behavior and the complementary nature itself are all grounded in CD.

It turns out that the qualitative phase transitions studied in CD reveal a great deal about how dynamic patterns of brain and behavior are self-organized. Phase transitions have been used extensively to identify state changes in complex systems, providing a crucial scientific window into the underlying dynamics involved. Phase transitions occur at 'stability thresholds', where quantitative changes in 'control parameters' lead to pattern instability and the emergence of new, qualitatively different dynamic patterns. And guess what? The complexity of a system's behavior is drastically reduced near the onset of transitions. Know it or not, you are very sensitive to phase transitions of your own brain~mind and in the world around you. You use quantitatively induced qualitative change as a way to switch your behavior and simplify the unceasing quantitative flood of input impinging upon your sensorium, enhancing your ability to survive and prosper!

J. A. S. Kelso and D. A. Engstrøm, *The Squiggle Sense*,
https://doi.org/10.1007/978-3-031-59369-7_29

qualitative~quantitative

~

Only qualitative, only quantitative, both qualitative and
quantitative, qualitative changing to quantitative, quantitative
changing to qualitative, between qualitative and
quantitative, neither qualitative nor
quantitative, beyond qualitative
and quantitative...

The Metastabilian says: Nothing lasts forever. While qualitative and quantitative aspects of my life and awareness are stable enough to perceive and identify, they are dynamic, and subject to change. The complementary nature of qualitative~quantitative perception~action grounded in CD is indispensable for my very viability and sentience. My *squiggle sense* is my sentient, intentional perception~action based on that grounding. The more I know about the Complementary Code of CD, the better I will understand the qualitative~quantitative perception~action at the core of my reality and behavior. This is one of the many treasures to be gained in the quest of my *squiggle sense...*

Related squiggles: synthesis~analysis, one~many, perception~measurement, ...

30 Of Modeler and Model

 A modeler is someone who makes theoretical descriptions of systems or processes in order to understand them and be able to predict how they develop—collinsdictionary.com

The complementary nature and your *squiggle sense* are both grounded in the science of coordination dynamics (CD). This means that a significant, nontrivial connection between them has been established not only as a result of ideas, metaphors, opinions and speculation, but due to hypothesis-tested, theory-driven scientific experimentation, vetted and published in hundreds of articles in peer-reviewed scientific journals. Such a vast knowledge base can be daunting for anyone to try to grasp. To explore and comprehend it seriously, you need a way to compress all this knowledge and help you navigate it. One way of doing that in modern scientific and technological areas of research is called modeling. Scientific models often provide 'big picture' perspectives that can be used to explain, predict, plan and test new hypotheses. In CD, the mathematical model comes in the form of a set of nonlinearly coupled, nonlinear limit cycle oscillators called the "Extended Haken-Kelso-Bunz (HKB) Model" which, because it was tested so thoroughly, came to be called a "Coordination Law".

The original HKB model was formulated in 1985 to account for novel experimental observations on human bimanual coordination that revealed universal features of self-organization like multistability, phase transitions and hysteresis. Since then, it's been used to successfully model coordination in different kinds of systems and different levels of description. Many elaborations and extensions of the basic HKB model led to the discovery of general principles and mechanisms of coordination that underlie a broad spectrum of behaviors at different scales of observation: between and within moving limbs; between limb movements and tactile, visual and auditory stimuli; both spontaneous and intentional social interactions with others; between humans and digital avatars and even between humans and other species (think of horse and rider).

The Coordination Law provides insights into *you*—your agency, movement, perception, learning, *squiggle* sense, and so much more. Of course, the basic equation itself may seem less captivating and compelling if you don't happen to be mathematically inclined. Yet it still can be revealing if you know what its symbols stand for (see illustration). Moreover, the model can be readily 'translated' into the *squiggles* of CD. Your *squiggle sense* can comprehend the model via *squiggles*! In the remaining *squiggle* frames, you can learn about the Coordination Law via the *squiggles* of the complementary code without the necessity of learning all the math. Still, it's important to be aware of the model equation and the role it has played in the scientific grounding of your *squiggle sense* in CD, and seeing into the heart of CD itself!

J. A. S. Kelso and D. A. Engstrøm, *The Squiggle Sense*,
https://doi.org/10.1007/978-3-031-59369-7_30

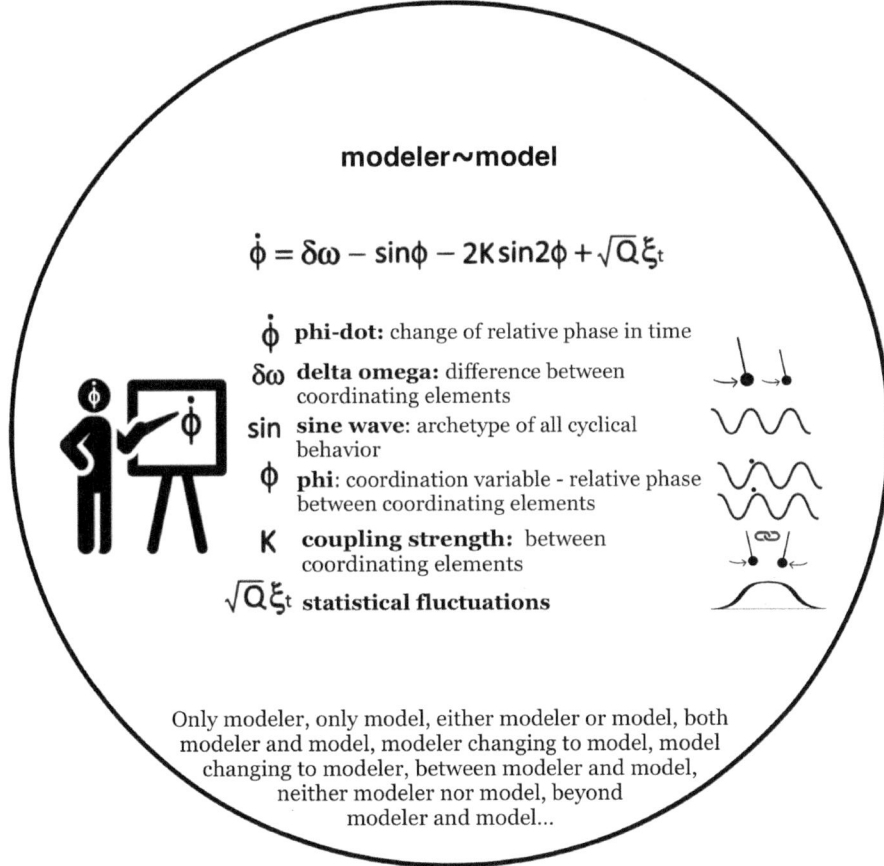

modeler~model

$$\dot{\phi} = \delta\omega - \sin\phi - 2K\sin2\phi + \sqrt{Q}\,\xi_t$$

$\dot{\phi}$ **phi-dot:** change of relative phase in time

$\delta\omega$ **delta omega:** difference between coordinating elements

$\sin$ **sine wave:** archetype of all cyclical behavior

ϕ **phi:** coordination variable - relative phase between coordinating elements

K **coupling strength:** between coordinating elements

$\sqrt{Q}\,\xi_t$ **statistical fluctuations**

Only modeler, only model, either modeler or model, both modeler and model, modeler changing to model, model changing to modeler, between modeler and model, neither modeler nor model, beyond modeler and model...

The Metastabilian says: I'm a sentient being, a modeler. My model of the world affects my ideas, my movements, my behavior, my life. The Coordination Law of CD is a source of my model of life and also the means by which I can model at all! This general model of coordination is directly relevant to what I am and everything I do. Amazingly enough, the CD Law leads to explanations of how I can make models of my world, which I then can use to navigate, adapt, survive and evolve. And best of all, I can learn about CD via the *squiggles* of its complementary code!

Related squiggles: designer~design, theory~experiment, simulation~reality...

31 Of Coordination Variable and Control Parameter

 Just as Newton did not need to know anything about the sun and planets except their motions, so dynamicists in cognitive science generally provide models that stand quite independently of implementation details—Tim van Gelder

As you now begin to play the "*Squiggle Game*" to help you learn the Coordination Law via the complementary nature of its *squiggles*, appreciate that its model equation wasn't always around, nor easy to produce. The ability to successfully capture real coordination behavior was hard won, an arduous reconciliation of theory and experiments, trial and error, observation and analysis of actual coordinating systems, like human beings coordinating movements. Happily, after all that work, it turns out that both the model and its *squiggle sense* stand quite independently of implementation details of actual systems, levels and contexts whose behavior the model captures. Indeed, the dynamics of relative phase, at the root of coordination, proves to be universal, transcending specific qualities and behavior of elements, components and processes from which it emerges.

In fact, the universality of relative phase is your entry point into the heart of the CD Law, via a foundational *squiggle* of its complementary code, coordination variable~-control parameter. In the model equation, the Greek symbol *phi* (ϕ) stands for relative phase, measured in degrees or radians. *Phi* is the equation's coordination variable (also traditionally called 'collective variable' and 'order parameter')— Coordination: collective, ordering, self-organizing phase relationships between elements moving together; variable: *phi* varies in time—evolving, changing, the essence of coordination. Scientists studying CD measure coordination variables experimentally. Several different factors can influence them, and much research and development in CD goes in to determining what they are.

Another crucial concept is a control parameter, the generic term for any adjustable factor able to affect the on-going dynamics of coordination variables in a lawful, predictable manner. For example, in CD experiments, continuously changing frequency causes established coordination patterns to reach stability thresholds, such instability leading to qualitative changes called phase transitions. This means one coordination pattern changes, usually switching abruptly, to another. Ironically, the very existence of phase transitions resulted in *phi's* identification as a coordination variable and conversely, movement frequency as a control parameter. Coordination variables like *phi* (ϕ), *squiggle* with control parameters like frequency. Theoretically, you can't have one without the other. Their codefining, inextricable, complementary, (nonlinear) dynamics are at the core of CD and the complementary nature.

J. A. S. Kelso and D. A. Engstrøm, *The Squiggle Sense*,
https://doi.org/10.1007/978-3-031-59369-7_31

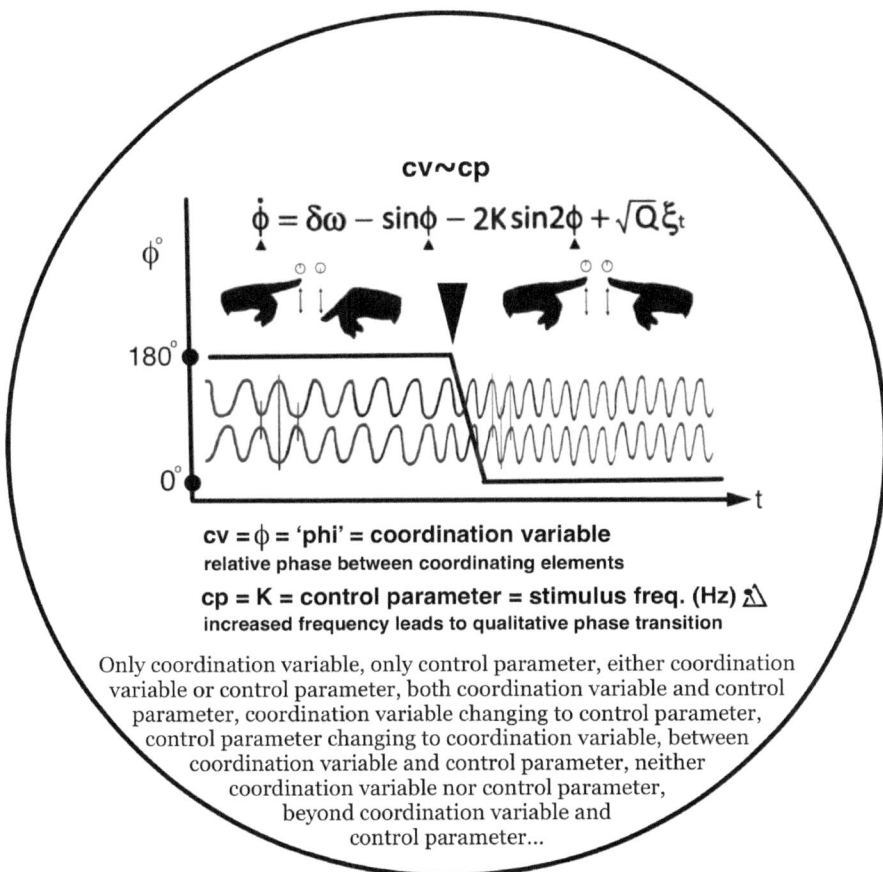

cv~cp

$$\dot{\phi} = \delta\omega - \sin\phi - 2K\sin2\phi + \sqrt{Q}\,\xi_t$$

cv = ϕ = 'phi' = **coordination variable**
relative phase between coordinating elements

cp = K = **control parameter = stimulus freq. (Hz)**
increased frequency leads to qualitative phase transition

Only coordination variable, only control parameter, either coordination variable or control parameter, both coordination variable and control parameter, coordination variable changing to control parameter, control parameter changing to coordination variable, between coordination variable and control parameter, neither coordination variable nor control parameter, beyond coordination variable and control parameter...

The Metastabilian says: The heart of coordination of my life, sentience, awareness, language, even love, springs from the lawful yet flexible and contextual universal CD of the complementary nature. This provides me a vast, novel and mostly untapped potential to know and begin to comprehend the complementary nature and my *squiggle sense*. The CD Law model equation makes it possible to study and understand the CD that grounds and expresses the complementary nature in living systems at all scales, including my own daily life. The relative phase *phi* captures the dance of coordination patterns. And control parameters lead those coordination patterns through qualitative changes—astonishingly simple yet profoundly complex!

Related squiggles: qualitative~quantitative, persist~change, stability~instability...

32 Of Heterogeneity and Coupling Strength

In Tar Baby, the classic concept of the individual with a solid, coherent identity is eschewed for a model of identity which sees the individual as a kaleidoscope of heterogeneous impulses and desires, constructed from multiple forms of interaction with the world as a play of difference that cannot be completely comprehended—Toni Morrison

Heterogeneity and coupling strength are complementary aspects at the very core of Coordination Dynamics (CD). The complementary nature of heterogeneity~coupling strength directly affects and shapes the universal self-organizing behavior of dynamic patterns. This squiggle is so fundamental to the CD paradigm that each aspect is included as a term in the Coordination Law equation. How fundamental? Well, without some heterogeneity or differences between individual coordinating elements (components, parts, processes, impulses, desires, etc.) there can be no metastability, no metastable mode—no *squiggle sense*. And without any coupling between individual coordinating elements, there can be no coordination at all!

Indeed, coordinated patterns arise in a self-organized fashion as a result of nonlinear coupling between individual coordinating elements. This is true at all levels, whether it be chemical bonds at the molecular level, receptor binding at the synaptic level, network connectivity at the neural level, all the way to the interactions between a teacher and student at the social level, and so on. The term 'K' in the model equation stands for the coupling strength between individual coordinating elements. Coupling strength in the model is intimately tied to experimental control parameters that exert both quantitative and qualitative, specific and non-specific changes on the coordination variable, *phi* (ϕ). So, one of the key determiners of how the coordination variable changes in time, *phi-dot*, is by changes in coupling strength.

Another important term in the Coordination Law is delta omega ($\delta\omega$) which expresses the heterogeneity or difference between individual coordinating elements (cells, neurons, muscles, body parts, brain areas, people, virtual partners, visual, auditory, tactile stimuli, combinations of all of them). For example, to play a drum set successfully, you must coordinate the motion of your hands, arms and legs. These movements, their accompanying kinesthetic sensations and the sounds they produce are all nonlinearly coupled via your nervous system. But the shapes, sizes, weight and muscular composition of your hands, arms and legs are all different, never mind the neural circuitry! This heterogeneity affects your overall ability to keep the beat. Heterogeneity~coupling strength is a *squiggle* of the complementary code of CD with the amazing property that it is applicable to many situations. At the same time it is contextually sensitive to all system levels of the complementary nature.

J. A. S. Kelso and D. A. Engstrøm, *The Squiggle Sense*,
https://doi.org/10.1007/978-3-031-59369-7_32

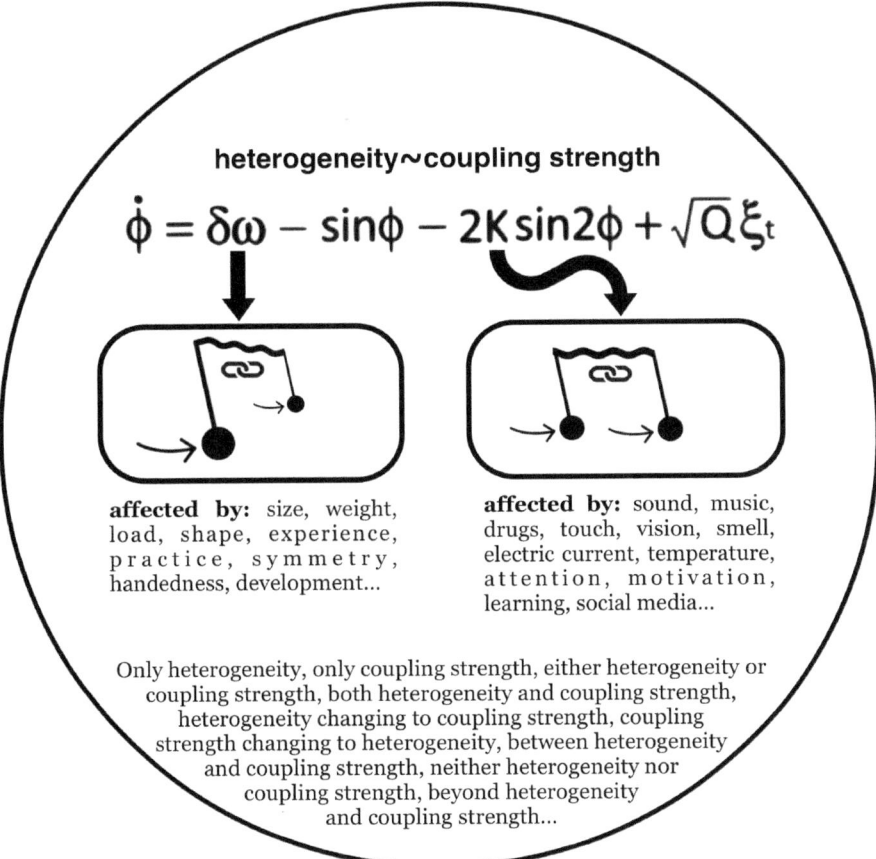

heterogeneity~coupling strength

$$\dot\phi = \delta\omega - \sin\phi - 2K\sin 2\phi + \sqrt{Q}\,\xi_t$$

affected by: size, weight, load, shape, experience, practice, symmetry, handedness, development...

affected by: sound, music, drugs, touch, vision, smell, electric current, temperature, attention, motivation, learning, social media...

Only heterogeneity, only coupling strength, either heterogeneity or coupling strength, both heterogeneity and coupling strength, heterogeneity changing to coupling strength, coupling strength changing to heterogeneity, between heterogeneity and coupling strength, neither heterogeneity nor coupling strength, beyond heterogeneity and coupling strength...

The Metastabilian says: So coherent states, state transitions and tendencies of my brain, behavior and countless other system~levels are captured by the same coordination law, which characterizes spatial~temporal phase relations of obviously very different structure~functions. Similarly, links between materially different structure~functions of myself and all other coordinations of the complementary nature appear to be governed by fundamental nonlinear coupling. It's insightful to realize that the complementary nature not only benefits from but requires heterogeneity.

Related squiggles: diversity~unity, individual differences~group, separate~together...

33 Of Deterministic and Stochastic

 Chance and necessity. Reality needs both—Hermann Haken

You are a sentient being, self-aware and perceptive, full of feelings, thoughts, memories, emotions, beliefs and opinions. Usually, these tend to be more or less stable entities, formed and developed over time. But life is also full of chance and imperfections—uncertainty, inconsistency, even error. How do you believe chance affects your daily life? Do you think that life is fundamentally deterministic, causally determined by preceding events and natural law, or is it essentially stochastic, statistical in nature, random, predominantly run by chance and probability? Do you think your preference affects your life and how you live it? Debates on such questions over the ages are legion. And yet, eons worth of thought and argument have generated more heat than light, and scant resolution.

Perhaps instead of resolutions, reconciliation is needed, a fresh perspective for progress to be made. Perhaps life is somehow essentially both. In this context, consider engaging your *squiggle sense* to explore the idea that life may well be essentially both deterministic and stochastic, that they are in fact inextricably related, complementary aspects, basically a *squiggle*. In that case, you already know one useful thing, that their complementary nature is grounded in Coordination Dynamics (CD). You can immediately wonder, for example, if choice and chance are somehow captured by the Coordination Law equation. Indeed they are. Together, the first three terms of the right hand side of the model equation are deterministic. Theoretically speaking, they serve to predict a system's dynamic patterns, that is, how a system will behave.

But as you know life always includes uncertainty. Chance is always lurking. Dynamic imperfections, random fluctuations, often called 'noise' or 'statistical effects', also exist. The last term on the right side of the equation is called a 'noise term.' It models the fact that all real systems are subject to fluctuations, the source of which lies both within the system itself and its environment. In CD, how a system behaves is always based upon both deterministic and stochastic processes. Deterministic~stochastic is a crucial *squiggle* of the Complementary Code of CD. All real systems, like you, have elements of both in all synergies and at all levels. Plan and accident, signal and noise, choice and chance, stability and variability *squiggle* right along with your thoughts and actions, desires and deeds. Decisions you make depend on both. Without them, life as you know it would neither be recognizable nor sustainable. 'Noise' provides an essential source of variability, without which selection cannot work, never mind skill.

© The Author(s), under exclusive license to Springer Nature Switzerland AG 2024
J. A. S. Kelso and D. A. Engstrøm, *The Squiggle Sense*,
https://doi.org/10.1007/978-3-031-59369-7_33

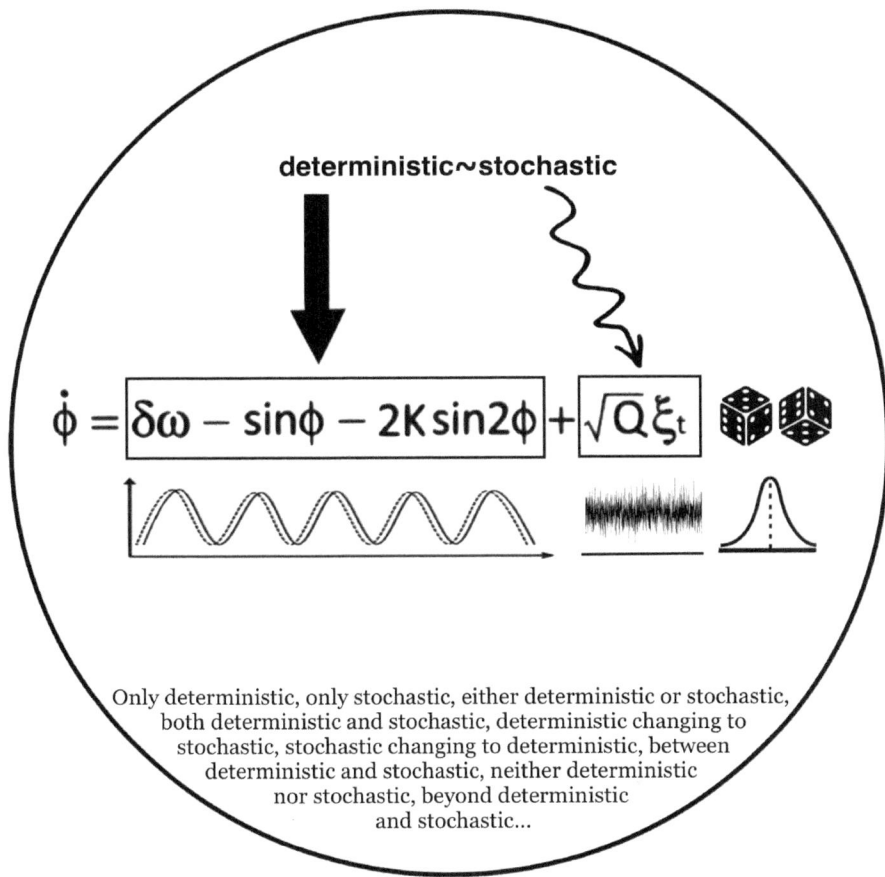

deterministic~stochastic

$$\dot{\phi} = \boxed{\delta\omega - \sin\phi - 2K\sin2\phi} + \boxed{\sqrt{Q}\,\xi_t}$$

Only deterministic, only stochastic, either deterministic or stochastic, both deterministic and stochastic, deterministic changing to stochastic, stochastic changing to deterministic, between deterministic and stochastic, neither deterministic nor stochastic, beyond deterministic and stochastic...

The Metastabilian says: I am a human being, I am sentient, aware, full of agency and intention. I live my life, make plans, follow my inspirations. My awakened *squiggle sense* helps me to appreciate that I'm a creature of the complementary nature, neither purely deterministic and predictable nor purely stochastic and probabilistic. I'm a mysterious creature, both determined and uncertain. Stochastic chance processes can lead to impediments and negative, unexpected outcomes, yet are also essential to the very vitality and humanity that defines and shapes my being~becoming, my self-awareness, my behavior, my humanity.

Related squiggles: of states~fluctuations, pattern~randomness, destiny~chance...

34 Of States and Transitions

Power ceases in the instant of repose; it resides in the moment of transition from a past to a new state, in the shooting of the gulf, in the darting to an aim—Ralph Waldo Emerson

Knowing that the Coordination Law equation exists and what all its terms mean is a crucial first step on your *squiggle* quest. But what comes next? How do you use it to help you better understand and appreciate the *squiggle sense* of yourself and world? A useful next step is to become acquainted with the graphs of the model, to learn what they reveal about the *squiggles* of the Complementary Code. The way scientists do this is 'feed' the model equation data and plot the outcomes as graphical visualizations. One such primary plot is of *phi-dot* versus *phi* (ϕ), which means: how phi is changing at the moment *versus* the current value of phi. Ironically, a common way to begin analyzing this kind of graph is to note where the relative phase, phi isn't changing. This happens wherever the solution of the equation for *phi-dot* equals zero. Those relative phases will persist until something in the equation is changed.

These zero points are called 'fixed points' of the equation. In physics, they correspond to states. Of course, state is a familiar word, as in physical states of matter, states of mind, energy states, even nation states. Which states persist (i.e. long enough to be observed) depends on their stability. Some states like the black dots in the picture are stable and attracting. Others, like the white dots are unstable and repelling. And of course the states themselves also change. Think of ice, water and steam, the different physical states or phases of matter. The 'persistence of states' *squiggles* with 'qualitative changes of state' in the coordinated *squiggle* dance of being~becoming. That the *phi-dot* versus *phi* plot provides a visualization of lawful persistence of coordinated states and transitions between them makes it an incredibly valuable tool. That it reveals more *squiggles* of the Complementary Code makes it even more so.

In both physics and Coordination Dynamics (CD), the moment of qualitative transition from one state to another state is called a phase transition, like the transitions at the freezing and boiling points of water. In CD though, stability thresholds are tipping points between qualitatively distinct patterns of coordination. While the idea that states have identifiable, predictable forms seems evident, the fact that transitions between states also do may seem less obvious or familiar. Actually, a system's behavior near transition points between two states is accompanied by certain predictable, universal effects that provide a window into underlying mechanisms of change. In fact, a number of signature phenomena surrounding phase transitions were discovered that led to the further development of CD. And guess what? They're all *squiggles!*

J. A. S. Kelso and D. A. Engstrøm, *The Squiggle Sense*,
https://doi.org/10.1007/978-3-031-59369-7_34

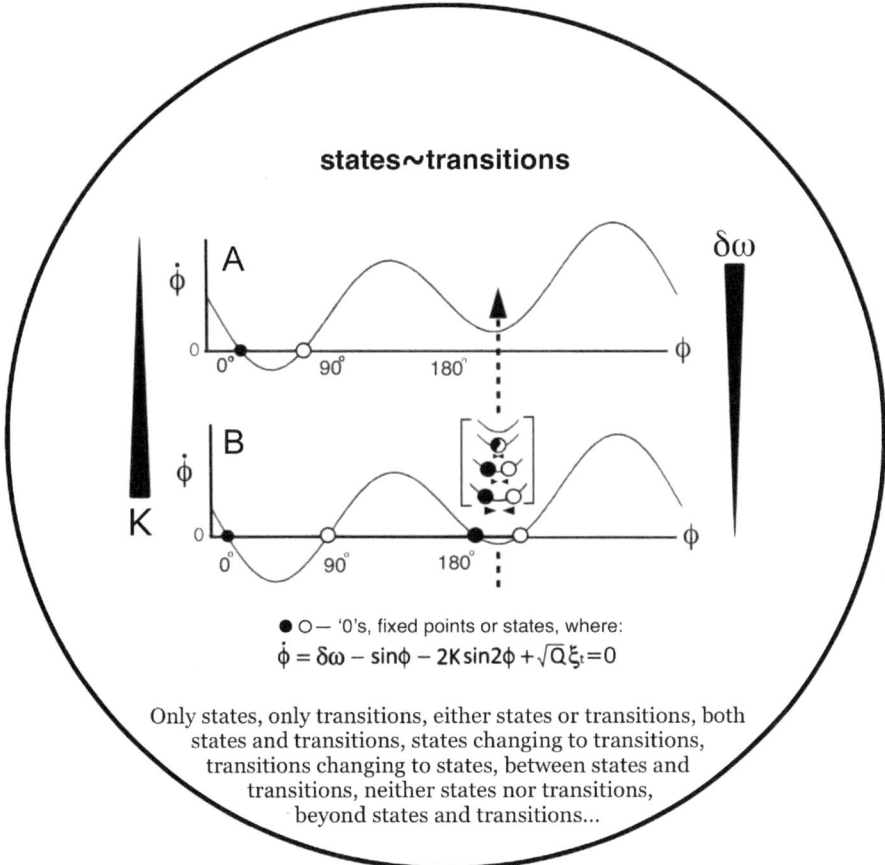

states~transitions

● ○— '0's, fixed points or states, where:

$$\dot{\phi} = \delta\omega - \sin\phi - 2K\sin2\phi + \sqrt{Q}\,\xi_t = 0$$

Only states, only transitions, either states or transitions, both
states and transitions, states changing to transitions,
transitions changing to states, between states and
transitions, neither states nor transitions,
beyond states and transitions...

The Metastabilian says: Here now, a way to visualize intuitions of my *squiggle sense*! In 'B' above, I notice that more than one dynamic *state* can exist at once. I can see that the zero points (fixed points, *states*) come in pairs, and wonder what that means? I can see in 'A' that some change in the equation has led to the *squiggly* graph being shifted, 'raised up' so that two of the four zero points, (fixed points, *states*) disappear—a phase transition or bifurcation has occurred! I know that such transitions of coordination patterns are happening all the time in the physical world, in my body, even in my thoughts and interactions with others. Can the Coordination Law equation allow me to visualize even more about my *squiggle sense* and the complementary nature? And what about the rest of the graph besides the 'zero points'? What other mysteries can be revealed and explained with such graphical visualization?

Related squiggles: persistence~change, structure~function, smooth~abrupt...

35 Of Attraction and Repulsion

Without contraries, no progression. Attraction and repulsion, reason and energy, love and hate, are necessary to human existence—William Blake

Of all the *squiggles* necessary to your human existence without which their could be no progression, attraction~repulsion surely ranks supreme. Attraction and repulsion are iconic expressions of physical polarity, as in bar magnets, from which comes the common experience that opposites attract and like poles repel. But attraction~repulsion in the dissipative, complex, synergetic systems of life that involve neurons, brains, behavior, minds, beliefs and opinions isn't as fixed as polarized bar magnets. It's more dynamic, flexible and contextual. In human behavior, opposite, contrary, polarized opinions rarely attract one another and often lead to conflict. "Like likes like" often runs the show, like 'like' buttons on social websites. In CD, attraction and repulsion emerge and disappear together due to the nonlinear dynamics of the Coordination Law.

Details of the CD of attraction~repulsion are revealed in the model equation plots, where attracting and repelling zero or 'fixed point' states come in pairs, often symbolized by filled and open circles. The black filled circles on the x-axis of the plots are attracting fixed point states. In Vector Flow plots (phi-dot vs. phi), when the slope near a fixed point is negative, changes in phi-dot are decelerating. Trajectories from any nearby initial conditions converge toward such 'fixed point attractors', where *phi-dot* $= 0$. The open circles on the x-axis of the plots are repelling fixed point states. When the slope around them is positive, changes in phi-dot are accelerating. Trajectories from any nearby initial conditions diverge away from the fixed point where *phi-dot* $= 0$, and will not return to it spontaneously. Such fixed points are called repellers.

A different but entirely parallel visualization of this behavior is provided by the Potential Well view—V(φ) versus φ. Its minima and maxima correspond more intuitively to attraction and repulsion. Imagine the fixed point as a ball at rest at the bottom of a well where it remains at rest. Push it, and it wiggles down and stops again at the fixed point as if attracted to it. Oppositely, a ball at rest at the peak of a hill (repeller) remains at rest, until nudged, then takes off down the side of the well as if repelled. The *squiggle* dance of attraction~repulsion is essential to CD. It's used to predict a synergetic system's behavior. The layout and location of attractors and repellers dictate the ongoing flow of the dynamics. Attractors, where a system's behaviors converge indicate stable coordination states. Repellers, where a system's behavior diverges, indicate intrinsically unstable states. And beyond where attracting and repelling fixed points disappear, metastable, complementary tendencies reign!

J. A. S. Kelso and D. A. Engstrøm, *The Squiggle Sense*,
https://doi.org/10.1007/978-3-031-59369-7_35

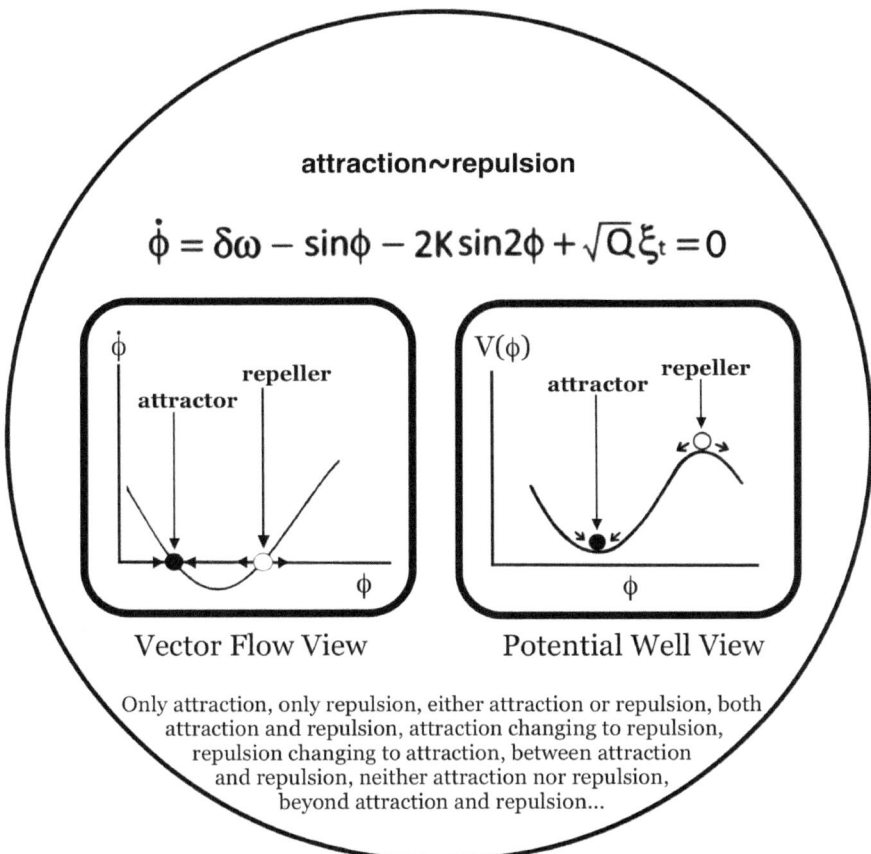

The Metastabilian says: It's a mystery and miracle that I'm somehow able to be both attracted and repelled by features of my life, which persist and change depending on the intricacies of the CD at play. It's incredible that a graph of a mathematical equation can capture the essence, the inner workings, the complementary nature of attraction~repulsion. The graphs help me visualize and conceptualize *the squiggle sense* of attraction~repulsion, like how attraction and repulsion emerge and disappear together! With this novel insight, I can suddenly imagine how to better avoid the extremism, intransigence, intolerance and hubris of polarized thinking.

Related squiggles: convergence~divergence, affinity~aversion, constraints~freedom...

36 Of Stability and Instability

Reflect frequently upon the instability of things, and how very fast the scenes of nature are shifted… Change is always and everywhere at work; it strikes through causes and effects, and leaves nothing fixed and permanent—Marcus Aurelius

Understanding the complementary nature can never only be about change and the instability of things. After all, things are things precisely because they persist, even if not permanently so. Stability, the resistance to change, strength to stand or endure, is a *sine qua non* of existence. A *squiggle sense* update of Aurelius's wisdom would be to say that stability and instability are always and everywhere at work, striking through causes and effects, leaving nothing permanently fixed—even change itself. Indeed, both the phenomena of stability and change in nature and the conceptual complementary nature of stability~instability are grounded in Coordination Dynamics (CD) as revealed by visualizations of the elementary Coordination Law.

To name some key examples: (1) Fixed point states of relative phase *phi* (φ) come in two complementary contrary types, stable attractors and unstable repellers. Note that in both coordination states, *phi-dot* $= 0$. But behavior of phi near the contrary fixed point states is drastically different. In the Potential Well plot you can see intuitively how values of phi will settle back to the fixed point of a stable attractor state, and how it will depart from an unstable repeller fixed point with the slightest nudge never to return spontaneously. (2) Stable and unstable fixed point states can appear and disappear together, depending on the parameters of the model. (3) Due to changes in model parameters like heterogeneity, coupling strength and statistical fluctuations of individual coordinating elements, a synergetic system may reach a stability threshold where a current state loses stability and a more stable state emerges spontaneously—this process is called a bifurcation or phase transition.

Analysis and reflection upon the stability~instability of states~transitions in CD led to the realization that instability is a universal, generic, dynamical mechanism underlying flexible switching between dynamic patterns in different systems, levels and contexts. Dynamical signatures of instability enabled discovery and identification of the key coordination variables that characterize nature's dynamic patterns. Two of these, 'enhancement of fluctuations' and 'critical slowing down', played key roles in the development of CD. And guess what? They *squiggle*! Without the *squiggle* dance of stability and instability, neither states, transitions between them, nor many other key features of CD such as multistability and metastability would have been discovered. Life, sentience, intention, learning, coordination all require instability just as much as they do stability!

J. A. S. Kelso and D. A. Engstrøm, *The Squiggle Sense*,
https://doi.org/10.1007/978-3-031-59369-7_36

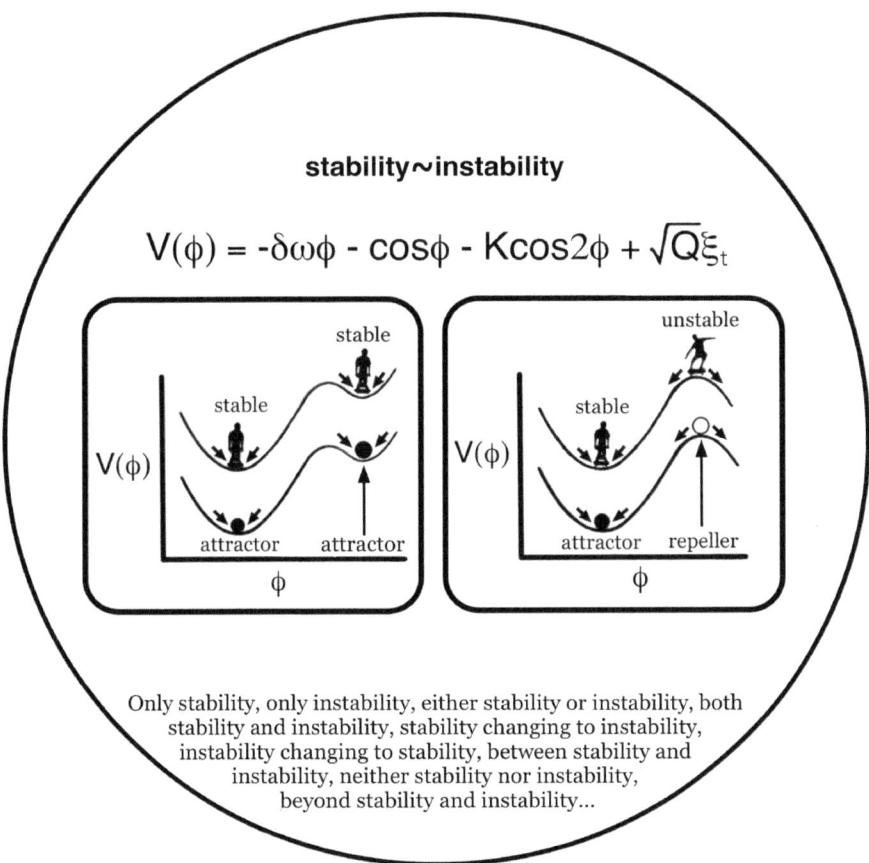

stability~instability

$$V(\phi) = -\delta\omega\phi - \cos\phi - K\cos2\phi + \sqrt{Q}\xi_t$$

Only stability, only instability, either stability or instability, both stability and instability, stability changing to instability, instability changing to stability, between stability and instability, neither stability nor instability, beyond stability and instability...

The Metastabilian says: I reflect frequently upon the stability~instability of things, how the scenes of my life shift depending on my frame of reference. Change~persistence is ever at work, striking through causes~effects, leaving nothing fixed and permanent for too long. My *squiggle sense* allows me to comprehend the *squiggle* dance of stability and instability of perception~action and my brain~mind.

Related squiggles: persistence~change, linear~nonlinear, smooth~abrupt...

37 Of E.O.F and C.S.D

Each system is trying to anticipate change in the environment—Kevin Kelly

As the Coordination Law successfully predicts, in open, synergetic, nonlinear dynamical systems like you, enhancement of fluctuations (e.o.f.) and critical slowing down (c.s.d.) are universal signatures of dynamic instability and predictors of change. Experiments show that they often precede dramatic, qualitative change—a phase transition is imminent! Random fluctuations are always present in dissipative dynamical systems. Far from thresholds of instability, fluctuations exist but their effects may not be pronounced or consequential. As the system approaches a phase transition, however, the effects of fluctuations are 'enhanced', their influence grows. Variability of the current patterned state increases as it loses stability and a new pattern self-organizes and reveals itself. A deep insight revealed by analysis of fluctuations is that loss of stability is neither instantaneous nor totally random. In certain scenarios, e.o.f. means it's possible to anticipate change!

Critical slowing down refers to a palpable increase in the time a dynamical system takes to recover from perturbations as it nears a phase transition. Like fluctuations, perturbations result in deviations from the system's current state, and take some time to recover from. In CD, critical slowing down is studied by systematically perturbing a coordination pattern, and measuring the time it takes for the pattern to return to normal. Near a qualitative phase transition, the Coordination Law predicts this "relaxation time" will increase. That is, as a system becomes less stable it takes longer and longer to recover from tiny perturbations, an anticipatory sign of impending change. Something is going to happen!

E.o.f and c.s.d. are complementary aspects of dynamic stability~instability that transcend specific details. They can signal an upcoming earthquake, an epileptic seizure or a heart attack, and can be viewed as universal aspects of certain kinds of phase transitions. That these quantitative complementary signatures of instability presage qualitative change is quite remarkable. You can see how the shape of the Coordination Law's Potential Well plot is systematically altered by the control parameter (K). E.o.f. and c.s.d. result from the 'attractive well' of the potential's minima becoming shallower and the effects of fluctuations and perturbations becoming greater as the system nears a tipping point or phase transition. *The squiggle sense* is to realize that control parameters can alter the shape of life's dynamical landscape and that small fluctuations and perturbations will always be around to test its stability~instability!

J. A. S. Kelso and D. A. Engstrøm, *The Squiggle Sense*,
https://doi.org/10.1007/978-3-031-59369-7_37

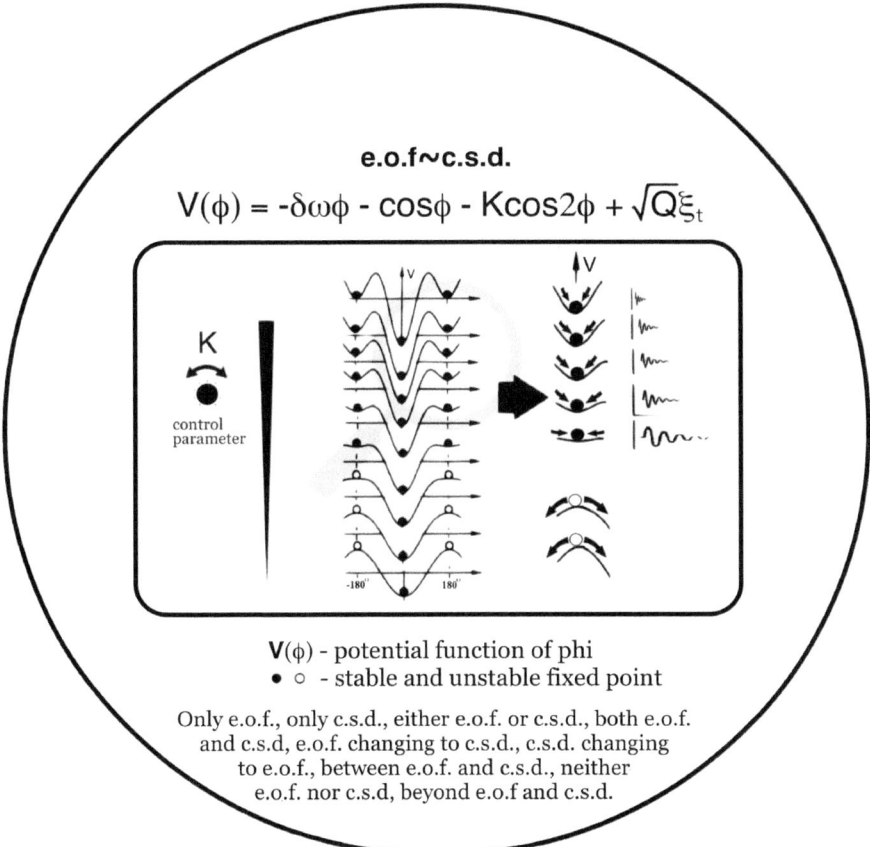

e.o.f~c.s.d.

$$V(\phi) = -\delta\omega\phi - \cos\phi - K\cos2\phi + \sqrt{Q}\xi_t$$

$V(\phi)$ - potential function of phi
● ○ - stable and unstable fixed point

Only e.o.f., only c.s.d., either e.o.f. or c.s.d., both e.o.f. and c.s.d, e.o.f. changing to c.s.d., c.s.d. changing to e.o.f., between e.o.f. and c.s.d., neither e.o.f. nor c.s.d, beyond e.o.f and c.s.d.

The Metastabilian says: When I'm far from threshold of some change that involves me, fluctuations and perturbations don't affect me much. But when they become critical, things become erratic, variable, hard to maintain. Suddenly I'm very sensitive to perturbations, a dog barking, an offhand remark, a bright light. I become distracted, find it hard to settle down, to relax, to focus. To know that these changes in my perception and action indicate a threshold is near is illuminating! I'm able to use these signs to anticipate impending change in myself and my environment. I can better appreciate that as frustrating, annoying, difficult as enhancement of fluctuations and critical slowing down can be, they're features of the same dynamics that allows me to change flexibly from one coordination pattern to another. My life and awareness depend on my sensitivity to them...

Related squiggles: anticipation~reaction, variability~stability, choice~instability...

38 Of Coordination Law and Dynamical Landscape

I had the landscape in my arms as I painted it. I had the landscape in my mind and shoulder and wrist—Helen Frankenthaler

The term dynamical landscape is a picturesque name for the 'big picture' view of Coordination Dynamics (CD). It's intended to conjure up an image of how all the phenomena modeled by the Coordination Law form a flexible, evolving terrain—a dynamical landscape. In other words, the flow of coordination behavior captured by the Coordination Law has an underlying, cohesive, predictable shape. Such dynamical landscapes of self-organizing, flexible, adaptable synergetic systems are affected by factors such as the heterogeneity of the individual coordinating elements and the coupling between them. And such factors change in time. Yet remarkably, they change in a way that can be understood and predicted by the Coordination Law.

Practically speaking, all the different graphical visualizations of the Coordination Law are really different perspectives of the same dynamical landscape, each in its own way providing valuable information and insights. But for those unaccustomed to them, the graphs can be as intimidating as the equations. Interpreting and understanding specific details of the graphical visualizations and how they capture and express nuances of the paradigm and phenomena of CD is undeniably challenging. Enter the big picture view of the dynamical landscape, which provides you a powerful conceptual image of the Coordination Law governing synergetic, pattern forming systems.

Imagine yourself as an informationally-coupled self-organizing system, whose evolving dynamical landscape follows the Coordination Law as you live your daily life. There is the dynamical landscape of your body, specifying the coordination among its moving parts. There is also the dynamical landscape coordinating your body and the world, the organism~environment, as well as your relation to other people, animals, tools and machines. And there is the dynamical landscape of your brain~mind. When you read and think about a *squiggle*, the dynamical landscape of your brain~mind shifts and your *squiggle sense* becomes engaged. Now, you can use the graphs of the CD law to analyze your own coordination behavior scientifically— though you probably won't. Still, being aware of the Coordination Law and how it shapes the dynamical landscape of synergetic systems is of great potential benefit and inspiration. The Coordination Law~dynamical landscape constitutes a paradigm shifting advance in knowledge and understanding of life's patterns available to everyone. It allows you to know yourself.

J. A. S. Kelso and D. A. Engstrøm, *The Squiggle Sense*,
https://doi.org/10.1007/978-3-031-59369-7_38

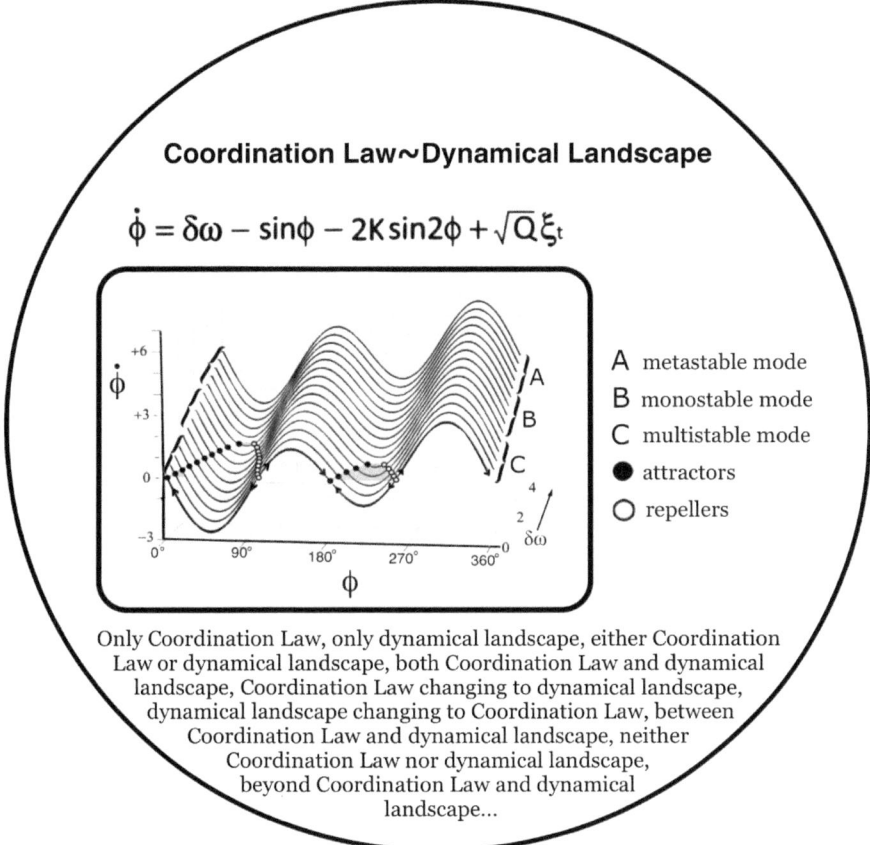

Coordination Law~Dynamical Landscape

$$\dot{\phi} = \delta\omega - \sin\phi - 2K\sin2\phi + \sqrt{Q}\,\xi_t$$

A metastable mode
B monostable mode
C multistable mode
● attractors
○ repellers

Only Coordination Law, only dynamical landscape, either Coordination Law or dynamical landscape, both Coordination Law and dynamical landscape, Coordination Law changing to dynamical landscape, dynamical landscape changing to Coordination Law, between Coordination Law and dynamical landscape, neither Coordination Law nor dynamical landscape, beyond Coordination Law and dynamical landscape...

The Metastabilian says: Life is fleeting, each passing moment filled with patterns upon patterns of coordination, within me, between myself, others and the world. It's a marvelous cacophony, so complex and chaotic in so many ways, coherent and simple in others. I think of all the transitions, shifts, changes of my mind, mood and movement that happen in a day, each task requiring a different feel, mode and dance to successfully accomplish it. Is there really any hope of understanding how it all works, how it all comes to be? Yes! By comparison, the equations of the CD Law and plots of its dynamical landscape aren't really so inaccessible to comprehend. That they can capture all these different coordinations of my life, and scientifically ground my *squiggle sense* which I use to comprehend the complementary nature, is mysterious, profound and inspiring.

Related squiggles: theory~practice, equation~simulation, map~territory...

39 Of Dynamical Landscape and Coordination Modes

 Most of an organism, most of the time is developing from one pattern to another, and not from homogeneity into a pattern—Alan Turing

The behavior of a living, synergetic system like you consists of a set of trajectories on a manifold that's diverse, contextual, and idiosyncratic—an evolving flow of different possible coordination patterns, all of which follow the Coordination Law. Behavior is nonlinear and changes from one pattern to another. It can go backwards, forwards and sideways, round and round, up and down... It's not just coordination patterns that emerge, persist and change in real time, but also the possible transitions between them. In Coordination Dynamics (CD), all possible coordination patterns that you can express, quantitative changes within them and qualitative transitions between them, can be imagined and visualized as a dynamical landscape whose shape depends on a variety of different factors or (in scientific lingo) control parameters. The *squiggle sense* of your dynamical landscape is revealed via visualizations and computer simulations of the Coordination Law.

A good way to see what's going on is by means of the "Parameter Space" view, which shows how different coordination modes persist and change as a function of the model's two control parameters, heterogeneity ($\delta\omega$) and coupling strength (K). All possible coordination patterns of the dynamical landscape fall into four different regions or archetypal coordination modes: (1) the uncoupled mode—coordinating elements move freely, independent and uncorrelated. No collective coordination states exist between them. (2) the metastable mode—coordinating elements are weakly coupled and differ in their intrinsic properties. Although no stable states exist, remnants or tendencies of stable states guide flexible coordination. (3) the monostable mode—coordinating elements are strongly coupled into a single stable coordination state. (4) the bistable or multistable mode—where two or more stable coordination states coexist.

In the Parameter Space view, arrows highlight different routes of change or travel between the four coordination modes of the dynamical landscape. Different combinations of parameter values can yield the same coordination modes. In some regions of parameter space, small changes in parameter values can lead to abrupt changes in modes. In others, large changes in parameter values produce no change at all. Life is nonlinear! Where you live in parameter space determines your behavior. Dynamical landscape~coordination modes is yet another foundational *squiggle* of the Complementary Code of CD. Its *squiggle* dance is the universal key to viable coordination, the full expression of the complementary nature.

J. A. S. Kelso and D. A. Engstrøm, *The Squiggle Sense*,
https://doi.org/10.1007/978-3-031-59369-7_39

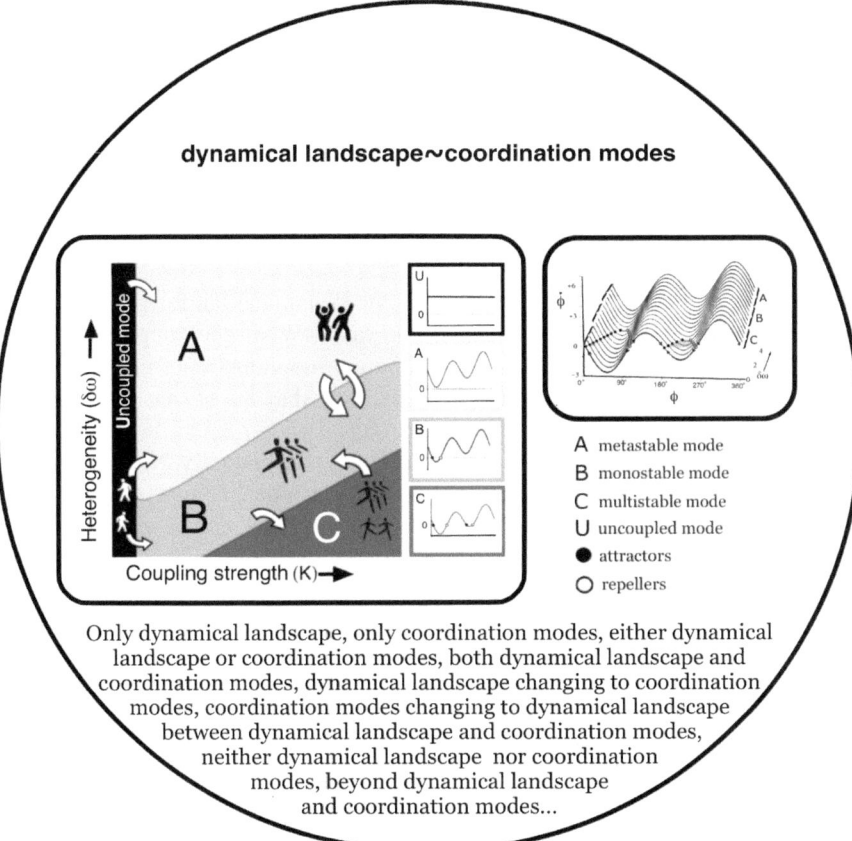

Only dynamical landscape, only coordination modes, either dynamical landscape or coordination modes, both dynamical landscape and coordination modes, dynamical landscape changing to coordination modes, coordination modes changing to dynamical landscape between dynamical landscape and coordination modes, neither dynamical landscape nor coordination modes, beyond dynamical landscape and coordination modes...

The Metastabilian says: I imagine my life's journey in terms of the dynamical progression around my own flexible, evolving roller coaster dynamical landscape and its four different archetypal coordination modes—all of them realizable changes of the coupling strength and heterogeneity of individual coordinating elements of different systems of my body, brain, mind and behavior. And therein lies my metastable mode that entails and is entailed by the complementary code, that special combination of coupling strength~heterogeneity that allows me to engage my *squiggle sense* of the complementary nature. And just as compelling are the other modes that aren't metastable. It's amazing to realize that my *squiggle sense* emerges~vanishes along with my metastable mode. I can see now how it can become hidden when I fixate on thoughts and deeds!

Related squiggles: cause~effect, potential~flow, smooth~abrupt...

40 Of Uncoupled and Coupled Modes

 Freedom and constraint are two aspects of the same necessity, which is to be what one is and no other—Antoine de Saint Exupery

The first of the four coordination modes in the dynamical landscape might seem a bit contradictory, though it's a mode all the same. When the coupling strength in the Coordination Law equation is zero, although individual coordinating elements may have the potential for collective coordination, they are at that moment uncoupled. They are free from one another, unbound and uncorrelated. Without coupling, they're free to do their own thing, to move and act independently. No self-organizing coordination between them and other elements can happen, no matter how similar or different their intrinsic properties might be. The mantra is, "no coupling, no coordination". The uncoupled mode allows for the free movement expression of individual elements, their autonomy.

If walking down the street someone bumps into you and you both keep walking—no coupling, no coordination. But if the person asks if you are okay and you make eye contact (coupling), a dialogue ensues (coupling) and coordination begins. You say your goodbyes, go your separate ways. Coordination between you and the person ceases—no coupling, no coordination. So the coupling vanishes again, though a memory of your interaction might not. Experiments in social CD show that remnants or memory of social interactions can linger in individuals even after they separate and are uncoupled. Without the uncoupled mode, truly independent movement isn't possible, nor would the onset and cessation of new coordination patterns be possible. On the other hand, viability requires coordination, which requires coupling. An uncoupled element, as free as it may be, won't last long without coupling of some kind to its "environment", whether natural or social. When that eventually happens, coordination begins anew.

The uncoupled and coupled modes of the dynamical landscape are complementary aspects. Wherever and whenever coupling and coordination is possible, uncoupling is also possible. This *squiggle* is as ubiquitous as coordination itself, and has many names. In neurosynaptic function, neurotransmitters bind with protein receptors (couple) at different affinities (coupling strengths), generate responses (coordinated actions), then release (uncouple, coupling strength diminishes to zero) and responses cease. The inextricable, complementary uncoupled~coupled modes of the universal dynamical landscape are fundamental, literally the beginning~ending of all coordinated dances. Together they're a *squiggle* essential to the complementary nature at all levels, from the molecular to the social.

J. A. S. Kelso and D. A. Engstrøm, *The Squiggle Sense*,
https://doi.org/10.1007/978-3-031-59369-7_40

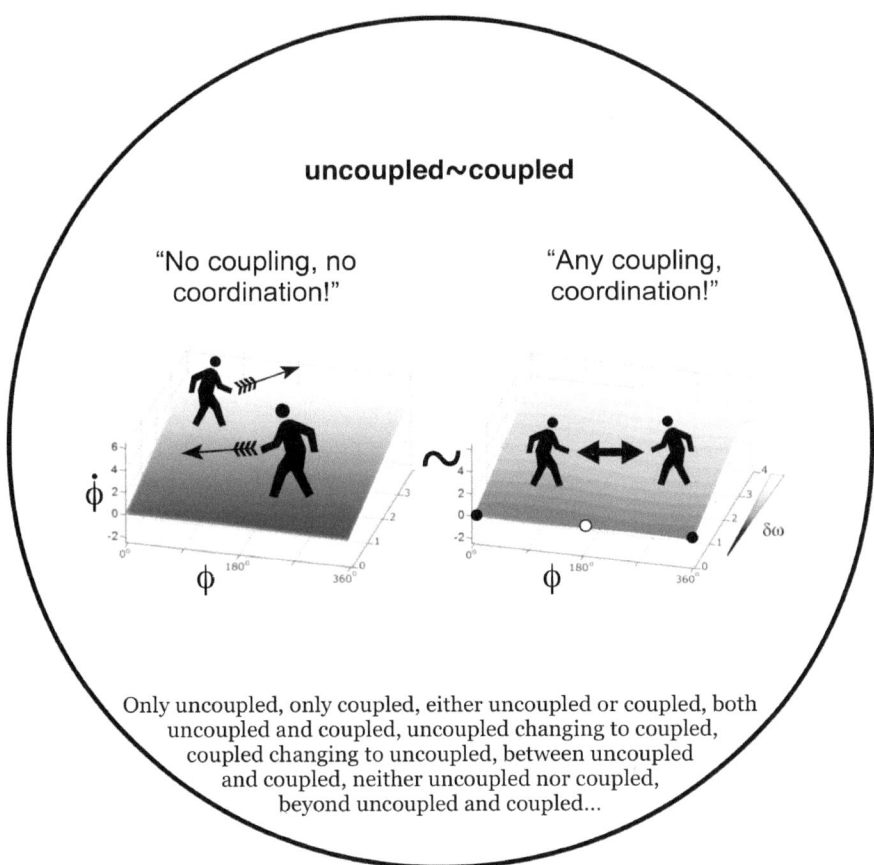

The Metastabilian says: Without coupling, the phase flow of the dynamical landscape is flat. No phasic relationships, no coordination patterns. But when just the tiniest bit of coupling is introduced, I notice the delicate wrinkle in the phase-plane surface. Along with it emerge fixed points of the Coordination Law at '0' and '180' degrees, in-phase and anti-phase. Birth of coordination involves creation of both a stable and an unstable fixed point, a stable attractor and an unstable repeller. One doesn't exist without the other. I notice that in this case of minimal coupling strength, if there is any difference between the coordinating elements ($\delta\omega$), both fixed points vanish, even though the wrinkle remains. (I wonder what that means?).

Related squiggles: unbound~bound, unlinked~linked, freedom~constraint...

41 Of Monostable and Bistable Modes

Multistability, the coexistence of several collective states for the same value of control parameters, is an essential property of biological coordination dynamics—J. A. Scott Kelso

Once there is any coupling between any individual coordinating elements, all the basic coupled coordination modes can emerge. The monostable and bistable modes have to do with stable states, represented graphically by the zero or fixed points of the Coordination Law equation. In the monostable mode, under current values of control parameters, only *one* stable coordination state or fixed point of the Coordination Law equation is present. It's associated with a close to zero phase relation called inphase or synchronized coordination. In the bistable mode, under current values of control parameters, two stable states or fixed points of the Coordination Law equation coexist. The first one is near zero relative phase corresponding to inphase or synchronization, and a second one near 180 degrees, corresponding to anti-phase or syncopation. Bistability is the simplest example of multistability—the existence of two or more states in a dynamical system for the same values of control parameters.

Notice that in-phase and anti-phase, synchronized and syncopated rhythms are polar contraries. In the bistable mode, equally valid polar alternatives coexist within the parameter space of the dynamical landscape. When control parameters like coupling strength are changed continuously by a scientist doing a laboratory experiment or, in the real world, by natural circumstances, the bistable mode can switch to the monostable mode, and vice versa. Such changes are indicated in the parameter space diagram as arrows crossing the boundaries of the bistable and monostable regions. In physics, qualitative switching from one mode to another is often called a phase transition. In mathematics, it's called a bifurcation.

The significance of the monostable mode and its main expression of synchrony is well established in nature. The bistable mode is a prerequisite for all duals and dualisms, polarization, dichotomy, binary explanations, dialectic, yin and yang—the switches and switching of brain and mental function. The complementary nature entails bistability by definition. In the context of monostability and bistability, the grounding of the complementary nature in Coordination Dynamics (CD) is clear. It explains how a single state can self-organize in the first place, how two stable states can coexist, and how transitions between them occur. The monostable and bistable modes of CD are a key *squiggle* of the Complementary Code, essential for coordination, life and awareness.

J. A. S. Kelso and D. A. Engstrøm, *The Squiggle Sense*,
https://doi.org/10.1007/978-3-031-59369-7_41

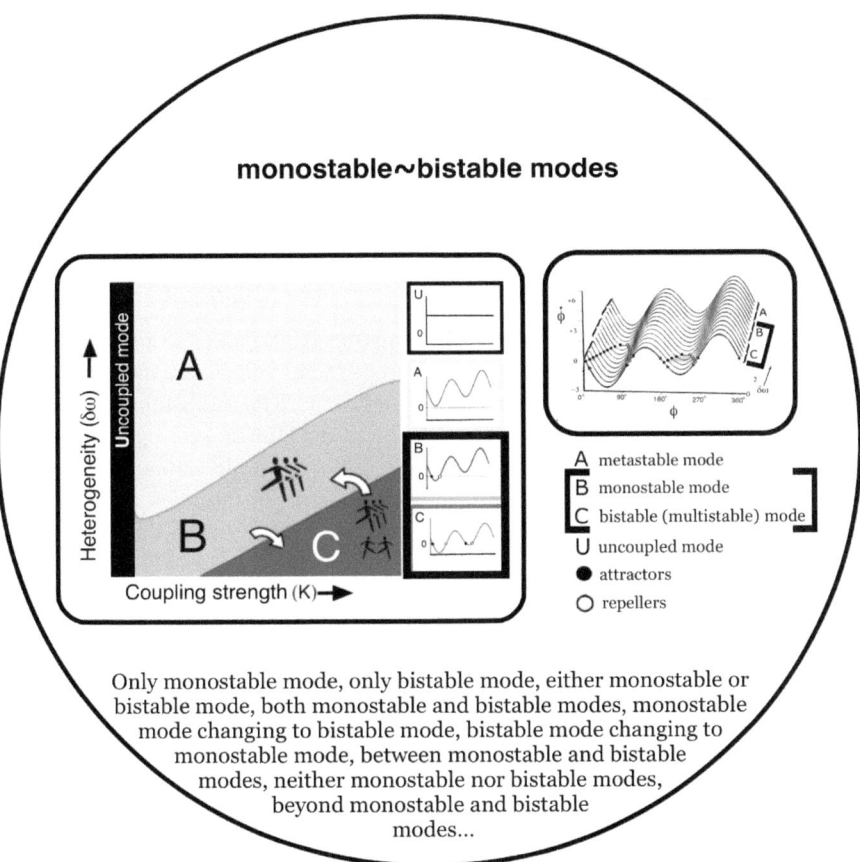

monostable~bistable modes

A metastable mode
B monostable mode
C bistable (multistable) mode
U uncoupled mode
● attractors
○ repellers

Only monostable mode, only bistable mode, either monostable or bistable mode, both monostable and bistable modes, monostable mode changing to bistable mode, bistable mode changing to monostable mode, between monostable and bistable modes, neither monostable nor bistable modes, beyond monostable and bistable modes...

The Metastabilian says: It's insightful to realize that all polarities, all complementary aspects, all squiggles entail bistability, even when the complementary aspects are the monostable and bistable modes themselves! Just considering these two coupled modes and the uncoupled mode of the dynamical landscape, it's already clear how monistic and dualistic policies, while certainly present and valid in their respective parameter regions...are limited and limiting when treated as mutually exclusive. The multistability of the complementary nature offers an elegant flexible compromise in the accommodation and assimilation of qualitatively different alternatives within the vast spectrum of possible coordination dances—even including free movement of uncoupled elements. And CD shows me how it works.

Related squiggles: one~two, unity~plurality, threshold~bifurcation...

42 Of Multistable and Metastable Modes

 (1) There cannot be an intermediate between two contradictions, but of one subject we must either affirm or deny any one predicate. (2) Now it is a mean between two vices, that which depends on excess and that which depends on defect...the vices respectively fall short of or exceed what is right in both passions and actions, while virtue both finds and chooses that which is intermediate—Aristotle

In the multistable mode of Coordination Dynamics (CD) two or more stable attractor states can be expressed for both the same and different values of control parameters. Which one is expressed depends on initial conditions. The multistable mode is characterized by relatively higher coupling strength and lower heterogeneity between coordinating elements than the monostable region. Multiple stable states are separated by unstable repellers, and threshold surpassing instabilities (bifurcations or phase transitions) are required to switch back and forth between them. Notice how taken together, all possible multistable and monostable modes only comprise about half of the dynamical landscape. The other half is occupied by the metastable mode, where heterogeneity between coordinating elements is relatively higher than in the other regimes and plays a crucial role.

In the phase space pictures of the metastable mode, the entire Coordination Law function lies above the x-axis, beyond the zero line. Amazingly, there are no fixed points or stable states at all for half the dynamical landscape! So *phi-dot*, the change of relative phase over time, never settles, yet the squiggly shape of the mathematical model remains. Even though multistability has vanished, remnants of previously multistable states remain. In CD, these dynamical remnants are called metastable tendencies. A main discovery of CD is that similar to the multistable mode, two or more complementary tendencies coexist and shape the ongoing coordination. The metastable mode displays attraction and repulsion without attractors or repellers. It's a both-and kind of complementarity. Both relative phase tendencies are simultaneously present, and you can move flexibly from one to another without the necessity of crossing energy thresholds required for nonlinear phase transitions.

The coexistence of multiple dynamic steady states and metastable mode tendencies in CD is stunning and unexpected. Even complementarity itself is complementary: two contrary means of achieving qualitatively coexistent coordination patterns are reconciled in the same mathematical model! Fixed contrary polar extrema represent bistable states, while the intermediate world between those poles, is metastable. Your *squiggle sense* is a product and expression of your metastable mode, the included middle, a wellspring of virtue and wisdom. Multistability~metastability is a remarkable *squiggle* of the Complementary Code of CD, essential to your brain~mind function, your very awareness and all your sentient potential.

J. A. S. Kelso and D. A. Engstrøm, *The Squiggle Sense*, https://doi.org/10.1007/978-3-031-59369-7_42

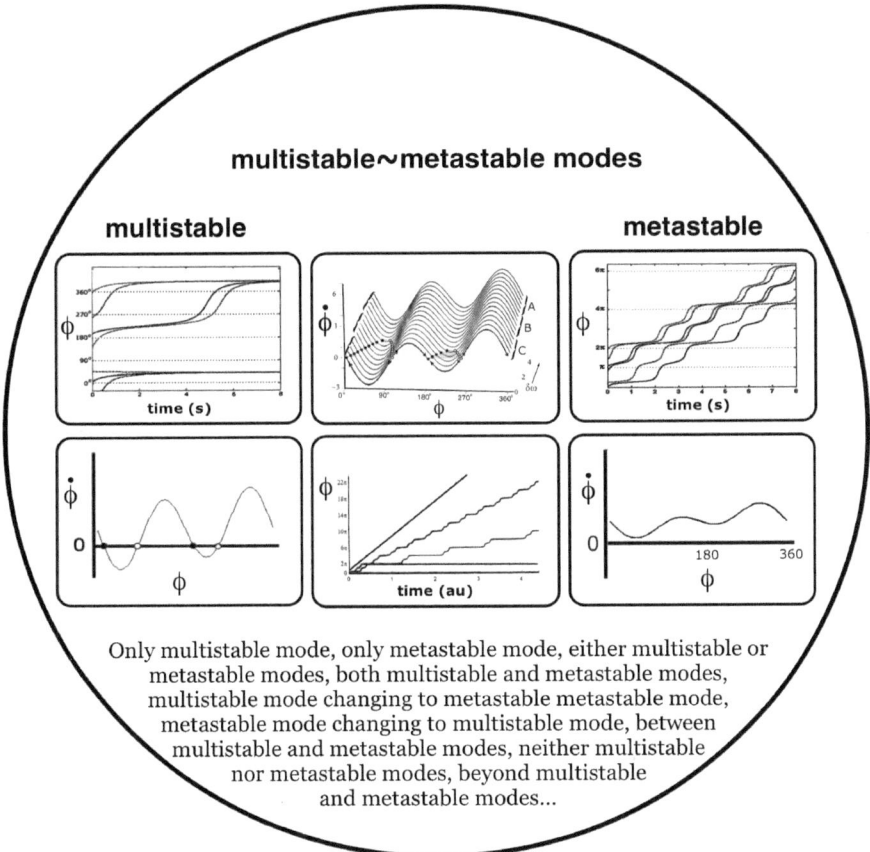

Only multistable mode, only metastable mode, either multistable or
metastable modes, both multistable and metastable modes,
multistable mode changing to metastable metastable mode,
metastable mode changing to multistable mode, between
multistable and metastable modes, neither multistable
nor metastable modes, beyond multistable
and metastable modes...

The Metastabilian says: My brain~mind embraces and entails both either/or complementarity of
multistable states and both-and complementarity of metastable tendencies. It is my complementary
nature. My dynamical landscape is a reconciliation of reconciliations! And as I reflect upon it
all, I have a flash of insight: the vast metastable mode composed of tendencies, preferences and
dispositions, isn't actually between contrary coexistent states at all. It's beyond them! And then
another: At this very moment of insight, my *squiggle sense* is engaged. In this very moment, I am
a metastabilian using metastability to reflect upon that metastability!

Related squiggles: states~tendencies, fixed~flexible, stability~novelty…

43 Of Integration and Segregation

 Synergy is cognate with metastable organization because, in the most general case, the synergic tendency of the involved degrees of freedom toward dependence (integration) is complemented by their anti-synergic tendency toward independence (segregation)—Michael T. Turvey

The complementary nature of integration and segregation plays an inherent, vital role in the synergic organization of living systems discovered, explored and modeled by Coordination Dynamics (CD). Synergies simultaneously reconcile their overall collective functions with the individual integrity of the coordinative elements (components, parts, members, processes, signals, thoughts…) that compose and form them. Dynamic, inextricable tendencies for individual coordinating elements to couple or bind together (integration), and for those same elements to retain and express their independence (segregation), coexist and function simultaneously. Such complementary integrative and segregative tendencies are exactly what is found in the metastable mode of CD. Like multistability and phase transitions, metastability is universal, relevant at all levels, including the very relevant level of your brain~mind.

Ample scientific evidence exists that metastability is essential for human brain~-mind~body~world function. Metastability reconciles the two main current and contrary theories of brain function: One is called functional segregation: it treats the brain as segregated self-governing regions, each localizable and capable of performing unique functions, such as perception, memory and movement. This is partially true. The other is called functional integration: it treats the brain as an integrated organ where the parts function together as a unitary whole. This is also partially true. The metastable mode of CD reconciles independent, individualized tendencies of specialized brain regions and interdependent, collective tendencies of those same regions to work together as a functional synergy.

Now, by definition, your *squiggle sense* is activated and engaged in the metastable mode of your brain~mind's coordination dynamics, at the very level of your self-awareness. It depends on this extraordinary 'both-and tendency complementarity' entailed by metastability. Think of it. Your ability to perceive, ponder and reconcile complementary contrarieties, the *squiggles* of you and your world, requires you to engage your metastable mode. Metastability is characterized by simultaneous complementary tendencies for integration and segregation of the neurons and neural ensembles of your brain~mind as those perceptions, thoughts and reconciliations are happening—the flow of consciousness itself. The complementary nature, your *squiggle sense* of it and your brain~mind's awareness of it, are all based on metastable CD!

J. A. S. Kelso and D. A. Engstrøm, *The Squiggle Sense*,
https://doi.org/10.1007/978-3-031-59369-7_43

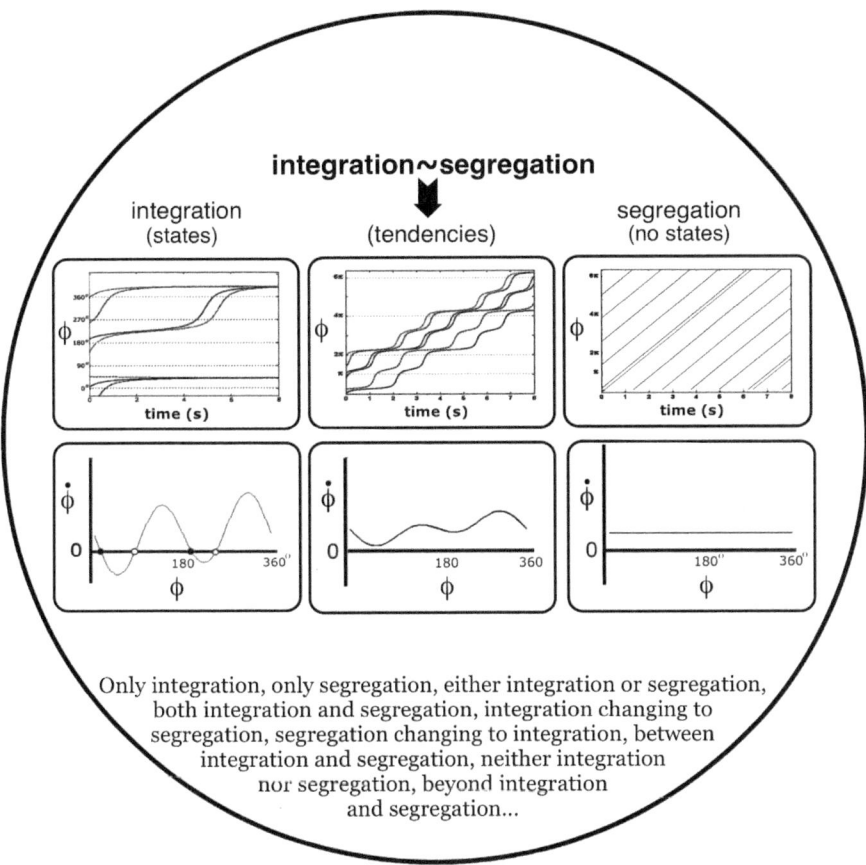

Only integration, only segregation, either integration or segregation, both integration and segregation, integration changing to segregation, segregation changing to integration, between integration and segregation, neither integration nor segregation, beyond integration and segregation...

The Metastabilian says: Kahlil Gibran once wrote, "Let there be spaces in your togetherness and let the winds of the heavens dance between you. Love one another but make not a bond of love: let it rather be a moving sea between the shores of your souls". The complementary tendencies for integration and segregation in the metastable mode of my brain~mind resonate perfectly with these profound, beautiful and wise words of the poet. From the view of CD, they are the poet in his metastable mode advising his readers to engage theirs. What an amazing thought: the complementary nature and metastability of love and wisdom!

Related squiggles: local~global, parts~whole, togetherness~apartness...

44 Of Dwelling and Escaping

 As we take, in fact, a general view of the wonderful stream of our consciousness, what strikes us first is this different pace of its parts. Like a bird's life, it seems to be made of an alternation of flights and perchings—William James

What does metastable integration~segregation look like in reality? Imagine that over some brief epoch in its ongoing metastable trajectory, the phase relation of a coordinating system continuously changes in time. In this epoch, even though the coupling strength between them is greater than zero, coordinating elements exhibit a tendency toward independent, segregated action, as if they were uncoupled, but not quite. As the escaping coordination variable phi (φ) approaches, say in-phase, it briefly dwells there, exhibiting a tendency for cooperative, integrated, coordination. Following this dwelling, the relative phase escapes again until it eventually converges toward and dwells near anti-phase, exhibiting a tendency again for integrated coordination. Its dwelling there soon ends and the phase relation escapes again, and so on. In the metastable mode, the system lives in a virtual sea of metastability. It never gets stuck.

This metastable dance of dwelling and escaping repeats as long as the system is in the metastable mode of its Coordination Dynamics (CD). In it, a segregative tendency to escape phase attraction coexists with an integrative tendency to dwell at preferred phase relations. The former provides the system with flexibility, the freedom to explore its full range of coordinative possibilities. The latter allows the system to maintain a certain degree of stability while doing so. So in the metastable mode, there's attraction~repulsion but no attractors or repellers. There are transient tendencies but no states. Metastability provides living systems the vital mix of flexibility and stability.

Escape velocities (how quickly a trajectory escapes dwelling), and dwell times (how long a system dwells near a preferred phase before escaping), are key measures of metastable coordination dynamics. They are a function of how strongly coordinating elements are coupled relative to how different they are from each other. The dwelling~escaping trajectories of the coordination variable *phi* over time reveal the actual evolution of coexisting integrative and segregative processes and tendencies. CD says that it is the metastable mode that underlies William James's powerful image of the stream of consciousness. Your brain~mind lives in a world of flights and perchings, of freedom and constraint, of escaping from and dwelling near bound states of coordination. In the context of Coordination Dynamics, metastability is the key to your life, awareness, the complementary nature and your *squiggle sense*!

© The Author(s), under exclusive license to Springer Nature Switzerland AG 2024
J. A. S. Kelso and D. A. Engstrøm, *The Squiggle Sense*,
https://doi.org/10.1007/978-3-031-59369-7_44

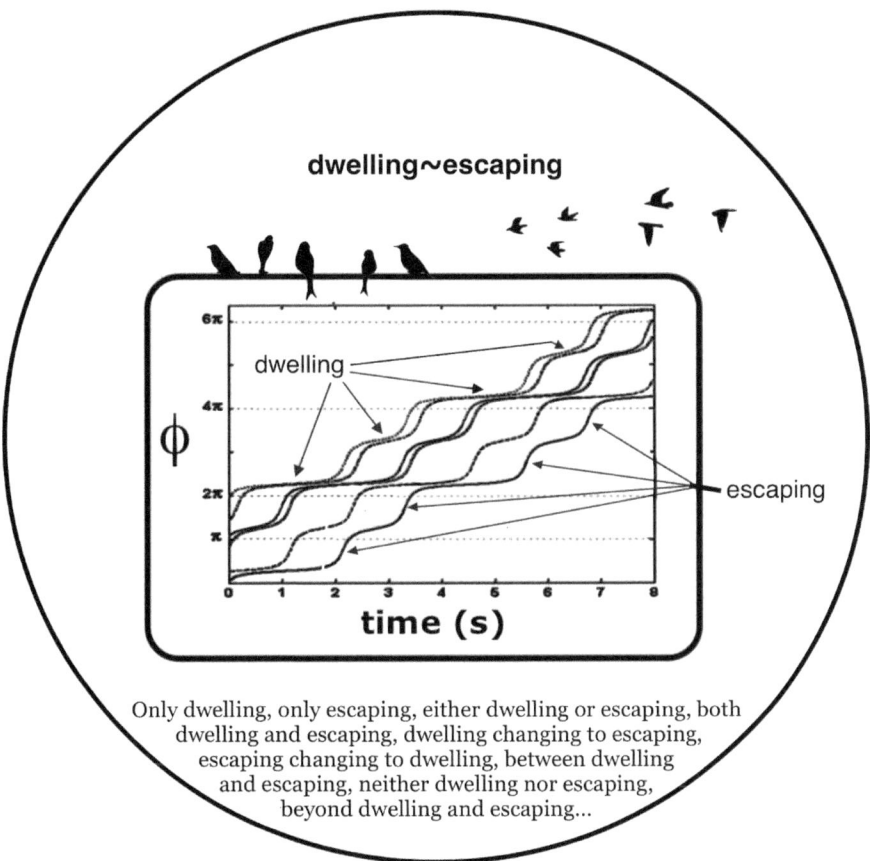

dwelling~escaping

Only dwelling, only escaping, either dwelling or escaping, both dwelling and escaping, dwelling changing to escaping, escaping changing to dwelling, between dwelling and escaping, neither dwelling nor escaping, beyond dwelling and escaping...

The Metastabilian says: How easy it is for me to imagine the dwelling and escape of my awareness, perception~actions, thoughts…all of my coordinated behavior. All this simultaneous integrating and segregating at all the levels of my existence. It's all a grand and mysterious coordination. My stream of consciousness, that is, the complementary flights and perchings of my conscious~unconscious, weave all the *squiggles* of my life into the synergetic collective~individual reality I call my 'self'. It's amazing to imagine how the metastable process of my imagination imagines the way it works. This is the complementary nature!

Related squiggles: persisting~changing, constraint~freedom, pausing~moving…

45 Of Symmetry and Broken Symmetry

The symmetries that are really important in nature are not the symmetry of things, but the symmetries of laws—Stephen Weinberg

Symmetry comes from a very old Greek word. Its original, everyday meaning is "a harmonious agreement in dimensions; due proportion". In the world of math and science, it has a more specific definition: "invariance of the form of an object or process under some transformation, like translation, reflection, rotation, scaling or time". Symmetries are crucial in nature and our understanding of it, where they're used to classify patterns. Indeed, the *squiggle sense* of symmetry~broken symmetry is the essence of metastability, of your metastable mode. To understand why, first realize that the full scientific name of the Coordination Law model is "the extended or broken symmetry form of the HKB model", the original model of Coordination Dynamics (CD) named after its architects, Haken, Kelso and Bunz.

Now, an assumption of the HKB model's original formulation is that individual coordinating elements are identical or nearly so. Their differences are considered negligible. In that case, the model is mathematically symmetric: It doesn't matter whether time flows forward or backward (2π-periodic) and is identical under left–right reflection, where positive and negative relative phase *phi* (φ) behaviors are equivalent. This well known, thoroughly tested model captures the behavior and transitions between monostable, bistable and uncoupled dynamical coordination modes very well. But when coordinating elements, parts and processes are identical, even if coupled, metastability can't happen. It's only when they are heterogeneous that metastable tendencies emerge. To accommodate this fact, the HKB model equations had to be extended to include symmetry breaking. This 'extended HKB model' or Coordination Law beautifully captures the multistable~metastable *squiggle* dance of heterogeneity~coupling.

From broken symmetry and the weakest of coupling, metastability emerges. In the model, the ($\delta\omega$) heterogeneity term bends and 'tilts' symmetrical potential attractor wells, which results in the ability of the system to both dwell in and escape basins of attraction. When heterogeneity vanishes ($\delta\omega = 0$), the equations return to the original, symmetrical HKB model: monostability, multistability, phase transitions and uncoupled modes, yes, but metastability, no. So what's really important is the complementary nature of symmetry~broken symmetry of the Coordination Law. Your awareness, your metastable brain~mind, the complementary nature and your *squiggle sense* all require broken symmetry in the CD. That is their foundation.

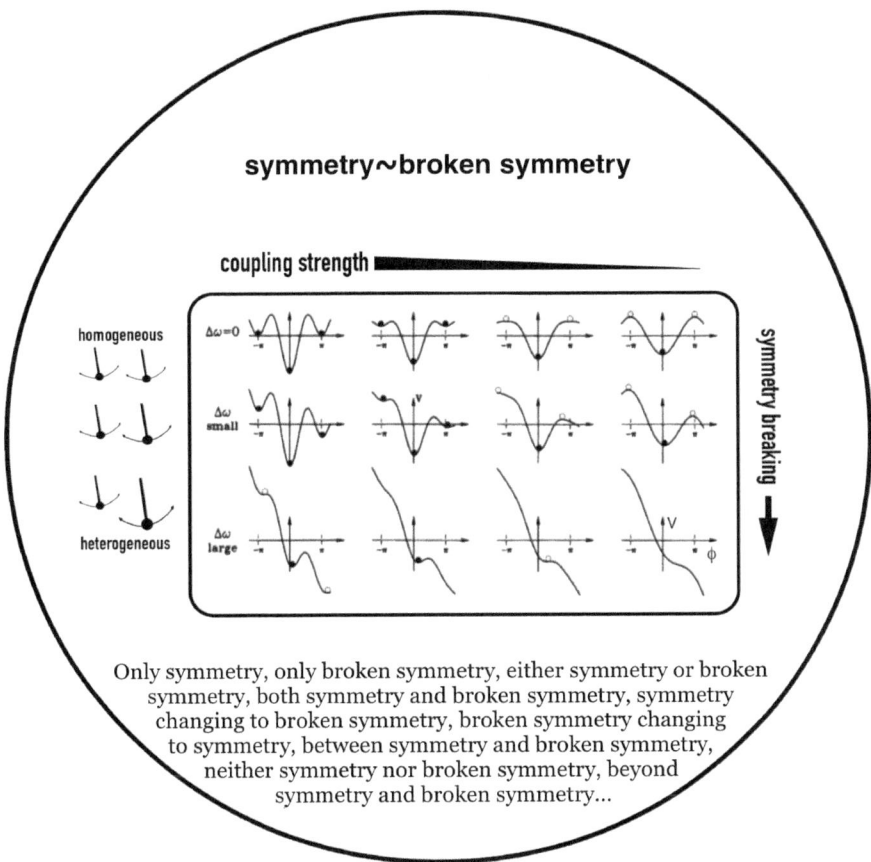

Only symmetry, only broken symmetry, either symmetry or broken symmetry, both symmetry and broken symmetry, symmetry changing to broken symmetry, broken symmetry changing to symmetry, between symmetry and broken symmetry, neither symmetry nor broken symmetry, beyond symmetry and broken symmetry...

The Metastabilian says: Like nonlinearity, the critical importance of broken symmetries to life takes getting used to conceptually. But when I engage my *squiggle sense* to reflect upon the complementary nature of symmetry and broken symmetry, I realize I've known about it all along. For it's what I actually do and experience. The synergies, broken symmetries, metastabilities and *squiggles*, as strange and unfamiliar as their names might be, provide me the life and self awareness that I actually experience, and that allow me to contemplate them. And now it's possible for me to reconcile my life experience with the grounded science and all the phenomena expressed by the Coordination Law. It opens up a whole world of new possibilities...

Related squiggles: homogeneity~heterogeneity, similarity~difference, states~tendencies...

46 Of Competition and Cooperation

The delicate and intricate pattern of competition and cooperation in the economic behavior of the hundreds of thousands of citizens of Stockholm offers a challenge to the economist that is perhaps as complex as the challenges of the physicist and the chemist—George Stigler

Like all human beings, you are very familiar with competition and *cooperation*, with every move you make, hand you shake, deal you make… All you do in your life, how you make a living, survive, work, play, what's been and what's to be, are sculpted from them. They are classic, epic complementary aspects at the very core of coordination and your existence. Your behavior, thought, perception~action, learning~memory, brain~body function, emotions and the world you live in, all depend on delicate and intricate patterns of competition and cooperation of individual~collective synergies. No wonder the interplay of cooperation and competition lies at the heart of the Coordination Law. Life itself consists of the *squiggle* dances of their complementary nature, the universal, inextricable, Coordination Dynamics of these two key complementary aspects.

The term delta omega ($\delta\omega$) expresses how heterogeneity or differences between intrinsic properties of coordinating elements compete for dominance. Such competition alters the dynamical landscape, the collective, self-organizing coordination patterns of nonlinearly coupled elements. The term expressing cooperation in the Coordination Law is $2K\sin2\varphi$, where K stands for how strongly the elements are coupled. The *squiggle sense* of competition~cooperation indicates that mutually exclusive policies that favor one and dismiss the other will eventually fail. Without the natural competition inherent among heterogeneous coordinating elements, metastability cannot emerge. Without cooperation of individual coordinating elements, self-organized coordination and the emergence of self-awareness is not possible.

In adaptation and learning, many studies have shown that new information can cooperate and compete depending on its relation to a learner's current predispositions. Learning tends to be fast when new information cooperates with the learner's intrinsic dynamics, and slow and laborious when it competes. Transitions in adaptation and learning occur when competition between new and old information is reduced, giving rise to stabilization of learned patterns, a cooperative effect. Similar delicate, intricate complementary dances of competition and cooperation of coordinating elements and processes are manifested in countless other scenarios, in many different systems and on many different levels of organization. Such is the powerful and empowering *squiggle sense* of competition and cooperation, whose intimate connection with heterogeneity and coupling in the Coordination Law can hardly be overstated.

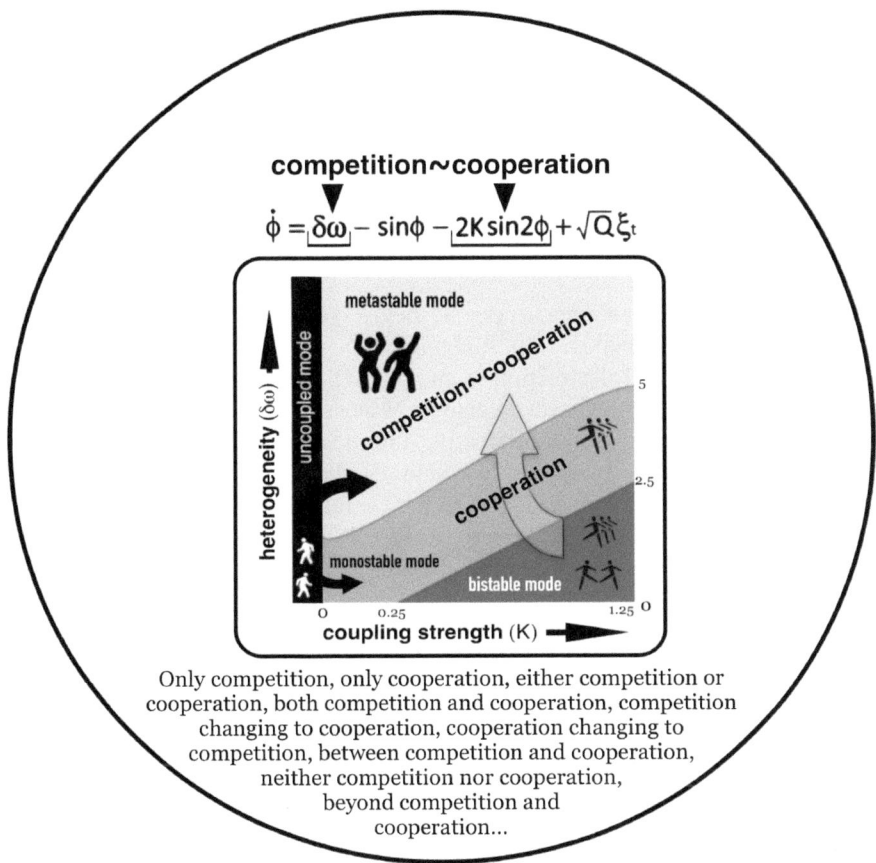

competition~cooperation

$$\dot{\phi} = \lfloor \delta\omega \rfloor - \sin\phi - \lfloor 2K\sin2\phi \rfloor + \sqrt{Q}\,\xi_t$$

Only competition, only cooperation, either competition or cooperation, both competition and cooperation, competition changing to cooperation, cooperation changing to competition, between competition and cooperation, neither competition nor cooperation, beyond competition and cooperation...

The Metastabilian says: With my *squiggle sense* engaged, I realize, accept and reconcile the inevitable, palpable, universal complementary nature of competition and cooperation in any and all system~levels. As amazing as that realization is, somehow it's not really that surprising. My body~mind has always known it. Yet in day-to-day life, often as not competition and cooperation are treated by so many as separate, mutually exclusive aspects, with one valued over the other as primary and fundamental. Then it hits me: without the natural *squiggle* dance of competition and cooperation in my brain~mind, body and behavior... the awareness necessary to choose competition over cooperation or vice-versa wouldn't even be possible!

Related squiggles: inter~intra, conflict~accord, sports~players...

47 Of Creation and Destruction of FI

 Creation and destruction are the two ends of the same moment. And everything between the creation and the next destruction is the journey of life—Amish Tripathi

As a sentient, self-organizing synergetic being, you survive and thrive amidst a dynamic flux of information. The *squiggle dance* of information is palpable. It's dynamical creation and destruction is an ongoing reconciliation of sensation and awareness that fills your life with diversity and nuance, inspiration and warning. But information can also be counterproductive, intimidating and misleading. There can be too much and too little, and both extremes can be detrimental. Now, considering the stupefying, omnipresent deluge of addictive, readily accessible online information available today, it's easy to forget that information has been crucial from the very beginning of sentient awareness, before the internet, computers, libraries, books, the printing press, predating civilization itself.

Indeed, information's original, essential, dynamic role is more often than not taken for granted. Ironically, that role is crucial to the uncountable day-to-day functions that make you human enough to ignore them, like your agency, perception~action, thought, intention and movement. Information that is meaningful and specific to a system's Coordination dynamics (CD) is called functional information (FI). In the context of the Coordination Law, coordination patterns and modes expressed in a given moment are a function of heterogeneity and coupling strength of individual coordinating elements, as well as fluctuations. To the extent that these basic control parameters are in turn affected, adjusted, or tuned by other sources both internal (e.g. from memory) and external (e.g. from the environment), information becomes functional. Functional information is capable of altering or modifying basic coordination patterns, delaying their appearance, stabilizing and destabilizing them. Functional information is an essential complementary aspect of Coordination Dynamics.

The creation~destruction of functional information occurs in the metastable mode of CD, exactly where your *squiggle sense* is active, the result of the twisting, streaming, *squiggle* dance of integrating~segregating tendencies. Among other sentient miracles, this enables you to have thoughts and actions in the first place, and for them to be creative, anticipatory, improvisational, and intelligent. The creation~destruction or 'flux' of functional information informs, modifies, and guides the coordination essential for your self-awareness, agency, and thinking. And all of it, the complementary nature, your brain~mind, *squiggle sense,* the creation~destruction of functional information—arise from the *squiggle* sea of metastability!

J. A. S. Kelso and D. A. Engstrøm, *The Squiggle Sense*,
https://doi.org/10.1007/978-3-031-59369-7_47

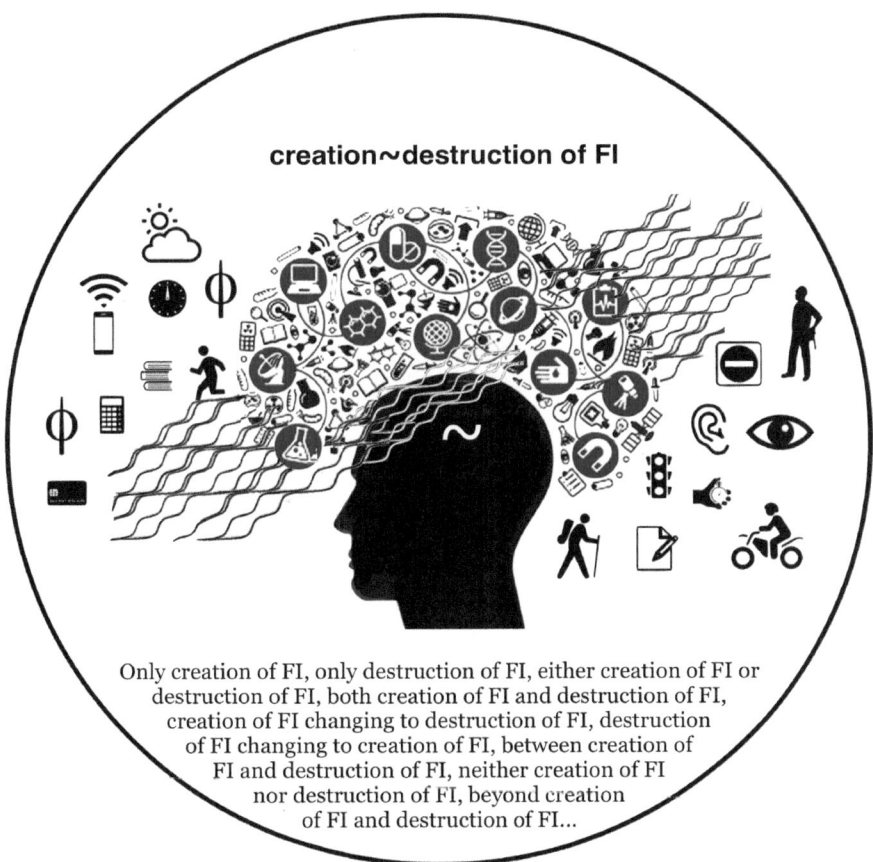

creation~destruction of FI

Only creation of FI, only destruction of FI, either creation of FI or destruction of FI, both creation of FI and destruction of FI, creation of FI changing to destruction of FI, destruction of FI changing to creation of FI, between creation of FI and destruction of FI, neither creation of FI nor destruction of FI, beyond creation of FI and destruction of FI...

The Metastabilian says: I'm a dissipative, dynamic self-organizing, synergetic system called a human being. As such, my very existence, behavior, health, survival all follow the Coordination Law. My dynamical landscape entails four universal coordination modes: uncoordinated, monostable, multistable and metastable. Of those, only the multistable and metastable modes are flexible enough to produce my higher sentient functions. But the crucial creation~destruction of functional information that informs, modifies, and enables my agency, the ability to guide my self-awareness, thought, coordination and *squiggle sense* occurs via my metastable mode. By realizing, understanding, appreciating and acting upon the idea that my metastable mode plays this essential role, I've become a metastabilian...

Related squiggles: emergence~disappearance, recruitment~annihilation, dwelling~escaping...

48 Of States and Tendencies

 What was so odd about the coexistence state is that two seemingly incompatible forms of behavior...were present in the same system at the same time—Yoshiki Kuramoto

You are a complex dynamic synergy of synergies: a wondrous spatial~temporal orchestration, an exquisite combination and composition of matter and energy, a coordination of levels organized as coexistent interwoven dynamical landscapes, doing uncountable multifaceted coordinations simultaneously. So it is with humanity: 8 billion individual human beings, each trying to survive and prosper in an ever changing natural and social environment. There's so much complexity happening at once, it seems impossible to comprehend. Yet, Coordination Dynamics (CD) and your *squiggle sense* can help you comprehend it. Remarkably, CD's dynamical landscapes contain multiple states, transitions and coexisting tendencies, dwelling and escaping, individuals expressing themselves as autonomous entities at the same time as cooperating together. Each synergy operates in its individual dynamical landscape as well as participating in other collective dynamical landscapes. Stable states and transient tendencies occur at the same time in the same system!

Coordinated behavior in which states and tendencies co-occur has been dubbed 'chimera' (after the hybrid creature in Greek mythology). Dynamic chimera are inevitable in the synergies of CD because the coupling among elements of synergies can be nonlocal as well as local, weak in some places and strong in others. A closer look at iconic CD experiments which studied the interplay of the coupling between the hands and sensory input from the environment revealed the empirical footprint of chimera— (mixed) dynamics. Results clearly showed that *both* multistable states and metastable tendencies are part and parcel of the basic repertoire of human behavior.

Dynamic chimera offer a way to study both integration and segregation at once, an important advantage when it comes to understanding complex systems like your brain~mind. The key point is that two seemingly contrary kinds of behavior, namely multistability and metastability, coexist in the same system at the same time. Chimera, thus, are a dynamic signature of the fundamental multistable~metastable *squiggle* that sits at the very heart of your *squiggle sense*. But hold on. You likely have noticed already that in chimera, metastability is just one aspect of the metastable~multistable *squiggle*. Could this mean that metastability, the very source of the *squiggle*, *squiggles* with itself?

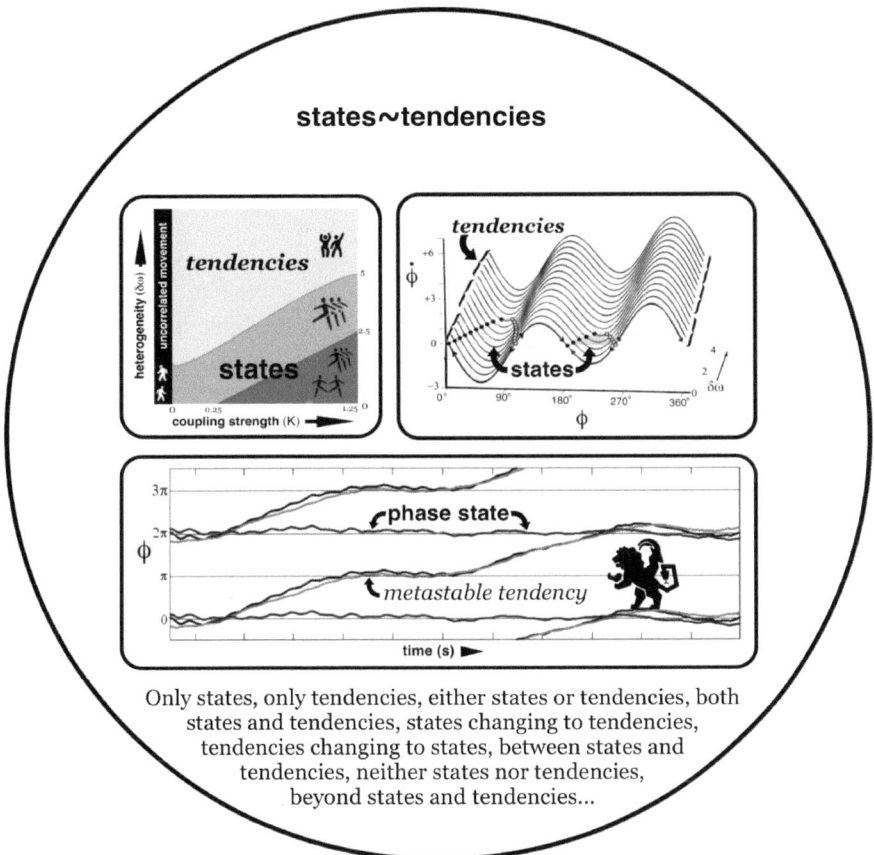

states~tendencies

Only states, only tendencies, either states or tendencies, both
states and tendencies, states changing to tendencies,
tendencies changing to states, between states and
tendencies, neither states nor tendencies,
beyond states and tendencies...

The Metastabilian says: I am a complex system, my brain~mind and body~mind is a synergy of synergies, an exquisitely heterogeneous combination and composition of levels, organs, subsystems, cells, organelles... all of which result in myself, an individual human being. I'm also one of 8 billion fellow human beings each unique and doing so many combinations of things. All of that complexity is not only accomplished by the capacity to move through and employ both states and tendencies of structure~function, but to employ them simultaneously on same and different levels. The phenomenon of dynamic chimeras demonstrates this amazing ability, which, after all, as mysterious as it is to comprehend...makes *squiggle sense*!

Related squiggles: hierarchy~heterarchy, multistability~metastability synergy~resonance...

49 Of Accommodation and Assimilation

 Every acquisition of accommodation becomes material for assimilation, but assimilation always resists new accommodations—Jean Piaget

At the psychological level, accommodation means a flexible tolerance of your brain~mind for novel dynamic coordination patterns. Assimilation means merging and reconciling novel input within a preexisting synergetic milieu, the evolving dynamical landscape or intrinsic dynamics. In Coordination Dynamics (CD), the *squiggle* dance of accommodation and assimilation is understood as the acquisition and stabilization of functional information. Your multistable modes and phase transitions between them are well-suited to assimilation and the pattern stabilization necessary for you to adapt, develop, learn and evolve. At the same time stability, or cohesion of pre-established synergies, naturally resists new accommodations, protecting you from informational overload.

To become a Metastabilian, you must somehow assimilate information you are even now accommodating about your *squiggle sense,* the complementary nature and the metastable CD of your brain~mind. If and when you do, you'll have adapted and altered your dynamical landscape to be more adept at intentionally engaging your *squiggle sense.* As a result, you should be better able to navigate pathways through your dynamical landscape and the multiplicity of coordination patterns expressed by them. Hopefully, you'll become more sensitive to the information created in your metastable mode and more readily perceive and act upon the Complementary Code and its endless dances of metastable tendencies. Ironically, this will also alter your ability to accommodate~assimilate new functional information!

Now, the CD paradigm predicts that the efficacy of accommodation and assimilation depends upon your intrinsic dynamics, the ongoing status of your dynamical landscape and the flux of information impinging on it. But guess what? You are actually predisposed to accommodate and assimilate functional information about your *squiggle sense,* metastable mode and Complementary Code! Metastability and multistability are active, indispensable modes of your dynamical landscape. Further, the complementary nature presents you uncountable contraries for your *squiggle sense* to perceive and act upon. Thus, you already have everything you need to proceed! But will you actually accommodate~assimilate the functional information presented in these *squiggle* frames in a way that changes you, enables you to evolve and advance, explore and discover, acquire new skills and better navigate your life's trajectory? This is your Metastabilian's challenge. It is your *Squiggle* quest!

J. A. S. Kelso and D. A. Engstrøm, *The Squiggle Sense,*
https://doi.org/10.1007/978-3-031-59369-7_49

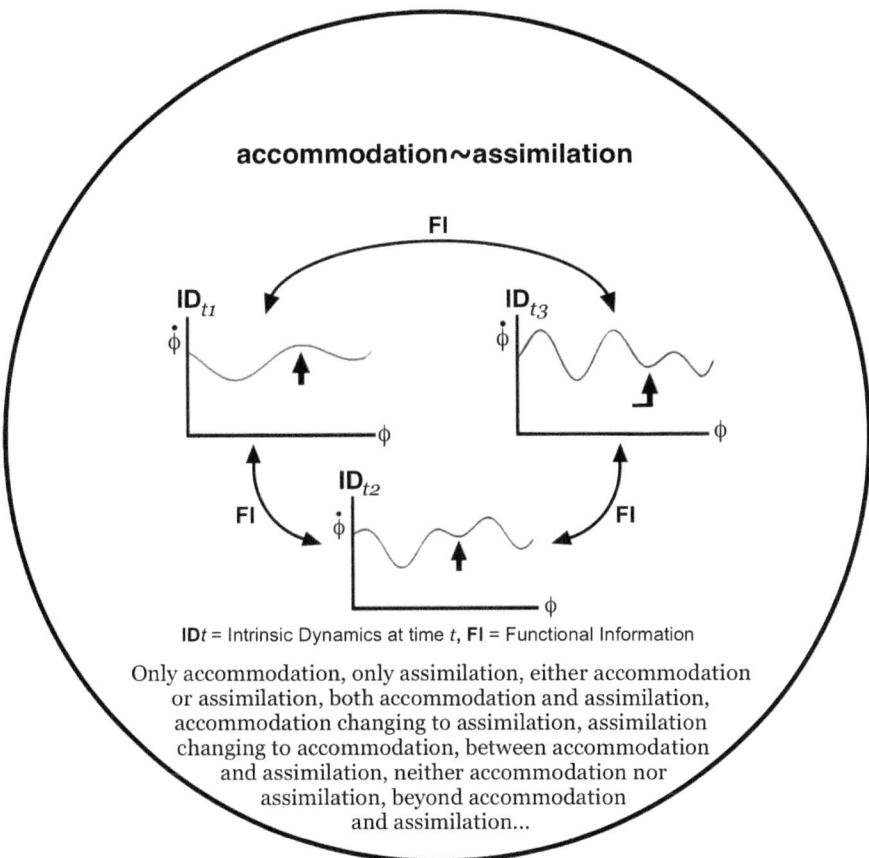

accommodation~assimilation

IDt = Intrinsic Dynamics at time t, FI = Functional Information

Only accommodation, only assimilation, either accommodation
or assimilation, both accommodation and assimilation,
accommodation changing to assimilation, assimilation
changing to accommodation, between accommodation
and assimilation, neither accommodation nor
assimilation, beyond accommodation
and assimilation...

The Metastabilian says: How amazing it is to be aware of, engage and wield my *squiggle sense* of the complementary nature. It leads me to so many novel insights and opportunities, discoveries and advances. It even helps me begin to comprehend how my brain~mind accommodates~assimilates creation~destruction of functional information. In the process of accommodating~assimilating functional information, my entire dynamical landscape changes flexibly but lawfully. As a Metastabilian, I intend, engage and guide those new complementary acquisitions of accommodation, and assimilate them with surprisingly little resistance. After all, they're welcome *squiggle* dances of my metastable~multistable brain~mind, a sentient, synergetic dynamical landscape!

Related squiggles: tolerance~adaptation, learning~memory, interaction~intra-action...

50 Of Metastabilian and Metastabilion

 The world is simple and complex, logical and weird, lawful and chaotic. Fundamental understanding does not resolve these dualities. Indeed, it highlights and deepens them. You can't do justice to physical reality without taking complementarity to heart—Frank Wilczek

Hopefully, the *Squiggle* frames in this book will help you engage your *squiggle sense* and take the complementary nature to heart. Their gist and root messages are: (1) Nature, including human nature, is essentially complementary. (2) Human brain~mind includes a *squiggle sense*, a sixth sense of the complementary nature (TCN). (3) The complementary nature, *squiggle sense* and brain~mind are grounded in Coordination Dynamics (CD). (4) The *squiggle sense* arises from and operates in the metastable mode of brain~mind CD, one of its four archetypal dynamical modes of coordination. (5) A Metastabilian is one who intentionally engages and wields their *squiggle sense* of TCN via their metastable mode of brain~mind. (6) In the metastable mode, functional information, meaningful to brain~mind CD, is created~destroyed.

As a Metastabilian, you use your *squiggle sense* of TCN~CD to discover, explore, invent, create and solve problems in your life pursuits and the world around you in as many contexts as possible, whenever and wherever you can. This activity, called The Metastabilian Movement, is already leading some pioneers to new insights in diverse fields of interest and expertise. As a Metastabilian, you can use your *squiggle sense* to join the *squiggle* quest, namely the pioneering effort to explore, study and comprehend Metastabilion—*"the set of any and all unique complementary metastable tendencies, or squiggles that comprise the complementary nature."* It's exciting to contemplate and anticipate the fundamental discoveries, inventions and innovations of the twenty-first century and beyond that will come from its exploration. One that's resulted in the *squiggle frames* of this volume is the discovery of the Complementary Code of CD. But there are untold more.

As a Metastabilian, your *squiggle sense* helps you move beyond fixated, divisive, polarized, intransigent, intolerant, hubristic thinking and actions that continue to plague humanity. You appreciate that no single polarized perspective is ever permanent, nor is it adequate to capture the reality of the world on its own. By shifting to your metastable mode, you are better at transcending fixation, making discoveries and innovations in all aspects and contexts of your life. The time is ripe for you to engage and wield your *squiggle sense,* to explore Metastabilion and produce and pursue the metastable trajectories of your own metastabilian movement. You might even use them as a platform on which to foster your individual enlightenment and social harmony grounded in TCN~CD. You could save the world!

J. A. S. Kelso and D. A. Engstrøm, *The Squiggle Sense*,
https://doi.org/10.1007/978-3-031-59369-7_50

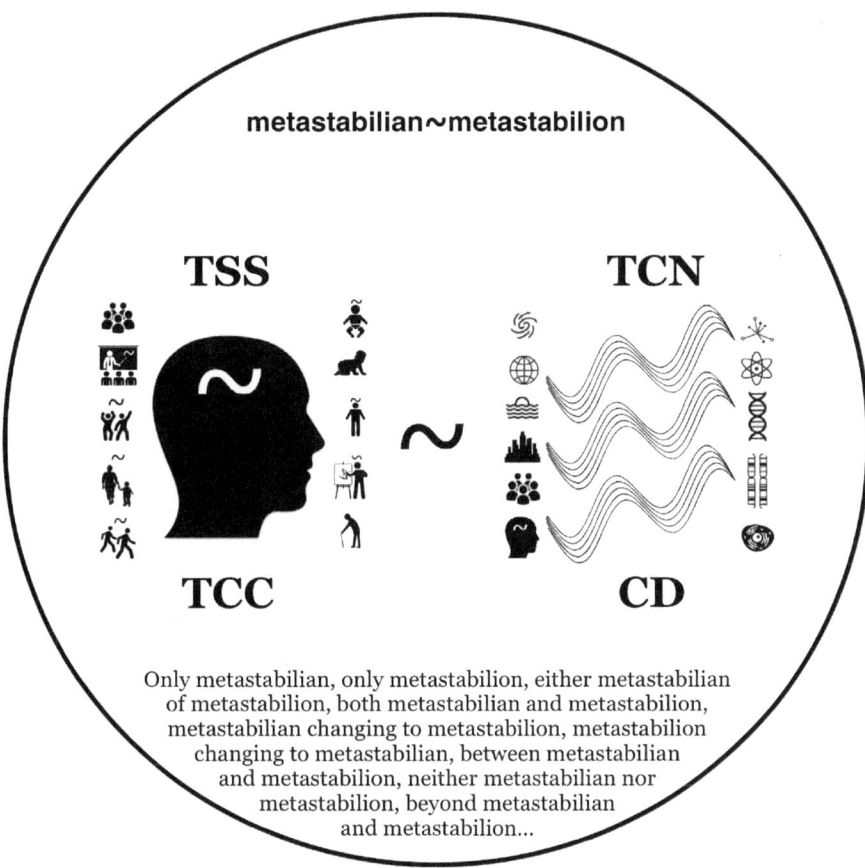

The Metastabilian says: It becomes ever more familiar, the onset of my metastable mode and *squiggle sense*. I sense the ebbing~flowing flight~perching, the relative phasic dances of the complementary nature. I experience the twinkling flux of functional information in my metastable mode. It heightens my awareness in so many different ways~means, piquing my curiosity~imagination and action. How vast is Metastabilion, this cosmic, metastable ocean of complementarity? Imagining it helps me transcend the limiting narrow-mindedness of polarized, either/or thinking~doing. Unfettered from the habit of dwelling in mutually exclusive duality, I advance and evolve in my thoughts, works and deeds. As a Metastabilian, I'm an intentional agent of my *squiggle sense* of the complementary nature, of my Metastabilian movement...

Related squiggles: metastability~complementarity, creation~discovery, TCN~TSS...

Coda—The Meaning of Mtsbwy

 It seems we have found a way at last to provide a scientific grounding for both polarization and reconciliation, which though discernible, differentiable aspects, are nonetheless inextricable.—J. A. Scott Kelso and David A. Engstrøm

Mtsbwy is an acronym that stands for "May the squiggle be with you". Mtsbwy is a Metastabilian greeting, a way to express your awareness, interest and appreciation of the *squiggle sense* of the complementary nature (TCN) to yourself and to others. You can think it, say it, write it and also 'sign' it simply by drawing the *squiggle* sign in the air. Mtsbwy means you are aware of TCN, your *Squiggle Sense* of it and the Coordination Dynamics (CD) that grounds it in science. Mtsbwy means you appreciate the significance of the Complementary Code of CD, those critical *squiggles* necessary for all levels of your synergetic being~becoming. Mtsbwy means you know about the Coordination Law equation, and the dynamical modes of dynamical landscapes—yours as well as other people's, groups and systems. Mtsbwy means you understand that functional information can sculpt dynamical landscapes and be sculpted by them, how incredibly flexible life in the complementary nature really is.

Mtsbwy in its delivery to others means you wish to share this novel perspective, outlook, knowledge~wisdom, in the hope that it will help you and those you say, write or sign it to transcend narrow-minded, limiting, either/or, "my way or the highway" thinking~behaving. Mtsbwy is a reminder to let go of common yet erroneous belief~dwelling in absolutes, permanent states, fixed points, rigid ideals. Saying, writing or signing Mtsbwy means you want to explore, navigate, appreciate and express the complementary nature of life via *the squiggle sense* of your metastable brain~mind. It means that you understand that you and your world are neither purely analytical nor empirical, individual nor collective, bound nor free. TCN grounded in CD is too dynamical, too flexible for such rigid idealizations.

Mtsbwy means you sense TCN of dualities and contradictions. Mtsbwy is a clarion call for all people to embrace, appreciate and wield their vital metastability, or in other words—become Metastabilian. Mtsbwy carries with it a fond and exciting hope, the prospect of an incalculable feat of civilization yet to be achieved in history, a creative~evolution of sentient advance, a novel way forward. Just as CD has been shown empirically and theoretically to transcend different kinds of things, processes, events, so the *squiggle* symbol is universal. *The Squiggle Sense* cuts across all languages, cultures, nationalities, religions. It is the gateway, a new universal language that embraces yet transcends contradiction, dichotomy, and polarities. And now we say to you as hopefully you will say to others—Mtsbwy! May the *squiggle* be with you!

J. A. S. Kelso and D. A. Engstrøm, *The Squiggle Sense*, https://doi.org/10.1007/978-3-031-59369-7

The Metastabilian says: My *squiggle sense* is awake. It gives me so many deep and novel insights into myself and my world. So what comes next? I have my companion, *The Squiggle Sense, Sixth Sense of The Complementary Nature and the Metastable Brain~mind* to inspire me to practice returning to my metastable mode as much as possible. But what more? One thing is sure—I'm not alone: The individual~collective complementary nature says so! There are more Metastabilians out there. Some know who they are, some don't. Yet. How can I meet and greet them? I can say "Mtsbwy! May the squiggle be with you!" and draw a *squiggle* in the air. What would it be like to introduce someone to their *squiggle sense* for the first time? What would it be like to meet other Metastabilians? What might come of sharing appreciation of the complementary nature between myself and other Metastabilians?

Related squiggles: knowledge~wisdom, self~other, individual~society…

Readings for the Squiggle Sense

 This is a brief, highly personalized reading list intended for those who wish to delve further into Coordination Dynamics, metastability, The Complementary Nature and related matters. Most of the material is in the form of Reviews/Encyclopedia Articles, though some key research articles are also included. The order is chronological. The work of many others, of course, is cited in these papers

Books

Kelso, J.A.S. (1995). *Dynamic Patterns: The Self Organization of Brain and Behavior*. Cambridge, MA: The MIT Press.
Tschacher, W., & J.P. Dauwalder (Eds.) (2003). *The Dynamical Systems Approach to Cognition: Concepts and Empirical Paradigms Based on Self-Organization, Embodiment and Coordination Dynamics*, Singapore: World Scientific.
Jirsa, V.K. & Kelso, J.A.S. (Eds.) (2004). *Coordination Dynamics: Issues and Trends* Vol. 1 Springer Series in *Understanding Complex Systems*, Berlin, Heidelberg.
Kelso, J.A.S., & Engstrøm, D. A. (2006). *The Complementary Nature,* Cambridge, MA: The MIT Press.
Fuchs, A., & Jirsa, V. K. (Eds.) (2008). *Coordination: Neural, Behavioral and Social Dynamics*, Springer Verlag, Berlin, Heidelberg.
Sheets-Johnstone, M. (2011). *The Primacy of Movement*. Amsterdam/Philadelphia: John Benjamins.
Fuchs, A. (2013). *Nonlinear dynamics in complex systems*. Springer, Heidelberg, New York.
Ally, M. C, (2017). *Ecology and Existence: Bringing Sartre to the Water's Edge*. Lexington Books, London.
Kelso, J.A.S. (Ed.) (2019). *Learning to Live Together: Promoting Global Harmony*. Springer Verlag, Berlin, Heidelberg.
Sheets-Johnstone, M. (2023). *The importance of evolution to understandings of human nature*. Brill, Leiden, Boston.

Reviews

Kelso, J.A.S., & Wallace, S.A. (1978). Conscious mechanisms in movement. In G.E. Stelmach (Ed.) *Information processing and motor control*. New York, Academic Press, pp. 79–116.
Kelso, J.A.S. (1981). Contrasting perspectives on order and regulation in movement. In A. Baddeley & J. Long (Eds.). *Attention and performance, IX*. Hillsdale, NJ: Erlbaum.
Kelso, J.A.S., & Tuller, B. (1984). A dynamical basis for action systems. In M.S. Gazzaniga (Ed.). *Handbook of Cognitive Neuroscience* (pp. 321–356). New York: Plenum.
Kelso, J.A.S., & Schöner, G. (1988). Self-organization of coordinative movement patterns. *Human Movement Science 7*, 27–46.
Schöner, G. & Kelso, J.A.S. (1988). Dynamic pattern generation in behavioral and neural systems. *Science, 239*, 1513–1520.
Kelso, J.A.S. (1990). Phase transitions: Foundations of behavior. In H. Haken & M. Stadler (Eds.) *Synergetics of cognition*. Springer Verlag, Berlin, pp. 249–268.
Turvey, M.T. (1990). Coordination. *American Psychologist*, August, 938–953.
DeGuzman, G.C. & Kelso, J.A.S. (1992/2018). The flexible dynamics of biological coordination: Living in the niche between order and disorder. In A.B. Baskin & J.E. Mittenthal (Eds.), *Principles of Organization of Organisms*. Routledge, New York, pp. 11–34.

Zanone, P.G., Kelso, J.A.S. & Jeka, J.J. (1993). Concepts and methods for a dynamical approach to behavioral coordination and change. In G.J.P. Salvelsbergh (Ed.), *The Development of Coordination in Infancy,* North Holland, Amsterdam. pp. 89–134.

Kelso, J.A.S., Ding, M., & Schöner, G. (1993). Dynamic pattern formation: A primer. In E. Thelen & L. Smith (Eds.) *Dynamic Approach to Development,* MIT Press, Cambridge, pp. 14–49.

Kelso, J.A.S. (1994). Elementary coordination dynamics. In S. Swinnen, H. Heuer, J. Massion, and P. Casaer (Eds.), *Interlimb Coordination: Neural, Dynamical, and Cognitive Constraints,* pp. 301–318. San Diego: Academic Press.

Kelso, J.A.S. (1994). The informational character of self-organized coordination dynamics. *Human Movement Science, 13,* 393–413.

Fuchs, A. & Kelso, J.A.S. (1994). A theoretical note on models of interlimb coordination. *Journal of Experimental Psychology: Human Perception and Performance, 20/5,* 1088–1097.

Kelso, J.A.S. & Haken, H. (1995). New laws to be expected in the organism: Synergetics of brain and behavior. In M. Murphy & L. O'Neill (Eds.) *What is Life? The Next 50 Years.* Cambridge University Press, pp. 137–160.

Kelso, J.A.S., Case, P., Holroyd, T., Horvath, E., Raczaszek, J., Tuller, B. & Ding, M. (1995). Multistability and metastability in perceptual and brain dynamics. In P. Kruse & M. Stadler (Eds.) *Ambiguity in Mind and Nature.* Heidelberg: Springer-Verlag, pp. 159–185.

Kelso. J.A.S. (1997). The other sciences of complexity. *Complexity,* 3, 7–8.

Tuller, B., & Kelso, J.A.S. (1998). Action theory and the production of speech. In J.L. Mey (Ed.). *Concise Encyclopedia of Pragmatics.* Oxford:Pergamon.

Kelso, J.A.S. (2000). Principles of dynamic pattern formation and change for a science of human behavior. In: Bergman, L.R., Cairns, R.B., Nilsson, L.-G., & Nystedt, L. *Developmental science and the holistic approach.* Mahwah, NJ: Erlbaum, pp. 63–83.

Bressler, S.L. & Kelso, J.A.S. (2001). Cortical co ordination dynamics and cognition. *Trends in Cognitive Sciences, 5,* 26–36.

Kelso. J.A.S. (2001). Metastable coordination dynamics of brain and behavior. *Brain and Neural Networks* (Japan) 8, 125–130.

Kelso, J.A.S. (2001). Self-organizing dynamical systems. In Smelser, N.J. & Baltes, P.B., (Eds. in Chief) *International Encyclopaedia of Social and Behavioral Sciences.* Amsterdam; Pergamon.

Oullier, O., & Kelso, J.A.S. (2006). Neuroeconomics and the metastable brain. *Trends in Cognitive Sciences,* 10, 363–364.

Kelso, J.A.S. & Tognoli, E. (2007). Toward a complementary neuroscience: Metastable coordination dynamics of the brain. In R.Kozma & L. Perlovsky (Eds.) *Neurodynamics of Cognition and Consciousness.* Springer, Heidelberg, pp. 39–60.

Turvey, M.T. (2007). Action and perception at the level of synergies. *Human Movement Science,* 26, 657–697.

Kelso, J.A.S. (2008). An essay on understanding the mind. *Ecological Psychology,* 20, 180–208.

Engstrøm, D. A. & Kelso, J.A.S. (2008). Coordination dynamics of the complementary nature. *Gestalt Theory,* 30, 121–134.

Raczaszek-Leonardi,J., & Kelso, J.A.S. (2008). Reconciling symbolic and dynamics aspects of language: Toward a dynamic psycholinguistics. *New Ideas in Psychology,* 26, 193–207.

Kelso, J.A.S. & Tognoli, E. (2009). Toward a Complementary Neuroscience: Metastable Coordination Dynamics of the Brain. In N. Murphy, G.F.R. Ellis, T. O'Connor, (Eds.) *Downward Causation and the Neurobiology of Free Will.* Springer: Heidelberg, pp. 103–126.

Kelso, J.A.S. (2009). Synergies: Atoms of brain and behavior. *Advances in Experimental Medicine and Biology, 629,* 83–91. [Also D. Sternad (Ed) *A multidisciplinary approach to motor control.* Springer, Heidelberg].

Kelso, J.A.S. (2009/2013). Coordination Dynamics. In R.A. Meyers (Ed.) *Encyclopedia of Complexity and System Science,* Springer: Heidelberg (pp. 1537–1564).

Fuchs, A., & Kelso, J.A.S. (2009/2013). Movement coordination. In R.A. Meyers (Ed.) *Encyclopedia of Complexity and System Science,* Springer: Heidelberg (pp. 5718–5736).

Oullier, O., & Kelso, J.A.S. (2009/2013). Social coordination from the perspective of coordination dynamics. In R.A. Meyers (Ed.) *Encyclopedia of Complexity and Systems Science,* Springer: Heidelberg (pp. 8198–8212).

Semetsky, I. (2010). Ecoliteracy and Dewey's educational philosophy: Implications for future leaders. *Foresight*, 12, 31–44.

Kostrubiec, V., Zanone, P.-G., Fuchs, A., & Kelso, J.A.S. (2012). Beyond the blank slate: Routes to learning new coordination patterns depend on the intrinsic dynamics of the learner—experimental evidence and theoretical model. *Frontiers in Human Neuroscience*, 6, 212 https://doi.org/10.3389/fnhum.2012.00222

Kelso, J.A.S. (2014). The dynamic brain in action: Coordinative structures, criticality and coordination dynamics. In D. Plenz & E. Niebur (Eds.) *Criticality in Neural Systems*, John Wiley & Sons, Mannheim, pp. 67–106.

Tognoli, E. & Kelso, J.A.S. (2014). The metastable brain. *Neuron*, 81, 35–48.

Kelso, J.A.S. (2016). On the self-organizing origins of agency. *Trends in Cognitive Sciences*, 20 (7), 490–499 https://doi.org/10.1016/j.tics.2016.04.004

Kelso, J.A.S., Stolk, E., & Portugali, J. (2016). Self-organization and urban design as a complementary pair. In Portugali, J. & Stolk, E., (Eds.) *Complexity, Cognition, Urban Planning and Design*, Springer, Heidelberg, pp. 43–53.

Balagué, N., Torrents, C., Hristovski, R., & Kelso, J.A.S. (2016). Sport science integration: An evolutionary synthesis. *European Journal of Sport Science*, https://doi.org/10.1080/17461391.2016.1198422.

Kelso, J.A.S. & Tognoli, E. (2017). Toward a complementary neuroscience: Metastable coordination dynamics of the brain. *Chaos & Complexity Letters, 11*, 141–162.

Kelso, J.A.S. (2018). Walls and Borders and Strangers on the Shore: On Learning to Live Together from the Perspective of the Science of Coordination and The Complementary Nature. In Kelso, J.A.S (Ed). *Learning to Live Together: Promoting Social Harmony*. Heidelberg: Springer, pp. 77–93.

Dumas, G., Lefebvre, A., Zhang, M., Tognoli, E., & Kelso, J.A.S. (2018). The human dynamic clamp: a probe for social coordination dynamics. In Mueller, S., et al (Eds) *Complexity and Synergetics*, Springer-Verlag, Heidelberg, pp. 317–333.

Tognoli, E., Zhang, M., Fuchs, A., Beetle, C.B., & Kelso, J.A.S. (2020) Coordination Dynamics: A foundation for understanding social behavior. *Frontiers in Human Neuroscience* https://doi.org/10.3389/fnhum.2020.00317.

Torrents, C., Balagué, N., Hristovski, R., Almarcha, M., & Kelso, J.A.S. (2021). Metastable coordination dynamics of collaborative activity in educational settings. *Sustainability*, 13, 2696. https://doi.org/10.3390/su13052696.

Hancock, F., Rosas, F.E., Zhang, M., Mediano, P.A.M., et al. (2023). Metastability demystified—the foundational past, the pragmatic present, and the potential future. https://doi.org/10.20944/preprints202307.1445.v1

Rossi, K. L., Budzinski, R.C., Medeiros, E.S., et al (2023). A unified framework of metastability in neuroscience. arXiv:2305.05328v1 [q-bio.NC] 9 May 2023

Kelso, J.A.S. (2023). Democracy deserves wisdom. In J.Portugali (Ed.) *The crisis of democracy in the age of cities*. Elgar, Cheltenham.

A Selection of Key Research Articles

Kelso, J.A.S., Southard, D., & Goodman, D. (1979). On the nature of human interlimb coordination. *Science, 203,* 1029–1031.

Kelso, J.A.S., Tuller, B., Bateson, E.V., & Fowler, C.A. (1984). Functionally specific articulatory cooperation following jaw perturbations during speech: Evidence for coordinative structures. *Journal of Experimental Psychology: Human Perception and Performance, 10,* 812–832.

Kelso, J.A.S. (1984). Phase transitions and critical behavior in human bimanual coordination. *American Journal of Physiology: Regulatory, Integrative and Comparative, 15,* R1000–R1004.

Haken, H., Kelso, J.A.S., & Bunz, H. (1985). A theoretical model of phase transitions in human hand movements. *Biological Cybernetics, 51,* 347–356.

Schöner, G., Haken, H., & Kelso, J.A.S. (1986). A stochastic theory of phase transitions in human hand movement. *Biological Cybernetics, 53,* 247–257.

Kelso, J.A.S., Scholz, J.P. & Schöner, G. (1986). Nonequilibrium phase transitions in coordinated biological motion: Critical fluctuations. *Physics Letters A, 118,* 279–284.

Scholz, J.P., Kelso, J.A.S. & Schöner, G. (1987). Nonequilibrium phase transitions in coordinated biological motion: Critical slowing down and switching time. *Physics Letters A, 123,* 390–394.

Kelso, J.A.S., DelColle, J. & Schöner, G. (1990). Action~Perception as a pattern formation process. In M. Jeannerod (Ed.), *Attention and Performance XIII,* Hillsdale, NJ: Erlbaum, pp. 139–169.

Kelso, J.A.S., Bressler, S.L., Buchanan, S., DeGuzman, G.C., Ding, M., Fuchs, A. & Holroyd, T. (1992). A phase transition in human brain and behavior. *Physics Letters A, 169,* 134–144.

Kelso, J.A.S. & Jeka, J.J. (1992). Symmetry breaking dynamics of human multilimb coordination. *Journal of Experimental Psychology: Human Perception and Performance, 18,* 3, 645–668.

Zanone, P.G. & Kelso, J.A.S. (1992). The evolution of behavioral attractors with learning: Nonequilibrium phase transitions. *Journal of Experimental Psychology: Human Perception and Performance, 18/2, 403–421.*

Zanone, P.G. & Kelso, J.A.S. (1997). The coordination dynamics of learning and transfer: Collective and component levels. *Journal of Experimental Psychology: Human Perception and Performance, 23,* 1454–1480

Friston, K.J. (1997). Transients, metastability and neuronal dynamics. *Neuroimage, 5,* 164–171.

Jirsa, V. K., Fuchs, A., & Kelso, J.A.S. (1998). Connecting cortical and behavioral dynamics: Bimanual coordination. *Neural Computation,* 10, 2019–2045.

Kelso, J.A.S., Fuchs, A., Holroyd, T., Lancaster, R., Cheyne, D., & Weinberg, H. (1998). Dynamic cortical activity in the human brain reveals motor equivalence. *Nature, 392,* 814–818.

Kelso, J.A.S. (2002). The complementary nature of coordination dynamics: Self-Organization and the origins of agency. *Journal of Nonlinear Phenomena in Complex Systems, 5,* 364–371.

Jantzen, K.J., Steinberg, F.L., & Kelso, J.A.S. (2004). Brain networks underlying timing behavior are influenced by prior context. *Proceedings of the National Academy of Sciences, 101,* 6815–6820.

Lagarde, J., Peham, C., Licke, T., & Kelso, J.A.S. (2005). Coordination dynamics of the horse~rider system. *Journal of Motor Behavior,* 37, 419–424.

Jirsa, V.K., & Kelso, J.A.S. (2006). The excitator as a minimal model for the coordination dynamics of discrete and rhythmic movements. *Journal of Motor Behavior,* 37, 35–51.

Tognoli, E., Lagarde, J., DeGuzman, G.C., & Kelso, J.A.S. (2007). The phi complex as a neuromarker of human social coordination. *Proceedings of the National Academy of Sciences, 104, 8190–8195 (*from the cover, *Scientific American Mind,* Aug./Sept., 2007).

Oullier, O., DeGuzman, G.C., Jantzen, K.J., Lagarde, J., & Kelso, J.A.S. (2008). Social coordination dynamics: Measuring human binding. *Social Neuroscience,* 3, 178–192.

Kelso, J.A.S., DeGuzman, G.C., Reveley, C., & Tognoli, E. (2009). Virtual partner interaction (VPI): Exploring novelbehaviors via coordination dynamics. *PLoSONE,* 4(6):e5749

Kelso, J.A.S. (2010). Instabilities and phase transitions in human brain and behavior. *Frontiers in Human Neuroscience* 4:23. https://doi.org/10.3389/fnhum.2010.00023

DeLuca,C., Jantzen, K.J., Comani, S., Bertollo, M., & Kelso, J.A.S. (2010). Striatal activity during intentional switching depends on pattern stability. *Journal of Neuroscience,* 30 (9), 3167–3174.

Riley, M.A., Richardson, M.J., Shockley, K., & Ramenzoni, V. C. (2011). Interpersonal synergies. *Frontiers in Psychology,* https://doi.org/10.3389/fpsyg.2011.00038

Wade, J.J., McDaid, L.J., Crunelli, V., & Kelso, J.A.S. (2011). Bidirectional coupling between astrocytes and neurons mediates learning and dynamic coordination in the brain. *PLoSONE,* 6.e29445

Kostrubiec, V., Zanone, P.-G., Fuchs, A., & Kelso, J.A.S. (2012). Beyond the blank slate: Routes to learning new coordination patterns depend on the intrinsic dynamics of the learner—experimental evidence and theoretical model. *Frontiers in Human Neuroscience,* 6, 212 https://doi.org/10.3389/fnhum.2012.00222.

Kelso, J.A.S. (2012). Multistability and metastability: Understanding dynamic coordination in the brain. *Phil. Trans. Royal Society B, 367,* 906–918.

Deco, G. & Jirsa, V. K. (2012). Ongoing cortical activity at rest: criticality, multistability, and ghost attractors. *J. Neurosci.32,* 3366–3375.

Kelso, J.A.S., Dumas, G., & Tognoli, E. (2013). Outline of a general theory of behavior and brain coordination. *Neural Networks,* 37, 120–131

Dumas, G., DeGuzman, G.C., Tognoli, E. & Kelso, J.A.S. (2014) The Human Dynamic Clamp as a paradigm for social interaction. *Proceedings of the National Academy of Sciences.* http://www.pnas.org/cgi/doi/10.1073/pnas.1407486111

Kelso, J.A.S., & Fuchs, A. (2016). The coordination dynamics of mobile conjugate reinforcement. *Biological Cybernetics,* 110 (1), 41–53.

Kelso, J.A.S. & Tognoli, E. (2017) Toward a complementary neuroscience: Metastable coordination dynamics of the brain. *Chaos & Complexity Letters, 11,* 141–162 [Special Issue on Neurodynamics: A Science in Transition, Essays Honoring Walter Freeman (F. Abrahams, Guest Editor)].

Zhang, M., Kelso, J.A.S., & Tognoli, E. (2018). Critical diversity: divided or united states of social coordination. *PLoSONE* https://doi.org/10.1371/journal.pone.0193843

Breakspear, M. (2017). Dynamic models of large-scale brain activity. *Nat. Neurosci.* 20, 340–352.

Nordham, C.A., Tognoli, E., Fuchs, A., & Kelso, J.A.S. (2018). How interpersonal coordination affects individual behavior (and vice-versa): Experimental analysis and adaptive HKB model of social memory. *Ecological Psychology* https://doi.org/10.1080/10407413.2018.1438196

Fuchs, A., & Kelso, J.A.S. (2018) Coordination Dynamics and Synergetics: From finger movements to brain patterns and ballet dancing. In Mueller, S., et al (Eds) *Complexity and Synergetics,* Springer-Verlag, Heidelberg, pp. 301–316.

Zhang, M., Beetle, C., Kelso, J.A.S., & Tognoli, E. (2019). Connecting empirical phenomena and theoretical models of biological coordination across scales. *J. Royal Society Interface* 16: 20190360. https://doi.org/10.1098/rsif.2019.0360

Alderson, T., Bokde, A., Kelso, J.A.S., Maguire, L., & Coyle, D. (2020). Metastable neural dynamics underlies cognitive performance across multiple behavioural paradigms. *Human Brain Mapping,* 41, 3212–3224. https://doi.org/10.1002/hbm.25009.

Kelso, J.A.S. (2021). Unifying large- and small-scale theories of coordination. *Entropy,* 23(5), 537. https://doi.org/10.3390/e23050537

Kelso, J.A.S. (2021). The Haken-Kelso-Bunz (HKB) Model: From Matter to Movement to Mind. *Biological Cybernetics,* 115 (4), 305–322. https://doi.org/10.1007/s00422-021-00890-w

Heggli, O.A., Konvalinka, I., Kringelbach, M.L., & Vuust, P. (2021). A metastable attractor model of self–other integration (MEAMSO) in rhythmic synchronization. *Phil. Trans. R. Soc. B* 376: 20200332. https://doi.org/10.1098/rstb.2020.0332

Kelso, J.A.S. (2022). On the Coordination Dynamics of (Animate) Moving Bodies. *Journal of Physics* (Complexity Section). https://doi.org/10.1088/2632-072X.

Sloan, A.T., Jones, N.A., & Kelso, J.A.S. (2023). Meaning from movement and stillness: Signatures of coordination dynamics reveal infant agency. *Proceedings of the National Academy of Sciences,* 120, e2306732120.